A Festschrift
in honor of
Vernon W. Hughes

Vernon W. Hughes

A Festschrift in honor of Vernon W. Hughes

Yale University 13 April 1991

edited by

Michael E. Zeller

Yale University

World Scientific
Singapore • New Jersey • London • Hong Kong

Published by

World Scientific Publishing Co. Pte. Ltd.

P O Box 128, Farrer Road, Singapore 9128

USA office: Suite 1B, 1060 Main Street, River Edge, NJ 07661

UK office: 73 Lynton Mead, Totteridge, London N20 8DH

The editor and publisher would like to thank the following publishers of the various journals and books for their assistance and permission to reproduce the selected reprints found in this volume:

American Institute of Physics (*J. Chem. Phys.*);
American Physical Society (*Phys. Rev., Phys. Rev. Lett.*);
CERN (*International Conference on Sector-Focused Cyclotrons and Meson Factories*);
Columbia University Press (*Quantum Electronics*);
Elsevier Science Publishers (*Phys. Lett., Nucl. Phys.*);
New York Academy of Sciences (*A Festschrift for I. I. Rabi*);
The Royal Swedish Academy of Sciences (*Physica Scripta*);
Springer-Verlag (*Z. Phys.*).

A FESTSCHRIFT IN HONOR OF VERNON W. HUGHES

ISBN 981-02-0703-4

Printed in Singapore by JBW Printers and Binders Pte. Ltd.

PREFACE

On May 28, 1991 Vernon Hughes celebrated his 70th birthday. To honor the occasion, and to take advantage of the opportunity to review the progress and status of a broad area of research, a symposium was held on the Yale campus on April 13. Approximately 200 physicists from around the world gathered to hear eight lectures by Vernon's colleagues. The topics ranged from reminiscences of early years of molecular beam resonance by Norman Ramsey to muon $g-2$ experiments, a subject of Vernon's current research, by Emilio Picasso. Also included were talks on polarized electrons by Wilhelm Raith and fundamental tests with polarized beams at high energies by Charles Prescott, a theory update on the muon $g-2$ and muonium hyperfine structure by Toichiro Kinoshita and a recollection of the development of muonium hyperfine interval measurements by Valentine Telegdi, a review of "Vernon Hughes' Worlds of Atoms" by Daniel Kleppner, and some thoughts about high precision measurements by Gerald Feinberg. The banquet speech was a remembrance of Vernon as a scientist and teacher of an international community by Gisbert zuPutlitz.

This book contains the written versions of the symposium lectures. We have also included comments by Kunitaka Kondo on Vernon's collaboration with Japanese physicists, remarks by Vernon at the symposium banquet, and selected publications.

The book is compiled in the hope that the physics community and other interested scholars can share in the enthusiasm for science that the attendees felt at the symposium. Each talk was related to an aspect of Vernon's work, but each speaker took his own tack in presenting his view of a particular area of physics and Vernon's role in it. Thus, from this volume the reader will not only learn of the physics explored by Vernon Hughes, but will get a glimpse of a group of eminent physicists and their approach to their science.

While the speakers were responsible for the outstanding content of the symposium, the responsibility for the successful organization and operation of the event was largely due to the administrative skills and diligent efforts of Nancy Soper and Jeanette DeFranco. Thanks are also due to Pat Fleming for compiling and typing these proceedings.

Michael E. Zeller

Participants in the Symposium

Bottom row left to right: C.S. Wu, G. Feinberg, T.D. Lee, W. Raith, D. Kleppner, N. Ramsey
Top row left to right: E. Picasso, V. Talegdi, T. Kinoshita, C. Prescott, V. Hughes, G. zu Putlitz

TABLE OF CONTENTS

Nuclear Physics

Muonium

Particle Physics

A Festschrift
in honor of
Vernon W. Hughes

Early Years of Molecular Beam Resonance

N. F. Ramsey
Lyman Physics Laboratory, Harvard University, Cambridge, MA 02138

Prehistory

Molecular beam resonance experiments originated from the fusion of two quite different fields: (1) Molecular beam experiments and (2) Speculation about space quantization when the axis of quantization is changed.

The first molecular beam experiment was that of Dunoyer[1] in 1911 to determine if atoms in a vacuum moved in straight lines. The techniques of molecular beams in subsequent years were greatly extended and improved by Stern,[2] Rabi[3] and many others.[4] Stern's[2] invention of the technique for deflecting beams of atoms by inhomogeneous magnetic fields were of special value in subsequent developments.

Darwin,[5] in 1928, speculated theoretically as to what would happen if an atom were initially space quantized in a magnetic field along one axis and the direction of the field were then changed. Phipps and Stern[6] and Frisch and Segre[7] did atomic experiments to observe the anticipated transitions between orientation states while Guttinger[8] and Majorana[9] further developed the theory of such transitions. There was partial, but not complete, agreement between theory and experiment until 1936 when Rabi[10] showed theoretically that the effects of the spins of the nuclei, as well as those of the electrons, had to be considered in weak magnetic fields where the nuclear and electron angular momenta were significantly coupled together. In that same year Gorter[11] published an account of his unsuccessful attempts with calorimetric techniques to detect nuclear transitions induced by an oscillatory magnetic field.

Magnetic Resonance Method

The problem of the transitions induced between orientation states again became important in another molecular beam context when Rabi[12] and his associates developed the powerful zero moment atomic beam method for measuring nuclear magnetic moments and found they could also measure the signs of the nuclear moments by inducing transitions between different orientation states of the atoms. This led Rabi carefully to consider the changes in orientation states of atoms that passed through successive regions of space with oppositely directed magnetic fields. Although the atoms had a modified Maxwellian velocity distribution, Rabi initially assumed, for simplicity, that they were all at a single velocity so that the changes in field were essentially oscillatory at a single frequency. As a result he obtained an exact solution and his brilliant 1937 paper[13] entitled "Space Quantization in a Gyrating Magnetic Field" provided the theoretical basis for all subsequent magnetic and electric resonance experiments. However, the equation was written in a form that obscured its resonance character and subsequently lost the resonance character when the results were averaged over all velocities and consequently over a variety of frequencies. Consequently, it was more than six months later, following a visit by Gorter,[14,15] before major efforts in Rabi's laboratory were directed toward the use of oscillatory fields and to the development of the now well known molecular beam magnetic resonance method.

The first successful observations of nuclear magnetic resonances were obtained at Columbia by Rabi, Zacharias, Millman and Kusch[15] in 1938 with LiCl. Soon thereafter, Kellogg, Rabi, Ramsey and Zacharias[16] observed nuclear magnetic resonances in H_2, D_2 and HD. Initially they were disappointed in observing a complex resonance pattern rather than the expected single resonance. But they soon realized that they were observing the radiofrequency spectrum of the molecule and the pattern was complicated by the internal magnetic and nuclear electric quadrupole interactions in the molecule in addition to the expected interactions of the magnetic moments with the external magnetic field. In 1940, Kusch, Millman and Rabi[17,18] extended this method to paramagnetic atoms, including observations of $\Delta F = \pm 1$ transitions where the relative orientations of the nuclear and electronic angular momenta are changed. The new molecular beam resonance methods were successfully used at Columbia to measure a number of nuclear, molecular and atomic properties, but this prolific activity was brought to an abrupt end in 1941 by World War II.

The speedy resumption of molecular and atomic beam magnetic resonance experiments at Columbia following the end of the war were highlighted by Lamb and Retherford's discovery of the Lamb shift[19] in the fine structure of metastable excited state of atomic hydrogen, by Rabi, Nafe and Nelson's discovery of the hydrogen anomalous hyperfine separation[20] and by Kusch and Foley's measurement of the anomalous magnetic moment of the electron.[21] These three discoveries provided the primary impetus for the development of the relativistic quantum electrodynamics. In addition, Rabi, Kusch, Feld, Lamb, Ramsey, Nierenberg, Foley, Mann, Bolef, Zeiger, Koenig, Prodell, Franken, Perl and Senitzky and others at Columbia used the magnetic resonance method to measure nuclear, atomic and electron magnetic moments and to study the magnetic and electric quadrupole interactions in numerous alkali atoms, diatomic molecules and alkali halides. Vernon Hughes, Tucker, Rhoderick and Perl measured[22] the magnetic moment of atomic ^{4}He in the excited metastable $^3S_1$ state.

Electric Resonance Method

Polar molecules were deflected with inhomogeneous electric fields in early experiments in Stern's laboratory[23] so, once the molecular beam magnetic resonance experiments were successful, it was apparent that electric resonance experiments could also be possible by utilizing inhomogeneous electric fields to produce the deflections and oscillatory electric fields to induce the resonance transitions. However, the development was greatly delayed by the interruption of World War II so that first successful electric resonance was that of Harold Hughes[24] in 1947 using CsF. The initial electric resonance experiment was soon followed by a number of others. With a more uniform electric field, Trischka[25] showed that what appeared to be a single line in the initial CsF experiments could be resolved into separate lines corresponding to different molecular vibrational states and that these lines at low electric fields could be further resolved into a hyperfine structure from which the interaction constants for the nuclear quadrupole and spin-rotational magnetic interactions could be determined. Vernon Hughes and Grabner[26] studied the electric resonance spectra of RbF and KF in detail. They determined the quadrupole interaction energy separately in each of the first four vibrational states of the molecule and showed a 4% variation between different states. They also observed for the first time a two quantum transition. Carlson, Lee, Fabricand and Rabi[27] found the surprising result that the Cl quadrupole interaction in KCl increases by more than 200% between the first and second vibrational states whereas the quadrupole interaction of K in the same molecule changes by only 1% between the same two states. Although the electric resonance method has the disadvantage of being inapplicable to atoms and homonuclear molecules, it has the great advantages of high deflecting power with correspondingly high beam intensity and the possibility of easily selecting specific rotational and vibrational states, which provides spectra which can be measured accurately and interpreted more easily. As a result these pioneering experiments at Columbia were the models for many subsequent experiment.

Conclusions

For most of the decade following Rabi's 1937 inventions of the molecular beam magnetic resonance method, Columbia had virtual monopolies on all kinds of electric resonance, magnetic resonance and molecular beam experiments, although nearly half of that time was interrupted by World War II. Despite the interruptions, the accomplishments of the laboratory were enormous in scientific discovery, in the development of new research methods and in the training of graduate and post doctoral students. By the early 1950's other resonance techniques, like NMR had been invented and new molecular beam resonance laboratories had been established, mostly by former students and associates of Rabi. These laboratories included Jerrold Zacharias's laboratory at MIT, Normal Ramsey's at Harvard, John Trischkas at

Syracuse, Vernon Hughes's at Yale, William Nierenberg's at Berkeley and Victor Cohen's at Brookhaven. By this time the "Early Years of Molecular Beam Resonance" had drawn to a close.

References

1. L. Dunoyer, *Le Radium* **8**, 142 (1911) and **10**, 400 (1913).
2. O. Stein, *Zeits. f. Physik* **7** 249 (1921) and **39**, 751 (1926).
3. I. I. Rabi and V. W. Cohen, *Phys. Rev.* **43**, 582 (1933).
4. N. F. Ramsey, *Molecular Beams*, Oxford Press (1956 and 1990).
5. C. Darwin, *Proc. Roy. Soc.* **117**, 258 (1927).
6. T. E. Oguoos and O. Stern, *Zeits. F. Physik* **73**, 185 (1931).
7. R. O. Frisch and E. Segre, *Zeits. F. Physik* **80**, 610 (1933).
8. P. Guttinger, *Zeits. f. Physik* **73**, 169 (1931).
9. E. Majaorana, *Nuovo Cimenta* **9**, 43 (1932).
10. I. I. Rabi, *Phys. Rev.* **49**, 324 (1936).
11. C. J. Gorter, *Physica* **3**, 503 and 995 (1936).
12. V. W. Cohen, S. Millman, M. Fox and I. I. Rabi, *Phys. Rev.* **46**, 713 (1934); **47**, 739 (1935) and **48**, 746 (1935).
13. I. I. Rabi, *Phys. Rev.* **51**, 652 (1937).
14. Rabi[15] referred to this visit in a footnote and Gorter 29 years later gave his recollections of the same visit in *Physics Today*, **33**, 25 (January 1967).
15. I. I. Rabi, J. R. Zacharias, S. Millman and P. Kusch, *Phys. Rev.* **53**, 318 (1938) and **55**, 526 (1938).
16. J. M. B. Kellogg, I. I. Rabi, N. F. Ramsey and J. R. Zacharias, *Phys. Rev.* **55**, 729 (1939); **57**, 728 (1939) and **57**, 677 (1940).
17. P. Kusch, S. Millman and I. I. Rabi, *Phys. Rev.*, **57** 765 (1940).
18. S. Millman and P. Kusch, *Phys. Rev.* **58**, 438 (1940).
19. W. E. Lamb, R. C. Retherford, S. Triebwasser and E. S. Dayhoff, *Phys. Rev.* **72**, 241 (1947); **79**, 549 (1950 and **89**, 106 (1953).
20. J. E. Nafe, E. B. Nelson and I. I. Rabi, *Phys. Rev.* **71**, 914 (1947) and **73**, 718 (1948).
21. P. Kusch and H. M. Foley, *Phys. Rev.* **72**, 1256 (1947) and **74**, 250 (1948).
22. V. Hughes, G. Tucker, E. Rhoderick and G. Weinreich, *Phys. Rev.* **91**, 828 (1953).
23. E.Wrede, I. Estermann and M. Wohlhill, *Zeits. f. Physik* **41**, 569 (1947 and *Zeits. f. Phsikal. Chem.* **B1**, 161 (1928).
24. H. K. Hughes, *Phys. Rev.* **76**, 1675 (1949).
25. J. W. Trischka, *Phys. Rev.* **74**, 718 (1948) and **76**, 1365 (1949).
26. V. Hughes and L. Grabner, *Phys. Rev.* **79**, 314 and 829 (1950).

Vernon Hughes' Worlds of Atoms

D. Kleppner
Department of Physics, Massachusetts Institute of Technology
Cambridge, MA 02139

Someone unfamiliar with the work of Vernon Hughes might regard the title of this talk as a trifle dramatic, but anyone who is familiar with his work will recognize that it is appropriate. A well meaning colleague, not overly gifted as an artist, (myself) attempted to illustrate Vernon Hughes' worlds of atoms with the results shown in Fig. 1. In case the iconography is not self evident, it is to be noted that the ascendant sun in the upper left is fueled by tests of fundamental principles. This sun illuminates a trio of three robust planets - "muonium", "positronium", and "helium". The object that bears some resemblance to a cup and saucer with wings is, in fact, a flying teacup of homey atoms and molecules. The fine structure constant threads the cosmos like a celestial messenger, and the g-factor anomaly of the muon can be seen soaring upward along a path of hope. Interlopers such as muonic helium and pionium are to be seen. A few other objects cannot quite be discerned, a visual rendering of the conventional apology for having overlooked anything.

Doing justice to such a body of work would monopolize these festivities and this volume. The 290 publications that extend from Hughes' earliest works on molecular beam electric resonance to his latest paper on the Lamb shift in muonium encompass seminal research in exotic atoms, new streams of fundamental tests, the creation of new technologies such as polarized particle sources, and experiments at energies that range form μ eV to GeV.

Consider, for example, Hughes's work on positronium. In a series of early papers[1–4] he studied the effects of electric and magnetic fields on positronium formation and quenching. These papers provided the foundations for using positronium as a testing ground for fundamental theory. He moved rapidly to these studies and measured the hyperfine structure of positronium, $\Delta\nu(Ps)$, a work carried out with Mader and Wu in 1957[5]. The paper states

> Positronium, the bound state of an electron and a positron, is an ideal system for a test of quantum electrodynamics because no particles foreign to the theory are present.

They determined $\Delta\nu(Ps)$ to a precision of 200 ppm. That early measurement did not seriously press theory, but over the years Hughes returned to the subject again

FUNDAMENTAL TESTS
PIONIUM
$g_\mu - 2$
MOLECULES
ATOMS
MUONIUM
& its
brothers &
sisters &
cousins &
aunts
POSITRONIUM
&
anti-
positronium
HELIUM
relativistic
theory
fine structure
hyperfine structure
anomaly
VERNON HUGHES' WORLDS OF ATOMS

and again, steadily improving the precision. The last measurement in 1984,[6] achieved an accuracy of 3.6 ppm. This exceeds the theoretical accuracy, leaving the ball in the theorists' court, at least for the moment. Meanwhile, the experimental arena has been spectacularly enlarged. In collaboration with a group at Heidelberg under zu Putlitz, Hughes participated in the discovery of pionium - the π^+e^- atom.[7] The spirit of this enterprise is summarized in the first paragraph, with words reminiscent of Hughes' earliest vision of the physics of exotic atoms:

> The observation of pionium completes the three-member family of atoms, consisting of the lepton-pairs (muonium, positronium), a baryon-lepton pair (hydrogen), and a meson-lepton pair (Pionium). Studying such exotic hydrogen isotopes is indispensable for testing precisely fundamental atomic theories.

The most far reaching of these studies of "exotic hydrogen isotopes" is Hughes' monumental work on muonium, a story that is told elsewhere in this volume. However, the theme of testing fundamental theories is recurrent in Hughes' work. Among these are his studies of the isotropy of space and charge neutrality.

The Isotropy of Space

According to Mach's Principle, the inertial properties of space are established by the existence of far-off mass. The underlying idea is most simply presented in Mach's commentary on Newton's rotating water-bucket experiment. The surface of water in a bucket that is set spinning forms a parabaloid as the fluid starts to rotate with the bucket. Mach pointed out that if the stars in the universe were somehow eliminated one by one, there would be no way to predict how the water would behave because rotation would have lost its meaning. Thus, the local properties of inertia must be related to the presence of mass elsewhere in the universe. Although the mass at great distances appears to be rather uniformly distributed, that is certainly not the case for nearby mass, for instance the mass in our galaxy. Thus, it is quite reasonable to expect that the inertial properties of space are anisotropic. One way this would manifest itself is by an anisotropy in mass: the inertial mass of an object could be different in different directions.

In a famous experiment[8] on the NMR spectrum of ^{7}Li, Hughes first demonstrated that one could use the Zeeman effect to study the isotropy of space, and set the first modern limit to a possible anisotropy. A similar experiment was reported shortly thereafter by R.V. Drever,[9] and the experiment has come to be called the Hughes-Drever experiment. With the advent of lasers and refinements in optical pumping techniques, the experiment has been refined. In the past few years several groups[10–12] have achieved a precision that sets a limit to any possible mass anisotropy of less than about 4×10^{-24}.

As with much of Hughes' work, the impact of the Hughes-Drever experiment is far greater than in the refinement of the limit of mass anisotropy. It has served as a driving force for refining the precision of NMR experiments, out of which has come, among other things, new limits to the electric dipole moment of an atom,[13] and new limits to possible nonlinearities in quantum mechanics.[14]

Charge Neutrality

The conservation of electric charge is among the strongest and most fundamental conservation laws in physics, but, as with all conservation laws, it ultimately rests on experiment. Closely related to charge neutrality is the assumption that the magnitudes of the charge of the electron and proton are equal. Zorn, Chamberlain and Hughes[15] set a limit to the neutrality of an alkali atom by searching for a deflection of the atomic beam as it passes through an electric field. The limit was $\delta q/q < 4 \times 10^{-19}$, and in a second generation of the experiment[16] the limit was improved to about 1×10^{-20}. The limit to sensitivity depends on the time the atom spends passing through the electric field. Other things being equal, the sensitivity varies inversely with the energy of the atomic beam. Recently, atomic beams have been cooled to the microkelvin regime. A number of groups are planning experiments to carry out a Hughes deflection experiment using ultra cold atoms. A new generation of charge neutrality experiments is in the works, one more example of the generative impact of Hughes' ideas.

Sub-Natural Linewidth and Quantum Cavity Electrodynamics

The connection between these two topics is not immediately apparent, but, as I shall explain, it exists.

The Lamb shift has played a continuing role as a test of basic atomic theory, but at the beginning of the 1960's there was an urgent need to confirm and improve the original measurements of Lamb and Retherford. The principal obstacle to precision is the short lifetime of the 2P state of hydrogen, which results in an extremely broad natural linewidth for the transition 2S→2P. Hughes pointed out[17] that one could observe a linewidth narrower than the natural linewidth by using the Ramsey method to observe selectively those atoms which live longer than the mean lifetime. The narrowing of the linewidth has its cost in loss of signal, for the linewidth decreases linearly with flight time, whereas the signal decreases exponentially when the flight time is longer than the natural lifetime. However, if the supply of 2S atoms is abundant, then the trade off can result in an important advantage. This principle was applied by F. M. Pipkin and his students and they substantially improved the accuracy of the Lamb shift.[18]

How this relates to the field of cavity quantum electrodynamics is as follows: Hughes' idea for achieving a subnatural linewidth, and its demonstration by Pipkin, stimulated me to think about other possible ways to reduce the linewidth in the Lamb shift measurement. Eventually this led to the speculation that one could avoid spontaneous emission from the 2P state by placing the atom in a cavity too small to support the mode of the radiated photon.[19] The idea was flawed for two reasons: first, because there was no way to create such a cavity at the wavelength of the Lyman-alpha transition, much less carry out such an experiment and, second, because it does not take one long to realize that the atom-cavity interaction would perturb the atom at the level of precision desired. So, the idea sat on the shelf for some time. However, techniques for working with highly excited atoms were developed during the next decade, and an experiment to demonstrate the inhibition of spontaneous emission became practical. We carried this out in my laboratory[20] in one of the earliest demonstrations of atom-cavity interactions, a subject that has since come to be called cavity quantum electrodynamics. This is but one example of how Hughes' ideas have helped to stimulate new areas of atomic physics. Here is another.

Magnetic Atoms

The origins of atomic physics are firmly embedded in astrophysics, and time is not weakening the connection. In 1974 Hughes published a paper with R. O. Mueller[21] that starts out

> Recent astrophysical evidence indicates the existence of intense magnetic field B of the order of $10^{12}G$ on pulsars, and in this connection the behavior of atoms has been discussed theoretically. From the atomic physics viewpoint, these intense magnetic fields provide a new atomic regime in which the magnetic interaction dominates the Coulomb interaction.

Although magnetic fields that are strong on the atomic scale are inaccessible in the laboratory, one can achieve the strong field regime by studying highly excited states of hydrogen, a point that is made in the paper's title: "Atomic Regime in Which the Magnetic Interaction Dominates the Coulomb Interaction for Highly Excited States of Hydrogen".

As far as I know the work did not extend beyond the paper, but once again a paper of Hughes was prescient in anticipating a new area of atomic physics. The essential idea is easily sketched: in a strongmagnetic field the important interaction is the diamagnetic interaction. This varies as $B^2\rho^2$, where ρ is the radius in cylindrical co-ordinates. The Coulomb interactions, of course, varies as 1/r. Because the characteristic dimension of a highly excited atom varies as n^2, where n is the principal quantum number, the ratio of magnetic energy to Coulomb energy varies as n^6 B^2. For large values of n, say n$\sim$100, the ratio can easily exceed unity in laboratory-sized fields.

We have pursued this line of research in my own laboratory for some time, and the problem has proven to be remarkably interesting. One area of interest stems from what might be called mainstream atomic physics, the sort of problem in the flying teacup of Figure 1. The diamagnetic hydrogen atom, as the problem has come to be called, is one of the simplest non-separable systems in atomic physics. Finding general solutions poses a challenge to atomic theory. The second stream of interest was completely unanticipated by us, and probably by Hughes, though the possibility of such a development is inherent in his calling attention to "a new atomic regime", for in new regimes one can expect new physics. This is the connection with nonlinear dynamics and disorderly motion. The classical motion of an electron around a proton in a magnetic field has been discovered to display a textbook example of a transition from orderly to disorderly motion as the energy is raised. The connections between disorderly classical motion and quantum mechanical behavior, the subject that has come to be known as "quantum chaos", is attracting wide interest. The diamagnetic hydrogen atom has come to be regarded as a paradigm for these studies. A good review of this subject is to be found in a recent volume of Comments in Atomic and Molecular Physics, [22] and in the references therein.

Helium

In spite of, or because of, being the simplest three-body system in atomic physics, helium is one of the most complicated of atoms. To classical mechanics the three-body problem remains largely an enigma. Quantum mechanics is only slightly kinder. Vernon Hughes early on staked out a path through this field, and as one might expect it is the path that leads through fundamental theory. He followed this path through a series of seventeen experimental and theoretical papers extending from 1953 through 1981. These studies focused on relativistic contributions to the magnetic moment in the $^3S_1$ state,[23–24] the fine structure interval of the 2^3P_{0-1} states,[25,26] and a few related topics such as the hyperfine structure of metastable helium-3.[27] The story is one of theory and experiment pushing each other through decades of refinements, with neither side winning, or perhaps with both sides winning. For the moment, the battle is a stand-off.

Nevertheless, the work has an unexpected ramification. New techniques for trapping and cooling atomic hydrogen should make it possible to measure the 1S→2S two-photon transition with a resolution that approaches the natural linewidth.[28] The atoms are confined by inhomogeneous magnetic fields. Non relativistically, the two-photon transition is field independent and the trap has no effect on the spectrum. Because of the relativistic correction to the g-factor, first measured by Hughes, there is a small but important first order Zeeman shift. One can find strategies for dealing with it, but it is noteworthy how this problem of purely academic interest has unexpectedly taken on practical importance.

And So On

This account does not do justice to Hughes' role in contemporary atomic physics. His reviews of fundamental atomic physics serve not only as the standard references but also the conscience of the field.[29–30] He has carried out innovative experiments on producing polarized electrons,[31–32] and applying them to studies of atomic scattering.[33–35] (The major scientific payoff for the polarized electron sources has been in particle physics, which is outside the territory of this talk.) He has studied atom-antiatom collision processes,[36] and has contributed to the determination of the fundamental constant.[37] The list goes on and on. All we can do is wonder that one person can accomplish so much, and that at a time in their careers when most scientists are slowing down, Vernon Hughes is moving ahead at full speed.

References

1. V. W. Hughes, S. Marder and C. S. Wu, *Phys. Rev.* **98**, 1840 (1955).
2. S. Marder, V. W. Hughes, C. S. Wu and W. Bennett, *Phys. Rev.* **103**, 1258 (1956).
3. W. B. Teutsch and V. W. Hughes, *Phys. Rev.* **103**, 1266 (1956).
4. V. W. Hughes, *J. Appl. Phys.* **28**, 16 (1957).
5. V. W. Hughes, S. Marder and C. S. Wu, *Phys. Rev.* **106**, 934 (1957).
6. M. W. Ritter, P. O. Egan, V. W. Hughes and K. A. Woodle, *Phys. Rev.* **30**, 1331 (1984).
7. H. -J. Mundinger, K. -P. Arnold, M. Gladisch, H. Hofmann, W. Jacobs, H. Orth, G. zuPutlitz, J. Rosenkranz, W. Schaefer, W. Schwarz, K. A. Woodle and V. W. Hughes, *Europhys. Lett.* **8**, 339 (1989).
8. V. W. Hughes, H. G. Robinson and V. Bertran-Lopex, *Phys. Rev. Lett.* **4**, 342 (1960).
9. R. P. W. Drever, *Philos. Mag.* **6**, 683 (1961).
10. J. D. Prestage, J. J. Drullinger, W. M. Itano and D. Wineland, *Phys. Rev. Lett.* **54** 2387 (1985).
11. S. K. Lamoreaux, J. P. Jacobs, B. Heckel, F. J. Raab and E. N. Fortson *Phys. Rev.* **A39**, 1082 (1989).
12. T. E. Chupp, R. J. Hoare, R. A. Loveman, E. Oteiza, J. M. Richardson, M. E. Wagshul and A. K. Thompson, *Phys. Rev. Lett.* **63** 1541 (1989).
13. S. K. Lamoreaux, J. P. Jacob, B. Heckel, F. J. Raab and E. N. Fortson, *Phys. Rev.* **A39**, 1082 (1989); S. A. Murthy, D. Krause, Jr., J. L. Li and L. Hunger, *Phys. Rev. Lett.* **63**, 965 (1989).
14. J. J. Bollinger, D. J. Heinzen, W. M. Itano, S. L. Gilbert and D. J. Wineland, *Phys. Rev. Lett.* **63**, 1031 (1989); T. E. Chupp, and R. J. Hoare, *Phys. Rev. Lett.* **64**, 2261 (1990); P. K. Majunder, B. J. Venema, S. K. Lamoreaux, B. Heckel and E. N. Fortson *Phys. Rev. Lett.* **65**, 2931 (1990).
15. J. C. Zorn, G. E. Chamberlain, V. W. Hughes, *Z. Phys.* **D12**, 12 (1963).
16. V. W. Hughes, L. J. Fraser, and E. R. Carlson, *Z. Phys.* **D10**, 145 (1988).
17. V. W. Hughes, *Quantum Electronics,* ed. C.H. Townes (Columbia Univ. Press, 1960), p. 582.

18. S. Lundeen and F. M. Pipkin, *Phys. Rev. Lett.* **46**, 232 (1981).
19. D. Kleppner, *Atomic Physics and Astrophysics,* ed. M. Chretien and E. Lipworth (Gordon and Breach, 1971), p. 49.
20. R. G. Hulet, E. S. Hilfer and D. Kleppner, *Phys. Rev. Lett.* **55**, 2137 (1985).
21. R. O. Mueller and V. W. Hughes, *Proc. Nat. Acad. Sci. USA* **71**, 3287 (1974).
22. *Comments in Atomic and Molecular Physics, XXV*, (1991), "Irregular Atomic Systems and Quantum Chaos", edited by J. C. Gay.
23. W. Perl and V. W. Hughes, *Phys. Rev.* **91**, 842 (1953); *Phys. Rev.* **89** 886 (1953).
24. B. E. Zundell and V. W. Hughes, *Phys. Lett.* **59A** 381 (1976).
25. A. Kponou, V. W. Hughes, C. E. Johnson, S. A. Lewis and F. M. J. Pichanick, *Phys. Rev. Lett.* **26**, 1613 (1971).
26. W. Frieze, E. A. Hinds, V. W. Hughes and F. M. J. Pichanick, *Phys. Rev.* **A24**, 279 (1981).
27. G. Weinreich and V. W. Hughes, *Phys. Rev.* **95**, 1451 (1954).
28. D. Kleppner, *The Hydrogen Atom,* ed. G. F. Bassani, M. Inguscio and T. W. Hansch, (Springer Verlag, Berlin, 1989) p. 68.
29. V. W. Hughes, *Atomic Physics 10,* ed. by H. Narumi, I. Shimamuro, (Elsevier Science Publishers, 1987), p. 1.
30. V. W. Hughes, *Precision Measurement and Fundamental Constants II*, ed. by B. N. Taylor and W. D. Phillips, (Natl. Bur. Stand. (U.S.), Spec. Publ. 617 (1984)) p. 237.
31. R. L. Long, Jr., W. Raith and V. W. Hughes, *Phys. Rev. Lett.* **15**, 1 (1965).
32. V. W. Hughes, R. L. Long, Jr., M. S. Lubell, M. Posner and W. Raith, *Phys. Rev.* **A5** 195 (1972).
33. M. J. Alguard, V. W. Hughes, M. S. Lubell and P. Wainwright, *Phys. Rev. Lett.* **39**, 344 (1977).
34. G. D. Fletcher, M. J. Alguard, T. J. Gay, V. W. Hughes, C. W. Tu, P. F. Wainwright, M. S. Lubell, and F. C. Tung, *Phys. Rev. Lett.* **48**, 1671 (1982).
35. G. D. Fletcher, M. J. Alguard, T. J. Gay, V. W. Hughes, P. F. Wainwright, M. S. Lubell and W. Raith, *Phys. Rev.* **A31**, 2854 (1985).
36. D. L. Morgan, Jr. and V. W. Hughes, *Phys. Rev.* **A7**, 1811 (1973).
37. V. W. Hughes, *A Festschrift for I. I. Rabi,* ed. by Lloyd Motz, (The New York Academy of Sciences) Series II, **38**, 62 (1977).

On Knowing Things Better And Better

Some Thoughts about Precise Physical Measurements

G. Feinberg

Department of Physics, Columbia University, New York, N.Y. 10027

One of the trademarks of Vernon Hughes' career as a physicist is that he has been a pioneer in carrying out precise measurements of a variety of physical quantities, and in extending previous limits of precision for such measurements. Examples of this work fill the list of publications. Here are just a few:

"Experimental Limit for the Electron-Proton Charge Difference" (1957)
"Upper Limit for the Anisotropy of Inertial Mass from Nuclear Resonance Experiments" (1960)
"New Value for the Fine Structure Constant α from Muonium Hyperfine Measurements" (1964)
"Precision Redetermination of the Hyperfine Interval of Positronium" (1967)
"Search for Muonium-Antimuonium Conversion" (1968)
"Parity Non-Conservation in Inelastic Electron Scattering" (1978)
"Experimental Test of Special Relativity from a High γ Electron g-2 Measurement" (1979)
"Limits on Neutrino Oscillations from Muon Decay Neutrinos" (1981)
"A Possible Higher Precision Measurement of the Muon g-2 Value" (1984)

It is obvious from this partial list of his publications that Vernon has been at the forefront of efforts in many areas of physics both to determine the limits of our knowledge and to increase the precision of those things that we already know.

I want to use this talk to discuss some philosophical considerations that I think are relevant to the quest for extreme precision in the measurement of fundamental quantities that Vernon and others have undertaken. While physicists are sometimes uncomfortable with philosophy, I think that philosophical analyses can have value in helping us to understand better what we are doing when we practice our science. Such analysis can make explicit some of the implicit assumptions of an area of scientific research, after which we can decide to what extent they are warranted.

The Rationale for Precision Measurements

I begin by asking whether indefinitely increasing precision of numerical quantities is a desirable aspect of knowledge, and if so, why this should be the case. Does it really matter to the progress of science if we know an additional decimal place of some observable physical quantity, such as the boiling point of neodymium nitrate? Several answers can and have been given to this rhetorical question. From a tactical point of view, there is always the possibility that an increase in precision in the measurement of any quantity will uncover some novel phenomenon whose study will lead to widespread changes in our understanding of matters that go far beyond the measurement itself. This is the reason that was offered in 1931 by F. Richtmeyer, who said,[1]

"Why should one wish to make measurements with ever increasing precision? Because the whole history of physics proves that a new discovery is likely to be found lurking in the next decimal place."

These words were perhaps meant as an ironic echo of the earlier and more notorious words of Michelson and Maxwell in the nineteenth century.[2]

"It is here that the science of measurement shows its importance—when quantitative results are more to be desired than qualitative work. An eminent physicist [Maxwell] has remarked that the future truths of physics are to be looked for in the sixth place of decimals."

There are certainly many examples that bear out these expectations. In the 19th century, it was recognized by Maxwell and Lorentz that the detection of the motion of the earth through the ether would require carrying out optical measurements to a precision of order $v^2/c^2 \sim 10^{-8}$, and the effort to do this culminated in the Michelson-Morley experiment, which of course showed no such effect. In the 1940s, measurements at Columbia by Kusch and Foley[3] of atomic Zeeman splittings, to an accuracy of 10^{-4}, this would revealed the electron anomalous magnetic moment. More recently, the attempt by Fitch, Cronin and coworkers[4] to extend the limit on CP conservation in the neutral kaon system beyond the 1% level led to the discovery of CP violation.

But I think that while the hope of discovering novelties is an important reason for the quest for greater numerical precision, it does not do complete justice to that quest. Even if novelties do not emerge from measurements with improved precision these measurements can satisfy an important scientific need.

Precise measurements of any quantity make an important contribution to creating the seamless web of knowledge that is the aspiration of science. One of the goals of science is to know as much as possible about everything in the natural world. This goal involves both the breadth and the depth of this knowledge. We want to know both the great unifying principles, such as the law of conservation of energy, and to know the detailed particulars about individual phenomena. Great precision in measurement contributes to both aspects of our quest. Most basically, it helps to shrink the range of our ignorance. In this regard, the totality of the numbers in the *Handbook of Chemistry and Physics*, or in the *Review of Particle Properties* are no lesser achievements of science than the equations of string theory.

Furthermore, the criterion of what constitutes a precision measurement changes over time. While at one time in the history of science the available level of technical skill may make a specific level of precision appear to be a natural resting place, scientists of a later time often find that this was just a temporary way station. Michelson's measurements of the speed of light, which represented the best that could be done in the early part of the twentieth century, are now rivaled in demonstrations done for first year physics students.

Let me mention another apsect of why additional precision is worth pursuing. One way to express this is by contrasting the achievement of an additional factor of ten in accuracy in measuring the muon g value with the achievement of calculating another million decimal places of $\pi^{1/e}$, or some other moderately interesting number. I think that most physicists, and perhaps even many mathematicians would agree that the former is a more significant accomplishment, but can we say why we think so?

In some cases, the reason is again tactical. In order to achieve higher precision in measurement, it is often necessary to invent new techniques. Thus the application of microwave techniques to spectroscopy directly following World War II led to improvements by several orders of magnitude in the measurement of some energy differences. For a more recent example, in the article on muonium by Vernon and Gisbert zu Putlitz,[5] in Tom Kinoshita's book *Quantum Electrodynamics*, there is a nice summary of the various techniques that have been used to improve the precision of the measurement of the muonium hyperfine interval over a period of twenty years by a factor of 10^4.

Sometimes, the new techniques needed for some precision measurement have wide applications to other measurements as in the development of microwave spectroscopy in the 1940s. When this happens, it seems to me as significant as finding a new theoretical principle that applies to a variety of physical situations. But even when they do not have such wide applications, the invention of new techniques for improving the precision of a measurement is a scientific advance that is as much worth celebrating as is the discovery or solution of a new equation.

Therefore, I think that one of the best reasons for doing specific precision measurements is that the techniques available at a given time allow them to be done. This rationale is not very different from George Mallory saying that he wanted to climb Mt. Everest, because it is there. Of course, this attitude must be constrained by the availability of resources as well as by the interests of the scientists who could do the measurements. But if a scientist gets an idea that allows for the measurement of any physical quantity to significantly greater precision, then carrying it out can be regarded as an important contribution to science, whatever the eventual consequences of the measurement. In other words, what I am saying is that carrying out precision measurements is its own reward, and need not be justified in terms of some supposedly higher calling.

What is Worth Measuring Precisely?

If we grant that greater precision of measurement is a worthy scientific goal, the next question that suggests itself concerns where to concentrate the efforts for such precision. The limited time and resources available to scientists require that we focus on a few areas for precise measurements. One obvious area for this focus is that of measurements of "fundamental" quantities, such as the properties of individual subatomic particles.

The Relevance of Theory to Precision Measurement

An argument seemingly contrary to Richtmeyer's which is sometimes put forward to justify such concentration on measurements of fundamental quantities is that there exist more reliable theoretical expectations for the results of such measurements. So, for example, the very precise series of measurements by the Seattle group of the gyromagnetic ratio of the electron is sometimes justified by saying that the result can be compared with the prediction of quantum electrodynamics, a prediction which, to the required accuracy is almost as difficult to extract as the measurement is to perform. The contrast between measurements which can be compared to a precise theoretical prediction, and measurements of such quantities as the proton g-value, for which no accurate theory is available, is sometimes expressed through the distinction between "precision tests" and "precision measurements."

It is true that when an accurate theoretical prediction is available for comparison with some precise measurement it is relatively easy to know when a problem exists, either with the measurement or with the theory. There have been some recent cases in which disagreements between theory and experiment in QED led to new theoretical analyses which superseded the previous ones and which now agree with the experiment. For example, this has occurred with the Lamb shift in hydrogen. However, agreement between theory and experiment is not always a reliable indication that both are correct to the accuracy at which they have been compared. Many of you remember the situation in the early 1950s, when theory and experiment involving the electron g-value were in agreement at the level of about one part in 10^5, but both were in fact incorrect at that level.

$(g\text{-}2)_{exp.} = 2\ (1.001\ 146 \pm .000\ 012)$ (mid 1950s)
$(g\text{-}2)_{theor.} = 2\ (1.001\ 145\ 4)$

The experiments at that time had been done by Kusch and coworkers,[6] the theory by Robert Karplus and Norman Kroll.[7] These numbers were soon corrected through theoretical work by Charles Sommerfield,[8] and experimental work by Peter Franken and Sidney Liebes Jr.[9] The experimental value of thirty years later is barely consistent with the earlier value, while the theoretical value is inconsistent with it.

$$(g\text{-}2)_{exp.} = 2\ (1.001\ 159\ 652\ 188 \pm 4 \times 10^{12}) \qquad \text{(late 1980s)}$$
$$(g\text{-}2)_{theory.} = 2\ (1.001\ 159\ 652\ 133 \pm 29).$$

The present experimental number comes from the "geonium" setup of Hans Dehmelt and his coworkers,[10] whereas the theoretical number is the work of Tom Kinoshita and his collaborators.[11] It will be interesting to see what the next thirty years will do to these present values.

Moreover, I think that while depth of theoretical understanding can be a useful guide to what measurements are worth pursuing, the extent to which this is the case can easily be overemphasized. There is in general no simple relation between the precision with which measurements can, have and should be done, and the accuracy of the theory that is used to understand those measurements. One of the triumphs of experimental physics was the measurement of the hyperfine structure interval in hydrogen to some twelve significant figures by Kleppner and Ramsey.[12] Even though the hydrogen atom is one of the simpler systems in nature, this measurement is about a million times more accurate than any available theory, a ratio that compares favorably with the corresponding ratios for most measurements on more complex systems than hydrogen. There is an inner logic to the development of techniques for precision measurement, and this development proceeds in a way that is basically independent of the development of theories of the associated phenomena. I believe that the subordination of measurement to theoretical understanding which has sometimes characterized the physics to the late twentieth century rcpresents a regrettable and perhaps ultimately disastrous reversal of the attitude with which modern science began in the seventeenth century, according to which experience was to be the ultimate source of all knowledge.

Indeed, I think that it can be argued, as was often done by I.I. Rabi, that many of the profoundest experimental discoveries are made when there is no theoretical guidance available about what to expect. Electromagnetism, radioactivity, superconductivity, fission, and the multiplicity of subatomic particle generations are but a few examples of experimental results that first emerged without significant input from theory. While these discoveries were all originally qualitative in nature, there are other examples, such as the quantum Hall effect, where great precision revealed that phenomena possessed unexpected features for which the theory emerged only afterwards, when it has emerged all.

The Virtues of Simple Systems

One place where theory is a useful qualitative guide to what measurements are worth pursuing to great precision is in the determination of which systems are sufficiently simple that the results obtained by measurements on them can be readily interpreted. For example, interest in the precise measurement of the hyperfine structure splitting in muonium is stimulated by the fact that muonium is as simple a two-body system as we can identify, so that its properties can be directly related to properties and interactions of the component leptons. The results for the muonium hyperfine interval allows the inference of other information of interest, whereas a measurement of similar accuracy for a lanthanum atom would not lead to such new information (at least I don't think it would!).

To the accuracy of present measurements, the hyperfine interval in muonium depends only on the masses and charges of the electron and muon.[13]

$$\Delta E_{HF,Theory} = 8/3\alpha^4 m_e{}^2/m_\mu F(m_e/m_\mu, \alpha)$$

where F $= (1 + m_e/m_\mu)^{-3}(1 + a_e + a_\mu + 1.5\alpha^2 - 3\alpha/\pi m_e/m_\mu\, ln(m_e/m_\mu) + \cdots)$, where a_e and a_μ are the g-2 values of the electron and muon.

The function F has been calculated up to terms of relative order α^3 and m_e/m_μ. At a slightly higher level of precision, which may soon be reached, the hyperfine interval also depends on the neutral-current induced weak interaction between muon and electron.[14]

At a higher level of accuracy, the theory of the muonium hyperfine interval involves properties of other particles, which are found in the loops that occur in radiative corrections. This mixing of the properties of different particles is already relevant to measurements of the muon g-value, which at the current level of accuracy is sensitive to hadronic scattering amplitudes. Since ΔE_{HF} contains a term proportional to a_μ, one can already see that the muonium hyperfine interval also involves other particles than the muon, electron and photon. However, the contributions of these other particles do not yet enter in at the present level of theoretical accuracy. Nevertheless, it is worth recognizing that at a sufficiently high level of precision the properties of all particles become entangled in the theoretical interpretation of any measurements.

This is one illustration of what someone has termed the decline of theoretical physics in the twentieth century. "In Newtonian mechanics the three body problem is insoluble; in relativistic mechanics the two body problem is insoluble; in quantum field theory the one body problem is insoluble, and in string theory the zero body problem is insoluble."

The corresponding numerical value of the muonium splitting is[13]

$$\Delta E_{HF,Th.} = 4\ 463\ 303.1 \pm 1.7\ kHz,$$

where the main error arises from uncertainty in the constants that enter into $\Delta E_{HF,Th}$.

This agrees well with the most precise measurements, done at LAMPF by the Yale-Heidelberg group[15] in 1982

$$\Delta E_{HF,Exp.} = 4\ 463\ 302.88 \pm .16\ kHz.$$

It is interesting that the muonium hyperfine splitting has been measured to better than 10^{-3} of the natural line width. In order to do this, an accurate theory of the line shape itself, under conditions in which the transition is perturbed by various time dependent external fields, is needed.[16] Methods for measuring energy differences much more accurately than the natural line width of the transition were first used long ago in atomic and molecular physics, where for the Lamb shift in hydrogen, the splitting has been measured to 1 part in 10^5 of the line width.[17] One method for making such precise measurements of energy splittings depends on having enough signal available to look at decays that occur long after the natural lifetime of the state.[5] I do not know if anyone has examined the limits of precision to which this line splitting procedure can be extended, or the accuracy of the theoretical analysis that goes into it. It is worth noting that also in the context of particle physics, there have been measurements of the mass of a state to an accuracy of 1% or less of its width. This has recently been done for the Z^0 at LEP.[18]

Simplicity of interpretation can also be found for other systems than subatomic particles. The improvement in our understanding of certain aspects of large, condensed matter systems has led to the recognition of circumstances in which the properties of such systems can also be accurately related to properties of the component particles, usually electrons. One example of this is the quantum Hall effect in certain two-dimensional structures, for which precise measurements of the Hall effect resistance has been further stimulated by the theory that precisely relates it to the electron charge and Planck's constant through the formula[19]

$$\mathrm{R} = \mathrm{h}/\mathrm{e}^2.$$

There are probably many other, yet unrecognized instances in which precisely measurable quantities involving macroscopic systems are related to simple properties of their components, and it seems likely that the search for and measurement of such quantities will be at the frontiers of future research.

Independence of the Environment

Another consideration that is relevant to whether ever greater precision should be sought for a specific physical quantity is the extent to which this quantity is determined intrinsically, as opposed to being affected by the environment of the system for which it is being measured. While physicists have regarded the speed of light in vacuum as being worth measuring with ever greater accuracy, until they finally elevated it into a unit of speed, the same has not been true about the speed of sound in air, for which the precision attained is much less. I think that one reason for this difference of emphasis, in addition to the wider role that light, as opposed to sound, plays in physical phenomena, is that the speed with which sound travels through air or any other medium ultimately depends on variable and non-reproducible features of the medium, so that extremely accurate measurements of it would amount to determining such quantities as the density of the air in a certain laboratory on a specific day of the year. While it might be argued that this measurement is also worth doing, it should be thought of less as a precision determination of the speed of sound as of a diagnostic tool for examining aspects of the laboratory environment.

The physical quantities which are most worth measuring with extreme precision are those that are insensitive to uncontrollable environmental factors, and so can be expected to give reproducible results at different times and locations. The properties of individual atoms and subatomic particles are prime examples of such quantities. The quantum Hall effect appears to share this property of being independent of the details of the material being measured.[20] However, it is worth noting that at the limits of precision now being reached for some quantities involving subatomic particles, the possible influence of the environment must at least be carefully examined. For example, both the radiative shifts of bound state energy levels and the g-values of particles can be influenced if the systems are contained in conducting cavities, through the change that such cavities induce in the allowed modes of the electromagnetic field. This effect has already been detected for atoms,[21] and eventually will become important for measurements on individual particles.

It is also worth noting that some properties of (very) large macroscopic systems, such as the rate of rotation of pulsars, are also insensitive to environmental influences and can be reproducibly measured to precisions comparable to those of the best measurements on subatomic systems.

Permanent Properties Versus Contingent Qualities

Another point relevant to what is worth measuring precisely involves the contrast between the measurement of permanent properties of systems, such as g-values, or differences of bound state energy levels, and the measurement of variable qualities, such as the momentum of a specific electron at a given time. Both our intuition and our practice suggest that it is the former measurements that are worth pursuing to extreme precision. So far as I know, nobody thinks that it is very interesting, for example, to determine with high precision the x-component of momentum of a particular free electron at one specific time. One reason for this view is that the measurements of variable qualities cannot easily be checked by other scientists and so do not fall under the usual criteria for verifiable results, at least those used by physicists. When precision measurements of variable qualities are done, it is usually as an incidental step in some further measurement, such as the determination of the mass of some particle that has decayed by emitting the electron. The measurement on the electron in question is just one bit of statistics used in the measurement that is ultimately being pursued. It is sometimes necessary to measure individual qualities very precisely in order to achieve modest precisions for the properties that are the real target of an experiment. For example, in the polarized electron experiment on which Vernon collaborated with Prescott and a group from SLAC, it was necessary to measure the scattering angles of individual electrons to a precision of 10^{-5} in order to determine the parity violating electron-quark weak interaction parameters to one significant figure.[22]

One rationale for this distinction between permanent properties and contingent qualities is that the former tend to be invariants while the latter depend on the observer. Anyone who measures the muon g-value or the muonium hyperfine interval will get the same value, whereas a component of momentum of a specific electron will depend on various contingencies of the laboratory in which the measurement is made. Physicists have not completely lost their interest in learning about features of the natural world which are objective, even though relativity and quantum mechanics have taught us that these objective features are far fewer than was previously believed.

A special example of this attitude is that physicists are usually interested in accurate measurements of quantities whose value does not change with time. In this attitude we differ somewhat from other scientists, such as astronomers, who are also interested in specific events, such as the supernova of 1987. One reason for this difference is surely that physicists rely on the capability of setting up similar situations over and again, in order to be able to take advantage of high statistics to decrease the error in their measurements. Of course, if we were interested in what happens for individual cases, this advantage would disappear.

What is a Fundamental Quantity?

In the course of preparing this talk I looked at some of the older articles on precision measurements of fundamental constants and was somewhat struck by the extent to which what is regarded as a fundamental constant has changed over time. Some indication of this change can be seen in Table 1, where I give the list of fundamental physical constants from the first article in *Reviews of Modern Physics*, published in 1929,[23] and compare it with those given in the latest such article, in *Reviews of Modern Physics* in 1987.[24] We see that for various reasons, more than half of the quantities that were regarded as fundamental constants in 1929 no longer qualify as such. Of course, there are many new quantities in the 1987 article, which I have not listed, such as the electron g-value. Nevertheless, it is clear that views of what is fundamental change over time.

There is an interesting consequence of recent developments in quantum field theory, especially of the notion of broken symmetries, and of the use of the renormalization group to calculate how quantities vary in theories without intrinsic scale parameters. Some properties that had previously been thought to be intrinsic to the subatomic particles are now thought of as contingent, both to the environment in the sense of the state of the universe, and to some details of the measurement itself, such as the energy at which it is carried out. For example, the fine structure constant α

Table I. Fundamental Constants Over Two Generations

1929 RMP[a]	1987 RMP[b]
speed of light	√
Newtonian Gravitational constant	√
Ratio of liter to cc	
Molecular volume of Ideal Gas	
Ratio of international to absolute units	
Atomic weights of light atoms	
Pressure of normal atmosphere	
Temperature of ice point	
Mechanical and Electrical Equivalent	
Faraday	√
electron charge	√
electron charge to mass ratio	(derived)
Planck's constant	√

[a] Ref. 23
[b] Ref. 24

is now considered to vary with the momentum squared of the photon, although this variation is relatively insignificant for the low energies that apply to QED precision measurements. The masses and mass ratios of particles, to the extent that they arise from spontaneous symmetry breaking, are functions of temperature and density, although again this variation is insignificant on the scale of present measurements. This variation in principle does not decrease the interest in measurements of these properties, but it does raise the question of whether there are more fundamental quantities that we should be trying to measure. It is not completely clear what fundamental intrinsic properties characterize the subatomic particles according to contemporary quantum field theory. In a theory such as massless QCD, with a single interaction and no intrinsic mass scale, neither the coupling constant nor the ratios of particle masses are fundamental quantities. Their character is more similar to such derived quantities in ordinary quantum mechanics as energy levels and atomic scattering lengths. In QCD without quark masses, the only fundamental quantity is the renormalization scale parameter, which is characteristic of the theory as a whole, rather than any of the particle states. In more complex theories, such as the SU(5) grand unified theory, there are other fundamental constants but again these constants are somewhat removed from what is directly measured.

It has sometimes been suggested that the quantities with which physics ultimately deals are dimensionless ratios, such as m_e/m_μ, or α. According to that view, it is such quantities upon which precise measurements should focus, whereas other, dimensional, quantities involve such contingencies as the specific piece of matter that has been chosen as the unit of mass, so that their measurement has less intrinsic significance. If this altitude were adopted, it would cut down the number of quantities that need to be measured precisely. Nevertheless, there are still many dimensionless ratios that can be constructed from physical quantities, such as the ratio of the heat conductivities of lanthanum, measured at two different temperatures. Some further constraints are needed if this idea is to be of much help in focusing our attention.

How Precisely Can Quantities be Measured?

Another question that is relevant to evaluating the virtues of scientific precision is whether numerical quantities can be known arbitrarily well, or whether there are definite limits in principle to the precision of such knowledge. Depending on the answer to this question we are either traveling an unending road, on which each milestone that we pass is but a record of where we have been, or we are approaching a definite goal whose achievement will mark the end of a specific part of the scientific enterprise.

Of course, the answer to the question of whether there are limits to precision in measurement varies from quantity to quantity and from theory to theory. Furthermore, the question must be considered separately for quantities whose values can vary continuously, such as the energy of a free particle, those such as angular momentum, whose value is quantized, and those which have a unique value, such as the fine structure constant. Quantum field theory, the deepest theoretical description that we presently possess, does not appear to place limits on the precision with which we can know individual physical quantities, such as a component of linear momentum, at one time, although it does restrict the precision of our simultaneous joint knowledge of several quantities represented by non-commuting operators, such as electric and magnetic field components.

One of the less celebrated, though eponymous virtues of quantum theory is that it assigns discrete, often integer values to many quantities which would vary continuously according to classical physics. Such quantization makes the measurement of such quantities to arbitrary precision considerably simpler. According to quantum mechanics the result of a measurement of such a quantity must always be one of the quantized values, not something in between. Provided that one accepts as correct this quantization of values and the usual measurement postulate, according to which the result of a measurement is always an eigenvalue of a quantum mechanical operator, it is only necessary to carry out the measurement to sufficient accuracy to distinguish between neighboring quantized values in order to know the quantity being measured to arbitrary precision.

It could be objected that the assumption of quantization is a theoretical gloss on observation, which strictly speaking should not be imposed. For example, we do not even know for sure from experiment that components of angular momentum in three dimensions are quantized. It would be interesting to carry out a high precision experimental test of this principle.

Furthermore, we do not definitely know the extent to which quantization applies in nature. There have been occasional suggestions that some quantities usually thought to be continuous, such as space-time coordinates, or the momenta of free particles are actually quantized, so that the same considerations concerning precision measurements apply to them as apply to components of angular momentum. That is, ultimately we would need only measure position or linear momentum to a precision sufficient to distinguish between neighboring quantized values. For example, if the universe is finite, with a radius approximately equal to the size of the presently observable universe, or about 10^{28} cm, then the linear momenta of photons are quantized in units of about 10^{-45} eV/c, which is of course much smaller than anything currently measurable.

On the other hand, we are learning from studies of "anyons" that for systems which can behave as if they are two dimensional the usual quantization of components of angular momentum need not apply. Such systems may contain objects which behave like particles whose angular momentum components can be any real number. Therefore, for such systems, the measurement of angular momentum components must be carried out to precision greater than integer multiples of $h/2\pi$ in order to determine their actual values. Some indication that systems of fractional angular momentum may be more than theorists' playthings has been found in analyses of the fractional quantum Hall effect and of superconductivity.[25]

How Reliable are Precision Measurements?

I mentioned previously that there have been instances where theory and precision measurements were in agreement, but both turned out to be somewhat in error. There have been many other cases in which the measured value of some quantity has varied over the years by considerably more than the estimates of error would have suggested was plausible. Here are tables of the "best" values of three fundamental quantities, h,c and α^{-1} with the associated estimates of error, as measured at different times.

It can be seen from these tables that for these quantities, there has over time been a systematic increase in the precision of the estimates and a systematic tendency to underestimate the errors. Of course the situation has been much worse for some results that do not qualify as precision measurements, such as those of the "Hubble constant," or of the original ρ parameter that occurs in muon decay, whose values have changed by many times the estimated errors of some early measurements. For example, the Hubble constant as originally measured by Hubble in 1929[26] was found to be 500 $\pm$ 60 km/sec/Mpc, whereas the present value[27] is thought to be 75 $\pm$ 25 km/sec/Mpc. For some of the quantities listed in my tables, there have been specific reasons for the changes in the value of the quantity being measured, as with the notorious error in the viscosity of air which led to Millikan's wrong estimate of e and h. In other cases the change does not seem to be the result of any specific error. Rather, there were a series of measurements over time, each of which overlapped the

Table II. The Change in Precision Measurements over Time
Planck's Constant in Units of 10^{-27} erg-sec.

Value	Date and Person
6.547 ± .011	1930–Millikan (based on measurement of e)[a]
6.625 17 ± .000 23	1955–Cohen and Dumond (review)[b]
6.626 076 ± .000 004	1991–Review of Particle Properties[c]

[a] Ref. 29
[b] Ref. 30
[c] Ref. 28

Speed of Light in Units of km/sec

Value	Date and Person
299990 ± 200	1874–Cornu[a]
299910 ± 50	1878–Michelson[a]
299901 ± 84	1900–Perriton[a]
299774 ± 6	1931–Michelson et al.[b]
299793 ± .3	1955–Cohen and Dumond (review)[c]
299792.458 ± 0	1991–definition

[a] Taken from Ref. 31
[b] Ref. 32
[c] Ref. 30

Inverse Fine Structure Constant (α^{-1})

Value	Data and Person
137.294 ± .11	1929–Birge (calculation from e,h,c)[a]
137.038 8 ± .000 6	1963–Cohen and Dumond (review)[b]
137.035 989 ± .000 006	1991–Review of Particle Properties[c]

[a] Ref. 23
[b] Ref. 30
[c] Ref. 28

Table III. History of Measurements of the Muonium Hyperfine Interval[a]

Year	Group	Result (MHz)
1961	Yale–Nevis	$5500 {+2900 \atop -1500}$
1962	Yale–Nevis	4 461.3 ± 2.0
1966	Yale–Nevis	4 463.18 ± .12
1970	Chicago	4 463.302 ± .009
1973	Chicago	4 463.304 0 ± .002
1977	Yale–Heidelberg	4 463.302 35 ± .000 52
1982	Yale–Heidelberg	4 463.302 88 ± .000 16

[a] All taken from reference 5

previous ones within the joint error estimate, but which gradually led to a new value that was well outside the earlier estimates of error.[28]

Just to show that this pattern is not universal, and that in some cases there have been a series of precision measurements for which the error estimates appear to have been realistic, I have copied a table of the history of measurements of the hyperfine frequency interval in muonium from the article[5] in Tom Kinoshita's QED book.

For a theorist perhaps the main lesson to be learned from this history is that we should be somewhat skeptical of the precise values and their associated errors that are attributed to fundamental constants at any one time, but that we can rely on the prospect that future experiments, by physicists such as Vernon who are dedicated to pushing the envelope of what can be measured, will surely produce values of considerably greater precision.

Acknowledgement

I would like to thank Dr. S. Lavine and Dr. V.P. Nair for helpful discussions on the topics discussed here.

References

1. This is quoted by R.N. Taylor in his article on Fundamental Constants in the *Encyclopedia of Physics*, 2nd edition, ed. R.G. Lerner and G.L. Trigg, (VCH Publishers, New York, 1991).
2. A.A. Michelson in the *University of Chicago Quarterly* for August 1894. The "eminent physicist" referred to by Michelson was J.C. Maxwell.
3. P. Kusch and H. Foley, *Phys. Rev.* **74**, 250 (1948).
4. J. Christenson et al., *Phys. Rev. Lett.* **13**, 138 (1964).
5. V.W. Hughes and G. zu Putlitz "Muonium" in *Quantum Electrodynamics*, ed. T. Kinoshita, (World Scientific, Singapore, 1990).

6. S. Koenig, A.G. Prodell and P. Kusch, *Phys. Rev.* **88**, 191 (1952).
7. R. Karplus and N. Kroll, *Phys. Rev.* **77**, 536 (1950).
8. C. Sommerfield, *Phys. Rev.* **107**, 328 (1957).
9. P. Franken and S. Liebes Jr., *Phys. Rev.* **104**, 1197 (1956).
10. See the article by R.S. Van Dyck Jr., et al., in *Atomic Physics 9*, ed. R.S. Van Dyck Jr. and N. Fortson, (World Scientific, Singapore, 1984).
11. T. Kinoshita in *Quantum Electrodynamics*, ed. T. Kinoshita, (World Scientific, Singapore, 1990).
12. S.B. Crampton, D. Kleppner and N.F. Ramsey, *Phys. Rev. Lett.* **11**, 338 (1963).
13. J.R. Sapirstein and D. Yennie in *Quantum Electrodynamics*, ed. T. Kinoshita, (World Scientific, Singapore, 1990).
14. M.A.B. Bég and G. Feinberg, *Phys. Rev. Lett.* **33**, 606 (1974).
15. F.G. Mariam et al., *Phys. Rev. Lett.* **49**, 993 (1982).
16. See ref. 5 for such a discussion concerning muonium.
17. See e.g., S.R. Lundeen and F.M. Pipkin, *Phys. Rev. Lett.* **46**, 232 (1981).
18. See e.g., Opal Collaboration, *Phys. Lett.* **B235**, 379 (1990).
19. See e.g., M. Buttiker, *Phys. Rev.* **B38**, 9375 (1988).
20. A. Hartland et al., *Phys. Rev. Lett.* **66**, 969 (1991).
21. R. Hulet et al., *Phys. Rev. Lett.* **55**, 2137 (1985).
22. C.Y. Prescott et al., *Phys. Lett.* **77B**, 347 (1978).
23. R. Birge, *Rev. Mod. Phys.* **1**, 1 (1929).
24. E.R. Cohen and B.W. Taylor, *Rev. Mod. Phys.* **59**, 1121 (1987).
25. See e.g., R.B. Laughlin, *Science* **242**, 525 (1988).
26. Edwin Hubble, *Proc. Nat. Acad. Sci.* **15**, 168 (1929).
27. Various methods of determining the Hubble constant, giving a range of values between 50–100 km/sec/Mpc arc discussed in *The Cosmological Distance Ladder* by M. Rowan-Robinson, (W.H. Freeman, New York, 1984).
28. For some examples of this, see the introduction to the Review of Particle Properties *Phys. Lett.* **B239**, 1 (1991).
29. R.A. Millikan, *Phys. Rev.* **35**, 1231 (1930).
30. E.R. Cohen et al., *Rev. Mod. Phys.* **27**, 363 (1955).
31. M. Gheury De Bray, *Nature* **120**, 602 (1927).
32. A.A. Michelson et al., *Ap. J.* **82**, 26 (1935).

Polarized Electrons

W. Raith
University of Bielefeld, Bielefeld, Germany

Polarized electrons have played a major role in Vernon Hughes' research for many years. First I will give you some background information on the state-of-the-art at the time when the Yale program was started and then describe how our work progressed, and what came out of it, directly and indirectly. Covered will be the research in which I or, later, people from Bielefeld participated. For the parity experiments with polarized electrons I refer to the talk of Charles Prescott.

The electron is a point-like elementary particle. Its main properties are the electric charge and the mass. Both properties are isotropic; they do not depend on direction. But just like the spinning earth has its North Pole, the electron has its spin which defines a certain direction and its magnetic moment pointing the other way. An ensemble of electrons is called "polarized" if the spins of those electrons point preferentially in one direction.

The electron was discovered at the end of last century in experiments with free particles. For decades only the isotropic properties charge and mass were known. In 1925 spin and magnetic moment were introduced in order to explain anomalies of atomic spectra. The history of this discovery is typical for the electron: As a free particle, its behavior is completely dominated by the interaction of its charge with electric and magnetic fields. Only for atomic electrons, which are charge-neutralized inside the atom, the effects related to spin and magnetic moment are easily observable. Therefore, it is expected to be difficult to investigate the anisotropic interactions of free electrons and to produce spin-polarized electron beams for that purpose.

Around 1960 only two ways of getting free polarized electrons were known: Mott scattering and β^- decay. Neither yielded beams useful for other experiments. In Mott scattering the electron is scattered through a large angle in the Coulomb field of a heavy nucleus. If the incident electrons are polarized perpendicular to the scattering plane, two detectors placed at scattering angles of $+\Theta$ and $-\Theta$ measure a "left-right" intensity asymmetry; if the incident electrons are unpolarized, the scattering electrons have a transverse polarization with opposite directions on the right and left side. Mott scattering is a good way of measuring the polarization of an electron beam, but the polarized beams produced this way are low in current and not very high in polarization. The same is true for the β^- decay electrons which are polarized because of the parity violation in weak interactions. Also, those electrons have an energy distribution with a width of several 100 keV which further reduces their usefulness.

More promising is it to begin with bound polarized electrons. There have been numerous proposals, starting as early as 1930, and many experimental attempts, most of them unpublished. Atomic electrons in a beam or a gas of paramagnetic atoms can be polarized by magnetic state selection or by optical pumping. One has to find a way of setting those electrons free, e.g., by photoionization or impact ionization, or of transferring the bound-electron polarization to free electrons by exchange. Ferromagnets are the only reservoirs of bound polarized electrons in nature. Some of the atomic electrons in a ferromagnet are polarized and their magnetic moments are responsible for the strong magnetic field. Of course, it is an intriguing idea to get those electrons out, somehow.

By now all of the above mentioned schemes have been tried and most of them yielded polarized electrons of measurable current and polarization. Up to 1960, however, none of those attempts had been successful. At that time optical pumping of atoms was still done with inefficient spectral lamps since tunable lasers did not yet exist. Ultra-high-vacuum physics was in it's infancy and, therefore, experiments with ferromagnets suffered from unclean surfaces. In 1961 H. Friedmann (Munich) reported a successful polarized-electron production. He polarized a potassium atomic beam by means of a Stern-Gerlach magnet, photoionized these atoms and performed Mott scattering with the ejected electrons. Unfortunately, soon thereafter he was killed in an accident. His work was not continued.

In 1963 Vernon decided to start the development of a polarized electron source specifically intended for high-energy electron-scattering experiments and hired me for that program. The resulting polarized electron source was later called PEGGY: PEG for Polarized Electron Gun and GY for femalization.

Time scale

1963-68 Work on the Yale prototype source[1,2] with R.L. Long, Jr., M.S. Lubell and M. Posner.

1970-73 Construction of PEGGY[3] and further developmental work with M.J. Alguard, G. Baum, J.E. Clendenin, J.S. Ladish, M.S. Lubell, and K.P. Schuler, first at Yale, later at the Stanford Linear Accelerator Center (SLAC).

1974-81 SLAC-PEGGY experiments with polarized targets by the Bielefeld-SLAC-Tsukuba-Yale collaboration[4–9] (groups in alphabetical order).

There was a 2-3 year time gap between the completion of the prototype source and the start of PEGGY construction. During that time Vernon tried to sell the idea of experiments with polarized electrons to the high-energy community. Fortunately for us the now-famous deep-inelastic scattering with unpolarized electrons evolved at

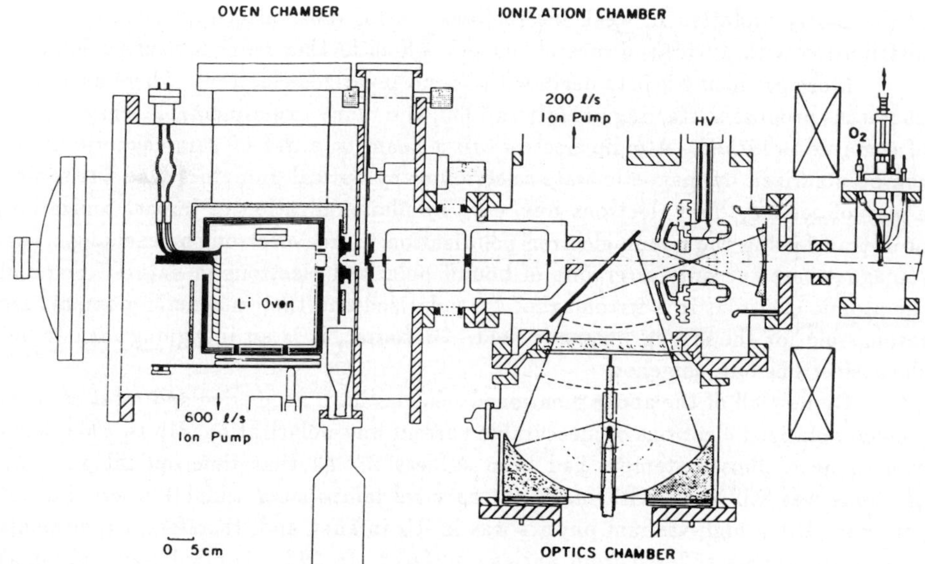

Figure 1: PEGGY cross section (from Ref. 3).

that time and led to interesting ideas for polarized-electron experiments. When our first high-energy experiment was approved we begin the construction of PEGGY at Yale. In 1972 Günter Baum and I left Yale for Bielefeld and that was the beginning of an international collaboration which was later joined by a group from Tsukuba.

A cross section of PEGGY is shown in Fig. 1. The major components (from left to right) are ^{6}Li atomic beam oven, permanent hexapole magnets, vortex-stabilized high-pressure flash lamp, uv optics, high voltage terminal, magnetic-field coil, electron optics.

After optimization of all components PEGGY yielded a 70 keV beam of up to 180 pulses per second with 10^9 electrons per pulse, a polarization of 85%, very stable operation and reasonably short "down times" for servicing.

PEGGY was the first polarized electron source developed for an experiment not related to the physics of the source itself. In the following years other sources were developed, most notably the GaAs source, now used almost exclusively, which was invented in 1975 in the group of H.C. Siegmann (Zürich). However, none of the later sources would have been better suited than PEGGY for our high-energy polarized-target experiments. Polarized targets suffer radiation damage from the electron beam and must be annealed repeatedly. We worked with 120 min of data taking followed by 45 min of target annealing. PEGGY delivered all the current the target could stand with a polarization twice as high as that of electron beams from GaAs sources.

Photo 1. The Stanford Linear Accelerator Center (SLAC). The big building in the foreground is Endstation A where our polarized proton target and the electron spectrometers were located.

Photo 1 shows the Stanford Linear Accelerator Center at that time. The high-energy end of the accelerator is a busy high-tech environment with many visiting scientists working there. PEGGY, however, had to go to the low-energy end, located two miles away (Photo 2) where not much action had been before we came. Photo 3 shows the real PEGGY, connected to the SLAC beam line, with two of her friends; the person behind Vernon is Günter Baum, now professor at Bielefeld.

Photo 2. The beginning of the 2-mile accelerator.

Photo 3. PEGGY installed at SLAC.

After our experiment on GeV Møller scattering[4] had shown that the electron polarization survives the acceleration, the planned experiments with the polarized-proton target began. Here I cannot describe the polarized-target technology. It is about as complicated as the polarized-source technology, involving very low temperatures, high magnetic fields, powerful microwaves and a tricky preparation of the target material.

The results of the deep-inelastic e-p scattering can be expressed as a function $A_1(\omega)$ with details shown in Eqns. 1-4.

$$A = \frac{d\sigma(\downarrow\uparrow) - d\sigma(\uparrow\uparrow)}{d\sigma(\downarrow\uparrow) + d\sigma(\downarrow\downarrow)} \tag{1}$$

$$A = D(A_1 + \eta_2) \approx DA_1 \tag{2}$$

$$\omega = \frac{2M_p(E - E\prime)}{Q^2} \tag{3}$$

$$x = \frac{1}{\omega} = \frac{quark\ momentum}{proton\ momentum} \tag{4}$$

After some data reduction we obtained A, the asymmetry for fully polarized particles. The parameter of theoretical interest is A_1, relating to the virtual photon. D and η are parameters known from the scattering kinematics. The Bjorken scaling parameter ω is given by the proton mass, M_p, the difference of the electron energy before and after the collision, E and $E\prime$, and the square of the four-momentum transfer, Q.

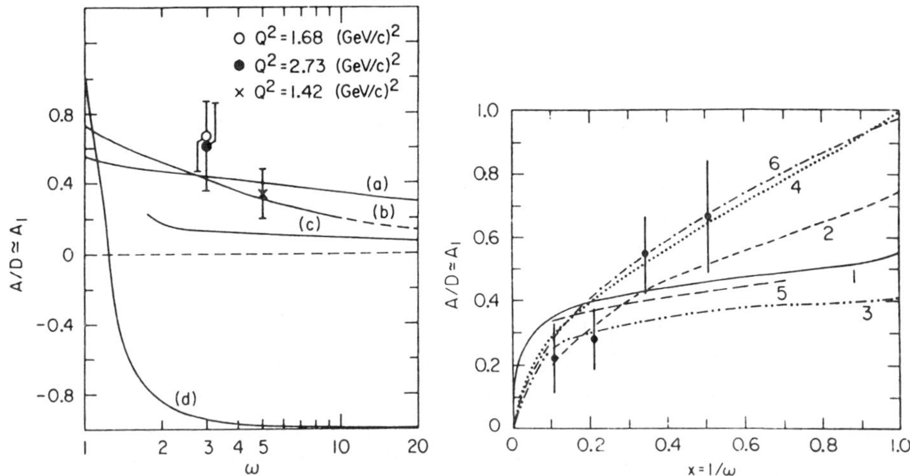

Figure 2: Experimental results compared with various theoretical models (from Ref. 6 and 7).

The results were compared with all the models available (Figure 2). Our 1976 paper [7] disqualified already two models, the 1978 paper [8] some more. Note that by that time it had become customary to plot versus the inverse scaling parameter, x, instead of ω.

Let me go back a few years: For getting the first high-energy experiment with polarized electrons and polarized target approved, a 1970 publication of Bjorken (Figure 3) was very helpful.

He recalled his 1966 sum rule which he himself had considered "worthless" because he had thought that verification was impossible. In this paper he estimated raw asymmetries on the order of 1% for polarization experiments with polarized electrons or muons. On the basis of the rather general assumption of quark structure for the hadronic electromagnetic and weak currents he had shown that the difference of the spin-structure function $g_1(x)$ for the proton and the neutron, integrated over the whole range of x can be related to the coupling constants of the β decay:

$$I_p - I_n = \int_0^1 [g_1^p(x) - g_1^n(x)]\, dx = \frac{1}{6} \mid \frac{g_A}{g_V} \mid (1 - \alpha_s/\pi) \tag{5}$$

Here I_p and I_n are abbreviations for the integral of the spin-structure function $g_1(x)$ over the whole range of x for the proton and neutron, respectively. On the right-hand side g_A and g_V are the axial-vector and vector coupling constants of neutron β decay. The QCD radiative corrections for finite Q^2 are represented by $\alpha_s[= 0.27(2)$ for $Q^2 = 10.7\ GeV^2]$. The value of the right hand side of Eq. 5 is well known:

$$I_p - I_n = 0.191\ (2) \tag{6}$$

PHYSICAL REVIEW D VOLUME 1, NUMBER 5 1 MARCH 1970

Inelastic Scattering of Polarized Leptons from Polarized Nucleons*

J. D. BJORKEN
Stanford Linear Accelerator Center, Stanford University, Stanford, California 94305

(Received 28 October 1969)

A previously derived sum rule, based on $U(6)\otimes U(6)$ equal-time commutation relations for the space components of the electromagnetic current, implies mean polarization asymmetries of greater than 20% throughout most of the inelastic continuum.

I. INTRODUCTION

SOME time ago, a high-energy sum rule involving electromagnetic scattering of longitudinally polarized leptons from polarized protons and neutrons was derived[1] and then dismissed as "worthless." However, it turns out to be interesting to reconsider that negative conclusion in light of the present experimental and theoretical situation.[2,3] We find that given "naive" quark-model equal-time commutation relations for the space components of the electromagnetic current[4] and reasonable estimates for the convergence of the related sum rule, there must be parallel-antiparallel asymmetry effects of greater than 20% over a large region of the "deep inelastic" continuum. It appears that the relevant experiments with electrons—or even muons—may be feasible.

. . . .

It appears to be possible to produce electron[13] or muon polarized beams which have nearly 100% longitudinal polarization. Polarized targets of ~4% polarization per nucleon are at present in use.[14] Therefore, nearly 1% raw asymmetries are predicted; this may well be within range of muon scattering as well as electron scattering experiments in the future.

[13] V. Hughes, M. Lubell, M. Posner, and W. Raith, in *Proceedings of the Sixth International Conference on High-Energy Accelerators* (unpublished).

[14] O. Chamberlain (private communication); M. Borghini, S. Mango, O. Runoffsson, and J. Vermeulen, in *Proceedings of the International Conference on Polarized Targets and Ion Sources, Saclay, France, 1966* (Centre d'Etudes Nucléaires de Saclay, Gif-sur-Yvette. France, 1967).

Figure 3: Title and text portions of stimulating publication.

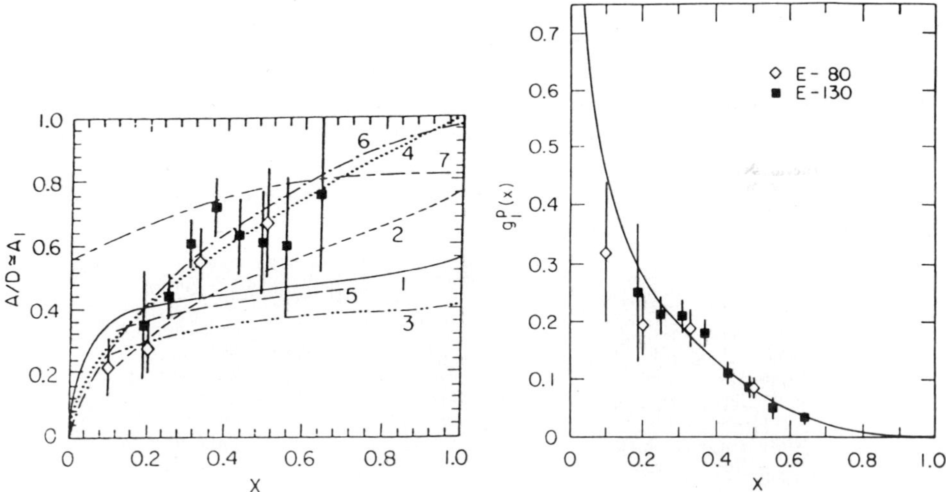

Figure 4: More results and a test of the sum rule (from Ref.9).

The left-hand side can be determined in polarization experiments. The spin-structure function for the proton (and analogously for the neutron) $g_1^P(x)$ is a product of several functions:

$$g_1^P(x) = \frac{A_1^P(x)F_2^P(x)}{2x(1+R)}. \tag{7}$$

$F_2^P(x)$ is a structure function known from unpolarized deep-inelastic experiments, x is a kinematical parameter and R is the ratio of longitudinal and transverse photo-absorption cross sections knows from other experiments. The spin-structure information is contained in $A_1^P(x)$, the asymmetry of Eq. 2.

However, for the Bjorken sum rule one needs also information about the spin structure of the neutron. In 1974 Ellis and Jaffe showed that with more specific assumptions about the nucleons the Bjorken sum rule can be broken up in one for the proton and one for the neutron.

$$I_P = \frac{1}{12} \mid \frac{g_A}{g_V} \mid \left[+(1-\frac{\alpha_s}{\pi}) + (Ellis - Jaffe\ term)\right] = 0.189(5) \tag{8}$$

$$I_n = \frac{1}{12} \mid \frac{g_A}{g_V} \mid \left[-(1-\frac{\alpha_s}{\pi}) + (Ellis - Jaffe\ term)\right] \approx 0 \tag{9}$$

The numerical value they obtained for the proton is practically equal to the Bjorken value, leaving something very close to zero for the neutron. In 1983 we published more A_1 results (Figure 4) and disqualified another theory (here #7). And we tried to check the sum rule. On the left diagram it looks as if more A_1-values for high x are needed in order to evaluate the integral. However, on the right one sees that the $g_1(x)$-values at high x are very small (the factor F_2/x brings them down). Actually needed are more A_1 values at <u>small</u> x. We made an extrapolation which appeared to be reasonable at that time and obtained a first value for the I_P which was consistent with the Ellis-Jaffe sum rule within the error margin.

The data shown in Figure 4 were the last ones which we obtained at SLAC. Since it is very difficult to get all participants together for an end-of-data-taking party, we celebrated many partial successes on the way with whoever happen to be present at the time. Photo 4 taken at such an occasion.

Photo 4. A success is being celebrated.

It was obvious, that we were right at the threshold of obtaining results of fundamental importance. Therefore, we proposed to take more and better proton data and to use a polarized deuteron target for obtaining first information on the spin structure of the neutron. An we had some ideas how PEGGY could be improved further. Unfortunately, at that time SLAC was determined to push the collider development with highest priority and rejected our proposal. Fortunately, that was not the very end of the story. But before I describe the next chapter, I will take you on a brief detour through atomic physics.

Spin-Offs in Atomic Physics

While we were waiting for approval and funds to construct PEGGY, we started a low-energy experiment on the scattering of polarized electrons ($e\uparrow$) from polarized hydrogen atoms ($H\uparrow$). In 1972, when Günter Baum and I went to Bielefeld, we utilized the Li beam technology which we had developed for PEGGY in order to perform low-energy $e\uparrow -Li\uparrow$ experiments. When Mike Lubell went to City College in 1981 he started a new, more elaborate experiment on $e\uparrow -H\uparrow$ scattering. Of

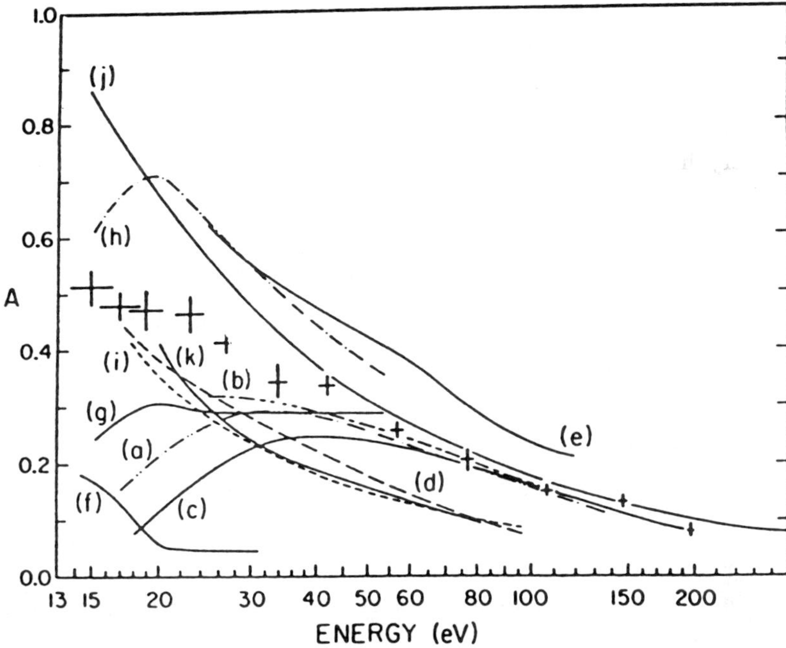

Figure 5: Ionization asymmetry (in the text called A_I) for $e \uparrow H \uparrow$ impact ionization (from Ref. 10)

the many results obtained with polarized electrons and polarized atoms I want to mention only three examples: One from Yale, one from City College, and one from Bielefeld, - all three related to impact ionization.

The Yale experiment of 1977 was the first "polarized-electron, polarized atom experiment" in atomic physics.[10] The result demonstrated that polarization experiments can provide new, unexpected information.

The lines labeled 'a' to 'k' are the predictions for the ionization asymmetry, A_I, as function of energy, E, of various theoretical approximations which deviate significantly in their A_I prediction but describe the cross sections for unpolarized electrons (not shown here) fairly well. Impact ionization is theoretically a very complicated process with three-body Coulomb interaction after the collision. All these theories contain phase factors determined by "judicious guessing".

Another interesting question is that of threshold behavior: Does $A_I(E)$ exhibit structure near threshold if the experimental energy resolution is adequate? According to the Wannier threshold theory the asymmetry should smoothly approach some threshold value. The Wannier theory had often been confirmed experimentally and is generally accepted. Only Aaron Temkin advocates a competing theory which predicts undulations in cross section <u>and</u> asymmetry as function of energy. The energy range

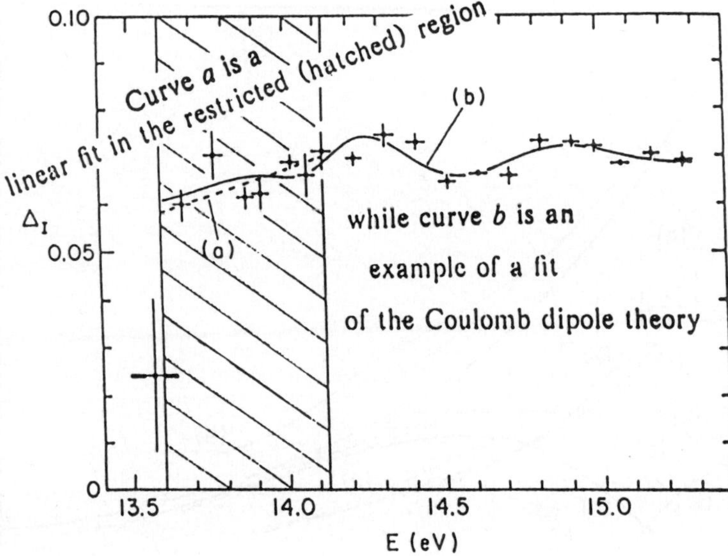

Figure 6: Values for the ionization asymmetry (in the text called A_I) near threshold (adapted from Ref. 11).

in which threshold laws are expected to hold decreases with increasing atomic number. Thus hydrogen is well suited for threshold studies. Figure 6 shows what Lubell and coworkers measured last year.[11] Lubell's result clearly favors Temkin over Wannier.

So far we considered angle-integrated ("total") ionization cross sections. The most crucial test of theory, however, is the so-called "triple-differential" cross section in which the directions of both outgoing electrons and the energy partition between them is measured. Typical is a polar diagram as shown in Figure 7. If one keeps the direction of electron "a" and the energy partition between the electrons "a" and "b" fixed, then the emission probability of electron "b" is described by such a polar diagram, exhibiting a "binary" peak (where the electron behaves like one with negligible binding) and a "recoil" peak (where the electron goes around the ion before taking off). We know from the 1977 Yale result that the angle-integrated asymmetry is substantial, about 0.5 near threshold. However, no theory predicted the triple-differential asymmetry. In Bielefeld we built an experiment where both outgoing electrons are energy analyzed and detected in coincidence (Figure 8). This is the first triple-differential ionization experiment with polarized incoming particles!

Figure 9 shows our first relative cross-section and asymmetry measurements which are just a few weeks old.[12] New calculations, which were stimulated by the knowledge about our experiment, have become available for comparison.

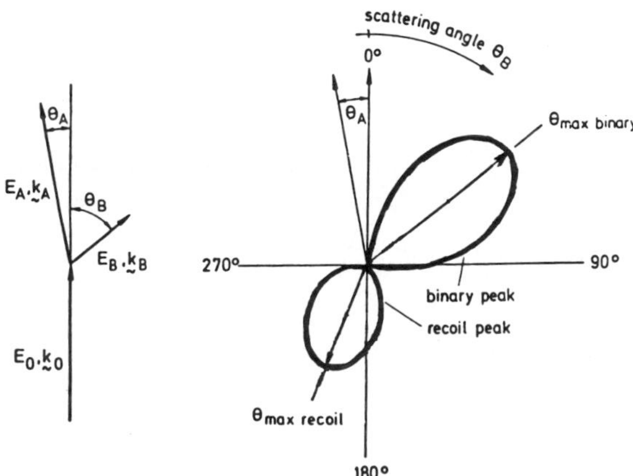

Figure 7: A polar diagram for low-energy triple-differential impact ionization.

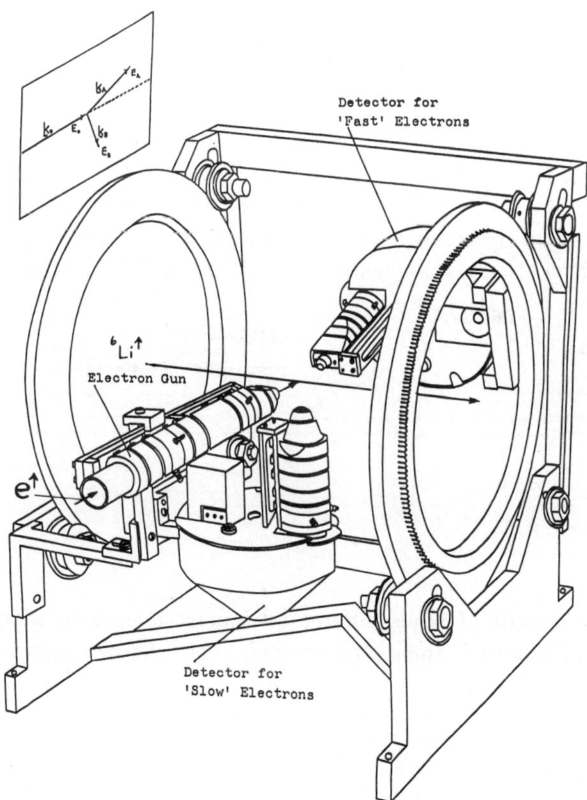

Figure 8: Bielefeld set-up for $e \uparrow Li \uparrow$ triple-differential ionization.

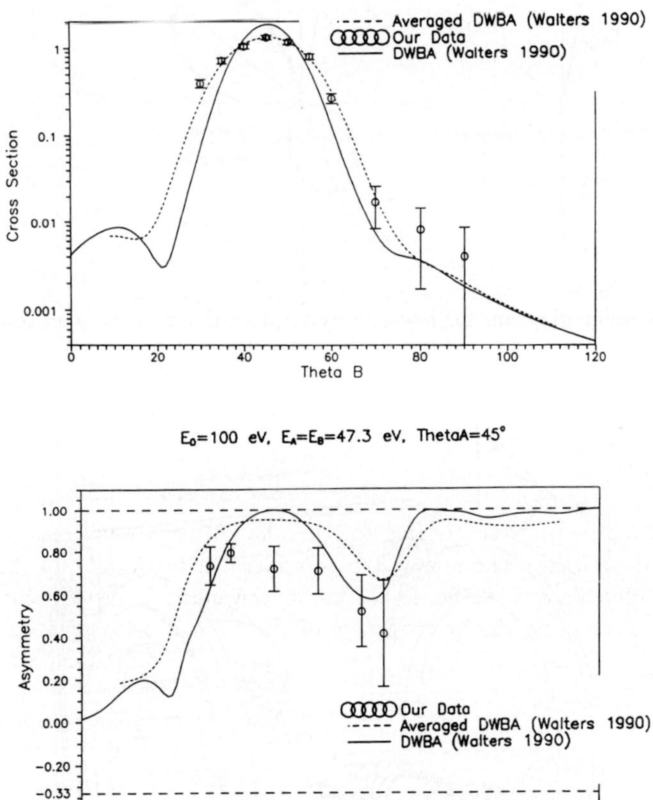

Figure 9: First results of triple-differential impact ionization with polarized particles and comparison with theory (from Ref. 12). Above: Cross Sections, Below: Asymmetries.

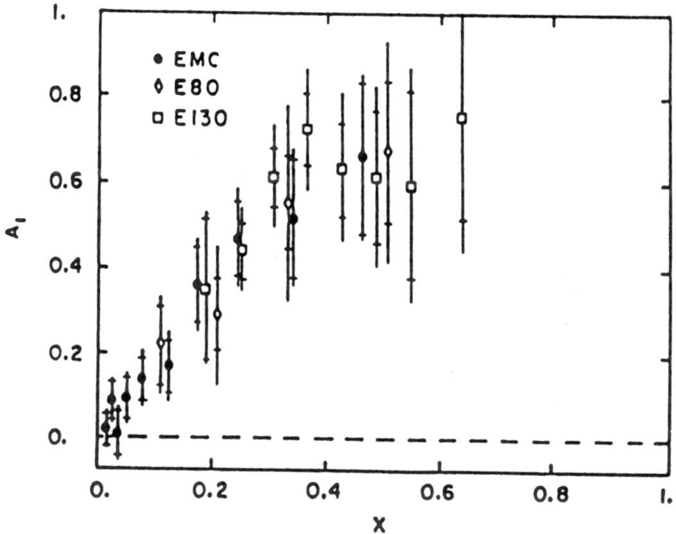

Figure 10: SLAC and CERN data for testing the Ellis-Jaffe sum rule (from Ref. 13).

Spin-Offs in High-Energy Physics

Now I am coming to the second chapter of the story about polarization experiments on deep-inelastic scattering. Bjorken had already envisioned polarization experiments with muons. The muons are polarized "at birth" due to the parity violation in the pion decay. As far as we know, the muon is just a heavy electron, well suited for continuing the investigation of the nuclear spin structure, which we could not finish at SLAC. At CERN the "European Muon Collaboration" (EMC) had performed experiments with unpolarized targets and planned to do polarization experiment. For the EMC polarization experiments people from Yale and Bielefeld joined this collaboration.

The EMC work[13,14] provided the asymmetry values for small x which were needed for a more crucial check of the sum rule (Figure 10). The sum-rule result derived from those combined SLAC-EMC data is inconsistent with the Ellis-Jaffe sum rule:

$$\begin{aligned} I_p^{exp} &= 0.126 \pm 0.010 \pm 0.015 \\ &\leq 0.151 \end{aligned} \tag{10}$$

$$\begin{aligned} I_p^{theor} &= 0.189\ (5) \\ &\geq 0.184 \end{aligned} \tag{11}$$

The theoretical consequences are considerable.

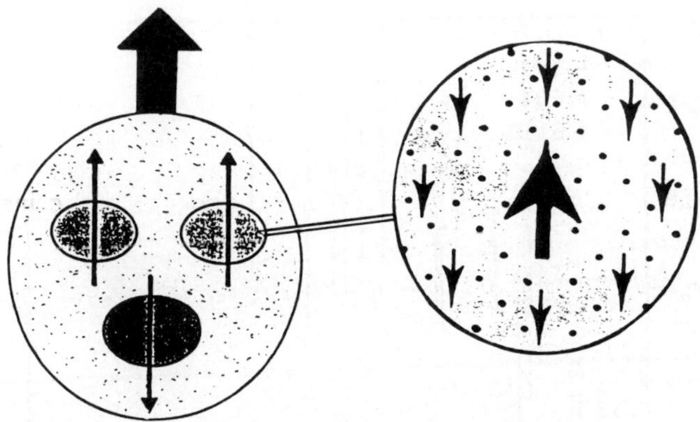

Figure 11: On the "spin crisis". The left picture shows a proton with its spin determined by the three valence quarks. The right picture illustrates the presumption that such a valence quark might have a net spin of zero because of the spin compensation by the polarized sea of quark-antiquark pairs.

Figure 11 shows two drawings adapted from popular theoretical articles.[15,16] In the old picture the proton consisted of three valence quarks surrounded by quark-antiquark pairs and gluons. It was firmly believed that the proton spin resulted from the valence-quark spin configuration. Now the valence quarks do not seem to contribute to the proton spin at all. The valence-quark spin could be compensated by the spin polarization of the surrounding quark-antiquark pairs as shown on the right of Figure 11. But where does the proton spin come from? From quark orbital angular momentum? Or from he gluons? The whole discussion has been called a "spin crisis". For the experimentalists whose work contributed to this result it is a most elating experience to see their theoretical colleagues so much disturbed by it. To confirm an existing theory is much less gratifying.

The remarkable sum-rule result of 1984/85 made measurements of the neutron spin structure a goal of highest priority. Is the Ellis-Jaffe sum rule for the neutron also violated, as one would assume if the more general Bjorken sum rule holds? If, however, the Bjorken sum rule were also violated, the theoretical consequences would be even more dramatic.

There are now two experiments approved and scheduled for neutron measurement.[17,18]

- CERN SMC (Spin Muon Collaboration) with a polarized deuteron target, a collaboration of about 130 people from 20 institutions put together by its spokesman Vernon Hughes,
- and SLAC experiment E-142 with polarized electrons and a polarized helium-3 target coordinated by Emlyn Hughes, scheduled for data-taking next year.

The SMC measurements with a polarized deuteron target are scheduled to begin now. The Bielefeld group is in charge of preparing the target material. Photo 5 shows how Gúnter Baum and his student are preparing little beads of paramagnetically doped deuterated butanol. Gúnter Baum sends his regards and congratulations from CERN. He would like to be here and celebrate with us, but they had to produce 400 000 beads for the CERN target in order to get the experiment going.

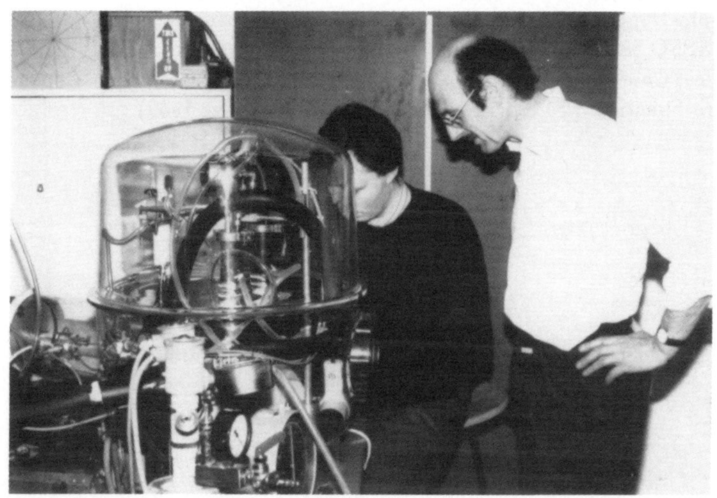

Photo 5: Preparation of the target material at Bielefeld.

References

1. R. L. Long, Jr., W. Raith and V. W. Hughes, *Phys. Rev. Lett.* **15**, 1 (1965).
2. V. W. Hughes, R. L. Long, Jr., M. S.Lubell, M. Posner and W. Raith, *Phys. Rev.* **A5**, 195 (1972).
3. M. J. Alguard, J. E. Clendenin, R. D. Ehrlich, V. W. Hughes, J. S. Ladish, M. S. Lubell, K. P. Schúler, G. Baum, W. Raith, R. H. Miller and W. Lysenko, *Nucl. Instr. & Meth.* **163**, 29 (1979).
4. P. S. Cooper et al., *Phys. Rev. Lett.* **34**, 1589 (1975).
5. M. J. Alguard et al., *Phys. Rev. Lett.* **37**, 1258 (1976).
6. M. J. Alguard et al., *Phys. Rev. Lett.* **37**, 1261 (1976).
7. M. J. Alguard et al., *Phys. Rev. Lett.* **41**, 70 (1978).
8. G. Baum et al., *Phys. Rev. Lett.* **45**, 2000 (1980).
9. G. Baum et al., *Phys. Rev. Lett.* **51**, 1135 (1983).
10. M. J. Alguard, V. W. Hughes, M. S. Lubell and P. F. Wainwright, *Phys. Rev. Lett.* **39**, 334 (1977).
11. X. Q. Guo et al., *Phys. Rev. Lett.* **65**, 1857 (1990).

12. L. Frost, G. Baum, W. Blask, P. Freienstein, S. Hesse and W. Raith, Abstract *XVII ICPEAC*, Brisbane, 1991.
13. V. W. Hughes, V. Papavassiliou, R. Piegaia, K. P. Schúler, and G. Baum, *Phys. Lett.* **B212**, 511 (1988).
14. J. Ashman et al., em Nucl. Phys. **B328**, 1 (1989).
15. F. Close, *Physics World* **3**, 29 (1989).
16. H. Fritzsch, *Phys. Bl.* **46**, 395 (1990).
17. V. W. Hughes, Spokesman of the Spin Muon Collaboration (SMC), "Measurement of the Spin-Dependent Structure Functions of the Neutron and the Proton", CERN/SPSC 88-47 SPSCP242 (22 December 1988).
18. E. Hughes, Coordinator, "A Proposal to Measure the Neutron Spin Dependent Structure Function", SLAC Proposal E-142 (13 October, 1989).

Fundamental Tests with Polarized Beams at High Energies

C. Y. Prescott
Stanford Linear Accelerator Center
Stanford University, Stanford, California 94309

Abstract

In 1970 Vernon Hughes proposed accelerating polarized electrons produced at low energies with a polarized electron source to high energies in the Stanford Linear Accelerator Center's linac. From this initial proposal, a broad program of fundamental physics grew, and continues to this day at SLAC and other laboratories around the world. The first measurements of proton spin structure and the demonstration of a parity non-conservation in inelastic electron scattering were major results. This talk looks at the physics resulting from that proposal, the history and sequence of events surrounding the experimental activities, and the technical developments that made these experiments possible.

Introduction

In the late 1960's experiments at SLAC in inelastic scattering of electrons from protons and from other nuclei, primarily deuterium, demonstrated a surprisingly large cross section and a feature of "scaling" which pointed toward point-like constituents within the proton. In the intervening time between those experiments and the present, we have developed a solid understanding of the nature of hadronic matter based on quark-gluon substructure, now embodied in the Standard Model. But in 1970, no such clear understanding existed. Vernon brought to SLAC a proposal to accelerate polarized electrons to high energies and to scatter them from polarized protons in the regions of kinematics probing deeply into the interior of the proton. Arguments for such experiments, both unpolarized and polarized, were strongly voiced by Feynman[1] and Bjorken[2] who were deeply involved in the theoretical and physical implications. Their published works developed the formal basis for theoretical ideas. In Bjorken's talks about the subject he discussed the concept of scaling and the implication for point-like constituents in the proton. With the advent of scaling behavior, demonstrated experimentally in the SLAC-MIT experiments[3], a physical interpretation was developed and advertised by Feynman[4]. The parton model, as Feynman's picture was called, explained deep inelastic results as an incoherent sum of scattering from constituent quarks. The quarks carry a fraction x of the total proton momentum, and the scattering of electrons, mediated by exchange of a virtual photon, is given by a sum over the probability distribution for the quark to have fractional momentum x, weighted by the quark charge squared. A particular version of the parton model was developed by Kuti and Weisskopf describing the parton as a composite object of three "valence" quarks and a "core" of $q\bar{q}$ pairs[5]. The parton model was not the only model used in attempting to explain the surprising results of the SLAC-MIT experiments. The vector-dominance model was one with considerable following. Resonance models and diffraction models were also concocted to explain the deep inelastic results.

The E80 Experiment

It was in this atmosphere of excitement, debate, and growing understanding of the structure of hadrons that Vernon came to SLAC in 1970 with his proposal to look at the spin-dependent part of the deep inelastic scattering process. The proposal was specifically to test for the asymmetries expected in the quark-parton models of the nucleon. The proposal consisted of three parts. The first was the development of a polarized electron gun for the injector section of the SLAC linac. This part required engineering a suitable device for linac injector operations based on principles demonstrated in a prototype device at Yale University. Secondly, he and his collaborators proposed to build a polarized target using techniques that were known to work at that time. This target would need to handle high currents from the electron beam in order to conduct the experiments within reasonable times. Thirdly, use of the existing end station spectrometers and counting house facilities was proposed, minimizing the need to develop the experimental detectors. Even with the use of existing facilities, the amount of work required to develop both a source and a target would strain the resources of SLAC and the Yale group. The experiment was given a name, E80, and was approved by the Program Advisory Committee at SLAC in June 1971.

The physics asymmetry called for in the theoretical discussions is defined as

$$A_1 = \frac{\tilde{\sigma}_{1/2} - \tilde{\sigma}_{3/2}}{\tilde{\sigma}_{1/2} + \tilde{\sigma}_{3/2}}$$

where $\tilde{\sigma}_{1/2}(\tilde{\sigma}_{3/2})$ refers to the cross section for virtual photons with longitudinal spin to be absorbed on a target with spin anti-parallel (parallel) to the incoming spin of the photon. This quantity is related to the experimental asymmetry

$$A_{exp} = \frac{\sigma_{1/2} - \sigma_{3/2}}{\sigma_{1/2} + \sigma_{3/2}}$$

where $\sigma_{1/2}(\sigma_{3/2})$ refers now to the cross section for scattering of polarized *electrons* on polarized protons with beam and target spins anti-parallel (parallel). It is related to A_1 by a QED factor D. D, the depolarization factor, gives the polarization for a

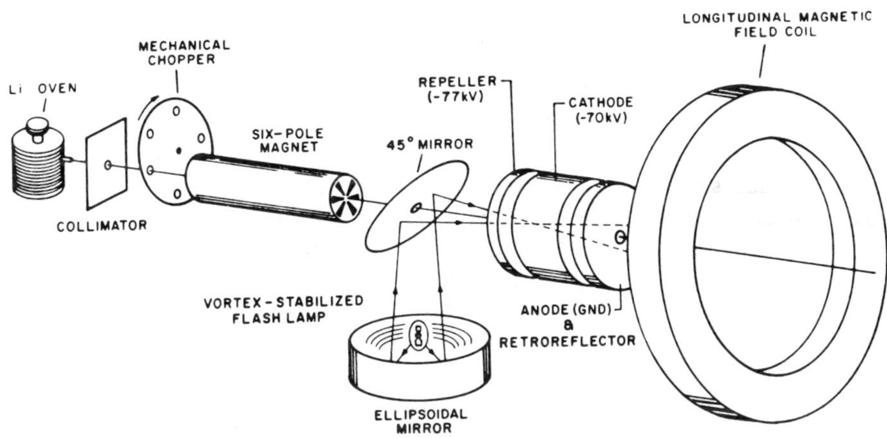

Figure 1: A schematic diagram of the PEGGY source showing the principal components: the atomic beam; the sextupole magnet; the ultraviolet optics; and the ionization region and electron optics.

vitual photon emanating from a 100% polarized incoming electron. The raw asymmetry measured in the experiment is given by $\Delta = P_e P_p f A_{exp}$ where P_e is the beam polarization, P_p is the free proton polarization, and f is the fraction of target nucleons which are polarized protons. The raw asymmetry has a value in the neighborhood of 0.01. The large asymmetries predicted for the physics turn into small effects in real experimental conditions, so great care would have to be taken to achieve statistical accuracies whilc maintaining control of all systematic effects.

The polarized electron gun (PEGGY) was developed by SLAC and Yale physicists and technical staff over a period of several years. It is shown schematically in Figure 1.[6] It operated on atomic physics principles, photoionization of state-selected ^{6}Li. The beam of Li was obtained from a Li oven, and passed through a sextupole magnet with geometry chosen so that one spin state was focussed and transmitted, while the opposite spin state was defocussed and lost on the walls and baffles. Passing into a region of longitudinal magnetic field and an electrostatic potential of -70 KeV, the neutral atoms were stripped of electrons by ultraviolet light from an argon flash lamp. The ejected electrons retained the high polarization (approximately 80-90%) and were directed by transport elements into the injector of the linac. Polarization of the electrons could be flipped by reversing the magnetic field. The source operated at 1 to 2 $\times 10^9$ electrons per 1.6 microsecond long linac pulse, at a rate of 180 pulses per second. The source achieved the highest intensities of any source available at the time.

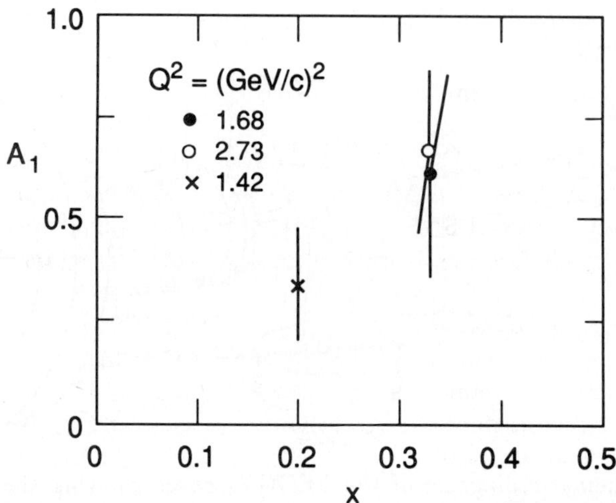

Figure 2: The derived physics asymmetries A_1 from E80 data, showing large positive values consistent with quark-parton predictions.

The polarized target consisted of a hydrocarbon material, 1-butanol, doped to about 5% with water saturated with porphyrexide. The target material was imbedded in a 5 Tesla magnetic field at 1 degree Kelvin and was pumped with radio frequency power at 140 GHz. Polarization of the free proton polarization reached 70%. The material suffered from radiation damage. The radiation dose of 4×10^{14} electrons per cm^2 reduced the polarization to $1/e$ of its maximum value. Annealing of the target material occurred when the target was warmed up to 130 degrees Celsius. Recooling and repolarizing to the maximum value was possible in a relatively short time. The target was annealed a number of times during the course of the experiment.

Figure 2 shows the results of E80. The large and positive asymmetries predicted by Bjorken and by Kuti and Weisskopf were first seen in E80. These results were published in August 1976.[7] By the time of this publication, the original deep inelastic scattering results had been augmented by neutrino deep inelastic scattering, which strongly supported the quark-parton model. In addition, the discovery of the charmonium system in $e^+ - e^-$ annihilation had established the existence of a fourth quark. The original issues that E80 proposed to study were widely accepted as valid. Nevertheless, the large and positive asymmetries of E80 were important confirmation of these ideas. The technical developments and the ideas that E80 developed were to become the beginning of a new program, the study of the spin structure of the proton.

The E80-Related Experiments

The E80 results were not limited to the study of spin-dependence in the deep-inelastic scattering process. The work on E80 was intertwined with another experimental program underway at SLAC, the search for parity non-conservation in deep-inelastic scattering. Before discussing the history of that program, I would like to continue the story of about E80 and its other results.

E80 in the course of data taking also looked at related processes, and extracted from these processes additional tests and limits. The first of these was polarized elastic *ep* scattering, where the asymmetry is the same as that defined for the inelastic regime. The process for elastic scattering of electrons by protons can be fully analyzed using measured values of the two proton form factors, G_m and G_e. The unpolarized cross section formula is given in terms of G_m^2 and G_e^2, while the asymmetries include terms linear in these factors. The measurement of these asymmetries allowed, first of all, a check of the experiment. The predicted asymmetries could be used to demonstrate that the experiment worked. Alternately, using the measured values, one could check that relative signs of G_m and G_e were as expected, namely positive. The measured value for A was 0.138 ± 0.031, which agreed with the prediction $A_{theory} = 0.112$ and ruled out the possibility of opposite signs.[8]

Another test of fundamental physics resulting from E80 data involved the Møller process, elastic polarized $e^- - e^-$ scattering. This process was used in E80 to measure the beam polarization. The polarized target electrons were obtained in a thin magnetic foil material, Supermendur. The foil was oriented nearly parallel to the incoming beam (i.e., the normal to the foil was nearly perpendicular to the incident beam direction) and the material was magnetized by a modest external field of about 100 gauss. The material, predominantly iron, supplied 2 polarized electrons out of the total of 26 surrounding the nucleus. The scattering process is purely a QED process with the spin-dependence readily calculable. Material effects and nuclear effects require that small corrections be understood. With this apparatus, one could study and measure accurately the longitudinal spin component of the incoming beam electrons. At high energies, the equation of motion for spin in a uniform magnetic field is given by

$$\Theta_{prec} = \gamma \left(\frac{g-2}{2} \right) \Theta_{bend}$$

where Θ_{bend} is the angle of bend for the momentum vector, i.e., the amount the momentum vector precesses in the magnetic field. The quantity Θ_{prec} represents the angle through which the spin precesses ahead of the momentum vector. At low energies $\Theta_{prec} \approx 0$, and the equation says that spin and momentum vector precess approximately the same amount. At high energies, the spin precesses well ahead of the momentum vector. Figure 3 shows the expected Θ_{prec} for the beam line at SLAC

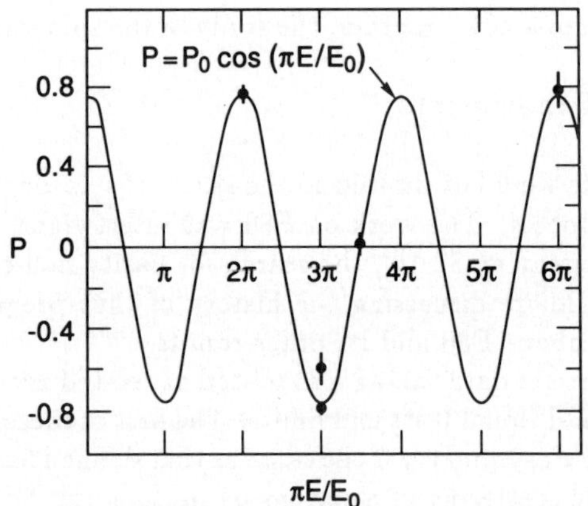

Figure 3: The spin precession for polarized electrons transported through the A-line at SLAC. The sinusoidal curve shows the expected energy dependence; the data are the measured values using the Møller scattering technique.

which has $\Theta_{bend} = 24.5°$. For every 3.237 GeV of beam energy, the spin precesses by an amount π in passing through the bend magnets. At 19.42 GeV, the total precession is 6π. The data points measured the longitudinal polarization at 4 separate energies.[9] These data are at relativistic γ-values up to 5×10^4, and provide an interesting check of special relativity in the extreme relativistic limit. In the paper, it is argued that Θ_{prec} is a combination of "kinematic" (i.e. Thomas precession) terms and "dynamic" (i.e., momentum) terms. The argument hypothesizes that the two special relativistic forms may involve different γ's. That is $\tilde{\gamma}$(kinematic)$\neq \gamma$(dynamic). One can write a form

$$\tilde{\gamma} = \gamma + C_1(\gamma - 1) + ...$$

where C_1 is hypothesized to be small. Given this form, then g-2 tests and E80 results could be compared. Using data at γ=29.2, the value for C_1 is $1.4 \pm 1.8 \times 10^{-8}$. Using $g-2$ data from e^- measurements at $\gamma = 1.2$, C_1 is $-2.6 \pm 1.8 \times 10^{-8}$. The corresponding value from E80 is $-1.0 \pm 8.0 \times 10^{-10}$ at $\gamma = 2.5 \times 10^4$. The high energies of the E80 data provided an improvement by an order-of-magnitude in this test, and furthermore carried the test to extreme relativistic motion.

The E130 Experiment and the Sum Rules

E80 opened a major program to investigate the spin structure for the purpose of a better understanding of the nucleon. The interest in pursuing this subject further stemmed from two sum rules, the Ellis-Jaffe sum rule, and the Bjorken sum rule. The Bjorken sum rule came first in 1966[10], and the Ellis-Jaffe sum rule in 1974[11]. I will discuss both of these, but I prefer to discuss the Ellis-Jaffe sum rule first, which was tested in E130. The Bjorken sum rule, regarded as a fundamental test of QCD, has not had an experimental test yet.

The Ellis-Jaffe sum rule (like the Bjorken sum rule) is derived from quark light- cone algebra, but with added assumptions about the strange quarks. Namely, the strange quarks are assumed to be paired with opposing spins, so do not contribute to the spin structure in the proton. The sum rule is an integral over x of the function $g_1(x)$ related to the physics asymmetry A_1 by

$$g_1(x) \approx \frac{A_1(x)F_2(x)}{2x(1+R)}$$

The function $g_1(x)$ is to be evaluated in the scaling region, high Q^2. Small terms arising from a second term $g_2(x)$ are neglected here. The Ellis- Jaffe sum rule is given by

$$proton: \quad \int_0^1 g_1^p(x)dx = \frac{1}{12}\left|\frac{g_A}{g_V}\right|\left[1+\frac{5}{3}\frac{3F/D-1}{F/D+1}\right] + O(\alpha_s) = 0.189 \pm 0.005$$

and

$$neutron: \quad \int_0^1 g_1^n(x)dx = \frac{1}{12}\left|\frac{g_A}{g_V}\right|\left[-1+\frac{5}{3}\frac{3F/D-1}{F/D+1}\right] + O(\alpha_s) = -.002 \pm 0.005$$

where $|\frac{g_A}{g_V}|$ is the ratio of axialvector to vector coupling in neutron beta decay $n \rightarrow pe\nu$, and F/D is the ratio of $SU(3)$ axial charges in the baryon octet. Measurement of the function $A_1(x)$ and $g_1(x)$ would be used to understand better the proton (and neutron) systems, and the integrals would verify our understanding of the origin of spin, coming from the quark substructure. E130 set out to explore this territory and to expand our knowledge of the nucleon.

E130 was proposed in 1976. The proposal was basically an upgrade of the existing E80 techniques. The PEGGY source intensity would be increased. The target would be improved to take into account lessons learned in E80, and to improve its operating efficiency. A new large acceptance spectrometer was proposed to increase counting rates. Overall, these improvements led to a factor of 40 improvement in rates over E80. The result of these improvements made possible the investigations of a wide range of kinematic points. Data from E130 were published in 1983.[12] These data are shown in Figure 4. With these data E130 demonstrated the scaling of A_1, which was expected if the unpolarized structure functions were taken as examples. But experimental proof of this important point was essential if these data were to be used to measure the Ellis-Jaffe sum rule. From E130 came the first evaluation of the integral. The value was

$$\int_0^1 g_1^p(x)dx = 0.17 \pm 0.05.$$

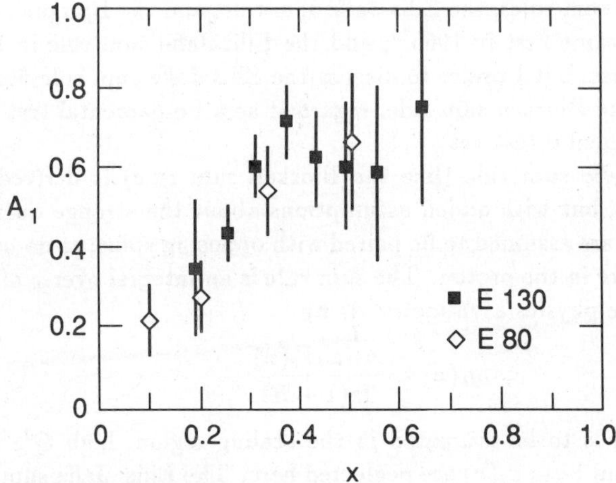

Figure 4: The physics asymmetry A_1 versus x from E130 and from E80. These data permitted the first evaluation of the proton Ellis-Jaffe sum rule.

This value agreed, within errors, with the predictions by Ellis and Jaffe. Because of the $1/x$ term in the $g_1(x)$ expression, low x values become important, relative to high x. To improve upon these measurements, more work was needed in the low x regime. By 1983, SLAC was busily planning its conversion to linear collider work, the SLC project. This activity would take many resources to carry out. Vernon and his collaborators submitted a new proposal to extend the deep inelastic spin structure measurements to low x and to the neutron. SLAC was unable to support this proposal, and sadly the opportunity for further information would have to wait. Vernon and his group turned to CERN to continue this program, using polarized muons at higher energies than available at SLAC. The proposal to use 200 GeV muons was ideal for the pursuit of spin structure at low x.

The EMC Experiment

The work at CERN is mostly beyond the scope of this talk, but is so important to the continuing history concerning spin structure, that I must mention it briefly. The EMC experiment took an exposure with a polarized proton target to measure A_1.[13] The data are shown in Figure 5. The figure includes data from E80 and E130 as well, and a curve as an example of a model which satisfies the Ellis-Jaffe sum rule. Figure 6 shows the integral of $g_1^p(x)$ as a function of the lower limit on the integral. The integral appears to extrapolate smoothly into $x = 0$ at a value of $0.114 \pm 0.012 \pm 0.026$, well below the predicted value 0.189. These results have stirred up considerable interest and controversy. For the integral to fail to agree with prediction, models of spin

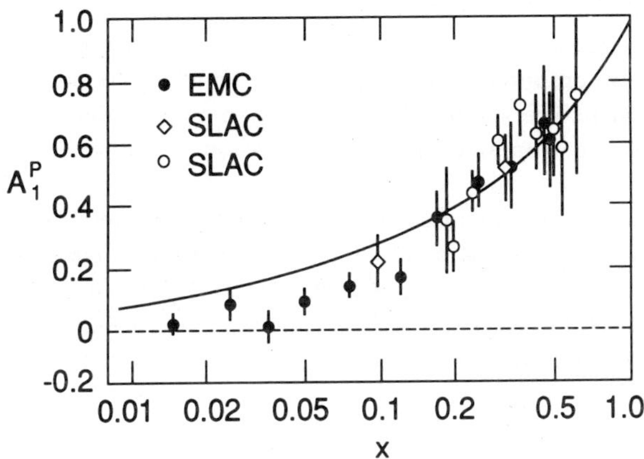

Figure 5: The proton asymmetry A_1^p from the CERN EMC experiment and from the SLAC E80 and E130 experiments. The curve is a model calculation from Ref. 27.

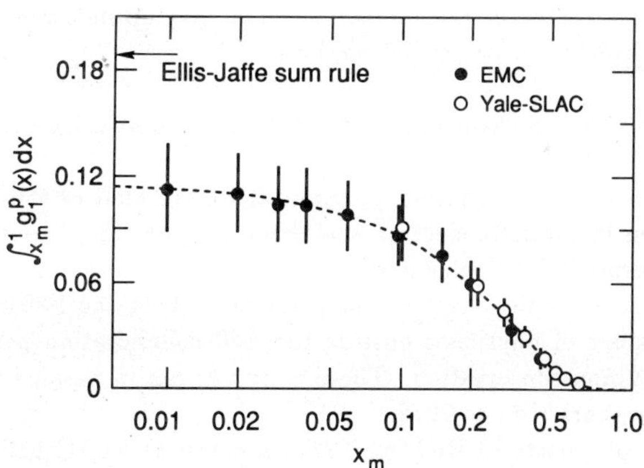

Figure 6: The integral of the spin-dependent structure function $g_1^p(x)$ from x_m to 1, where the lower limit x_m coincides with data points. The Ellis-Jaffe sum rule prediction is also indicated.

in the proton must be modified. An analysis by the EMC Collaboration shows the implications of the experimental results. The net quark spin appears to sum to 0! The contribution to spin from strange quarks is not zero, but appears to be negative! The net proton spin must be assigned to something else, perhaps gluons or orbiting partons. This apparently surprising result has led to the term "spin crisis", and many suggestions to modify our picture of the proton. Future experiments at CERN, SLAC, and DESY are proposed and being planned hopefully to clarify the situation. The neutron spin structure should significantly add to our picture. Furthermore, measurement of the neutron spin structure will allow the Bjorken sum rule to be evaluated.

The Bjorken sum rule was first derived in 1966, based on fundamental principles that have led to the formulation of QCD. It stems from quark light-cone algebra and isospin symmetry. It is part of the underpinnings of QCD. Bjorken first described this sum rule as "worthless" because at the time it appeared impossible to measure experimentally. Advances in technology and the growing importance of quarks in particle physics led Bjorken to rethink the issue of spin-dependence in deep inelastic electron scattering. He commented in 1969 that it was "interesting to reconsider in light of the present experimental and theoretical situation."[14] Feynman in his book on photon-hadron interactions in 1972 says "Its verification of failure would have a most decisive effect on the direction of future high energy theoretical physics." With the technologies that exist today, it is unthinkable that this fundamental test of QCD should go untested. This situation must be changed, and within the next several years, experiments here and abroad will move to solve this problem. It is also unthinkable that the Bjorken sum rule could fail. Experiments will have to be *very* good before the community of particle physicists would give up its current ideas and start to rethink its approach to strong interactions.

The Search for Parity Non-Conservation in Inelastic Electron Scattering

Let me now turn to another program at SLAC, that of the search for parity non-conservation in inelastic electron scattering. To do this, I must go back to the beginning of Vernon's involvement at SLAC and unravel the intertwined activities with E80. These activities involve some, but not all, of the E80 collaboration. In addition, a number of physicists outside the E80 collaboration participated in the search for parity non-conservation. These activities ran in parallel to the E80/E130 work during the years 1971 to 1978.

In 1970, physicists in Richard Taylor's group at SLAC had been discussing ideas on how to search for parity non-conservation in electromagnetic interactions. Vernon visited SLAC in the fall of 1970 to propose accelerating polarized electrons to high energies. That proposal was a crucial component in an approach to a parity non-conservation search. The incoming polarized beam could be used to look for parity non-conservation by comparing scattering rates for oppositely polarized longitudinal spin. If parity is a good symmetry in the scattering process, then beams of opposite longitudinal spin will scatter equally. A measure of unequal scattering is evidence for breakdown of the parity symmetry. Starting from this very simple, but key, idea we began to work on an experimental approach to search for parity non-conservation. In early spring of 1971, I visited Yale University to meet with Vernon and his group to discuss the possibility of mounting a parity experiment. The experiment to search for parity non-conservation with a polarized beam required an unpolarized target. The E80 collaboration had proposed a polarized target. In the meeting at Yale, Vernon supported the idea of pursuing a search for parity non-conservation, and we discussed three possible strategies: (i) to use the E80 data and by averaging over target polarization to obtain a sample of data from unpolarized material; (ii) to ask for an extension to E80 for the purpose of running an unpolarized target material; and (iii) to propose a new experiment. We agreed to pursue (i) and (iii).

Motivating a parity non-conservation search would prove to be somewhat difficult. The PEGGY source was limited to 1 to 2×10^9 electrons per linac pulse, and the counting rates in the spectrometers were low. Statistical accuracy and therefore sensitivity to parity non-conserving effects would be limited. It was clear that scenarios (i) and (iii) would not achieve sensitivity to weak-electromagnetic effects. Speculations on connections between the weak force, for which the charged current was known to be maximally parity violating, and the electromagnetic force could be found in literature. For example, Zel'dovich in 1957 discussed "Electromagnetic Interactions with Parity Violation,"[15] in which an intrinsic parity non-conserving piece was hypothesized. In 1958 he wrote a remarkable paper in which he discussed both scattering of longitudinally polarized electrons at high energy and rotation of the plane of linear polarization of light by atoms as experimental techniques to use in a search.[16] In spite of the speculation in literature about such effects, we did not find much help upon which to base a proposal. We developed the formalism by analogy to neutrino scattering, which had been fully discussed in the literature.[17] If parity conservation is relaxed in the electron scattering process, then a $W_3(\nu, Q^2)$ term appears in the cross section for polarized electrons. Current conservation requires such terms to vanish at $Q^2 = 0$, so we argued that such effects might appear at large Q^2 and may have escaped detection in prior low energy experiments such as studies of nuclear transitions. In 1972, the "E95" experiment was proposed and approved. It would use the PEGGY source and an unpolarized hydrogen target to search for the hypothesized parity non-conservation. By choosing an unpolarized target, the amount of target material considerably exceeded that of the polarized target of E80. Furthermore, E95 would select kinematic points more suitable to a search than was possible for E80. Thus a separate experiment was justifiable. But even with the improvements, E95 could not reach the very difficult goal of the weak-electromagnetic interference. In spite of this shortfall, we proceeded with considerable energy to learn how to look for parity non-conservation.

In 1976, E80 published a limit on the parity non-conserving asymmetry,

$$A = \frac{\sigma_L - \sigma_R}{\sigma_L + \sigma_R} \leq 5 \times 10^{-3}$$

at Q^2 values between 1.4 and 2.7 $(\mathrm{GeV/c})^2$.[7] The subscripts L and R refer to the left-handed and right-handed helicities of the incoming beam. In 1978, E95 published an improved asymmetry value $A = -3.9 \pm 8.4 \times 10^{-4}$ at a Q^2 near 1 $(\mathrm{GeV})^2$.[18]

Meanwhile, during the period between the time of the E95 proposal and the actual running, a new proposal was taking shape. This proposal was submitted and approved in 1974, and given the designation E122. It was specifically aimed at testing the Weinberg-Salam model of electroweak interactions. This proposal was based on new technologies and ideas to improve the beam intensity, to increase the detector acceptances, and to improve substantially the beam monitoring and control capabilities. The need for a much increased beam intensity over that of the PEGGY source was paramount. Without a significant improvement, statistical accuracy could not be achieved. Following the proposal of Garwin, Siegmann, and Pierce,[19,20] it was suggested that a solid state source, based on gallium arsenide illuminated with polarized laser light, could provide the required intensity. In addition, reversal of the electron spin at the source could be accomplished easily and without influence on other beam parameters. The gallium-arsenide source, driven by circularly polarized laser light, provided these rapid, easy, and systematic-free reversals. The E122 proposal also suggested constructing a large acceptance fixed-angle spectrometer to replace the more traditional end station spectrometers. Counting of scattered electrons at a sufficiently high rate to achieve statistical accuracies $A \leq 10^{-5}$ proved to be impossible by conventional digital counting techniques. A technique to count electrons by integrating over each 1.5 microsecond long beam pulse was proposed. The combined enhancements for E122 would provide greater than a 1000-fold increase in rates over the previous experiment, E95.

E122 was approved by the SLAC EPAC with the condition that a solid state laser-driven source could be developed as proposed. With this approval, work began in earnest to construct a gallium-arsenide source along the lines proposed by Garwin, Siegmann, and Pierce. Development and commissioning of such a source was to occupy the next three years at ETH Zürich and at SLAC.

While this development was underway, there was activity at other laboratories and universities. Progress in understanding neutral currents was strong, particularly in neutrino scattering. However progress in the electron sector of the theory was running into problems. Considerable activity to look for parity non-conservation in optical rotation experiments was occurring. By the summer of 1977, a year ahead of the experiment at SLAC, experimenters at Oxford and Washington had converged on results for optical rotation in atomic bismuth vapor. The results by the two groups, published in adjacent articles in Physical Review Letters, concluded that no optical rotation occurred at a level predicted by the Weinberg-Salam theory.[21,22] This reported null result was a serious problem for the minimal $SU(2) \times U(1)$ version in the Weinberg-Salam theory. Considerable discussion and debate ensued, particularly among particle and atomic physicists. The impact of these arguments on preparations for E122 were minor, but not negligible. The previous experiences of two searchs, E80 and E95, giving null results, plus in 1977 the reported null results from two experiments in atomic physics, generated a feeling that E122 could likewise be a null experiment. The consequences of a possible null experiment made the experimenters in E122 redouble the efforts to provide a clean experiment, since proving that a null result is valid would be much more difficult than if the experiment were to see a positive signal.

By late 1977 the gallium arsenide source had begun to work in the laboratory, and polarizations of 40 to 45% at high currents had been obtained. The original proposed beam currents were surpassed by factors of 10 to 100, providing far more polarized electrons than the accelerator could capture and accelerate. These high current electron beams were produced by moderate laser beams falling on gallium arsenide crystal surfaces coated with cesium and oxygen, thus photoemitting electrons. Polarization of the electrons required circular polarization of the illuminating laser light. The control of polarization of the electrons was accomplished by a linear polarizer followed by a Pockels cell to achieve circular polarization. A Pockels cell requires a precise voltage to make it optically active at full quarter wave retardation, and reversing the voltage reverses the sign of the retardation. Hence the circular polarization of the laser, and the longitudinal polarization of the photoemitted electrons is readily reversible. For operation with a pulsed linac like the SLAC linac operating at 180 pulses per second, reversal of the electron polarization between pulses was easy. It was decided that for E122, the pattern of reversals from pulse to pulse would be random, with the sign of the polarization in each pulsc to be transmitted to the computer to be logged with the data for each machine pulse. From time-to-time (several times per day), the plane of linear polarization of the laser light was manually rotated by 90 degrees. This resulted in changing the sign of the circular polarization emerging from the Pockels cell relative to the voltage applied. Figure 7 shows the optical system at the source of polarized electrons.

The source operated at full beam intensity for the accelerator in a test in December 1977, and the first physics run was scheduled for Spring 1978. Checkout and commissioning of the experiment began in February 1978, after extensive preparations. During this period of tune-up, a report from Novosibirsk appeared in our hands.[23] This publication reported observation of optical rotation in bismuth, using the same optical transition (648 nm) for which Oxford had originally reported a null result. It is difficult today to evaluated this paper. The results from Novosibirsk have not been confirmed by later work. Subsequent measurements by experiments and refined work on the theory have converged to an accepted value several standard deviations away, approximately 1/2 of that reported by the Novosibirsk group. These extremely small effects in atoms are susceptible to systematic errors that are difficult

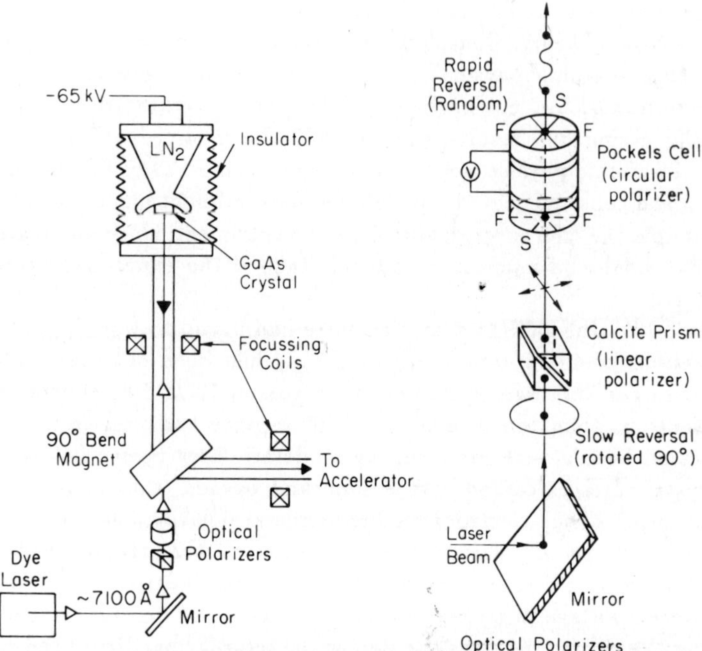

Figure 7: A schematic of the gallium-arsenide source and a blow-up of the active optical elements. Rapid and slow reversals of the electron spin was controlled by the circular polarization of the laser light

to pin down, and the early atomic parity non-conservation experiments seemed to have difficulty with the control of these small effects.

The E122 experiment was beginning to come in with data soon, in April 1978. The signal for parity non-conservation was evident rather quickly. Runs lasting less than an hour would show significant spin dependent asymmetries of the general size expected for weak-electromagnetic interferences in the Weinberg-Salam theory. But there remained the need to prove that systematic errors were not present. Figure 8 shows an example of a sequence of runs for the April period. The asymmetry follows the solid line, reflecting the orientation of the calcite prism, the linear polarizer. Figure 9 combines these data and adds in a point at 45 degrees (a null measurement). The curve illustrates the expected cosine modulation of the asymmetry. The null point is particularly important, since small systematic effects which could creep in, would be seen in this point as an offset from zero. The 45 degree point agrees with zero, and indicates lack of systematic errors at the level of the indicated statistical errors. In Figure 9 each point is double, indicated by the two symbols. These points were obtained simultaneously in a gas Cerenkov counter and a lead-glass shower counter placed immediately behind. The electrons which passed into the spectrometer acceptance first passed through the gas counter, were counted, then passed into a lead-glass counter and were again counted. The two counters agreed to high accuracy, as indicated. The counters proved to be a good consistency check, but were highly cor-

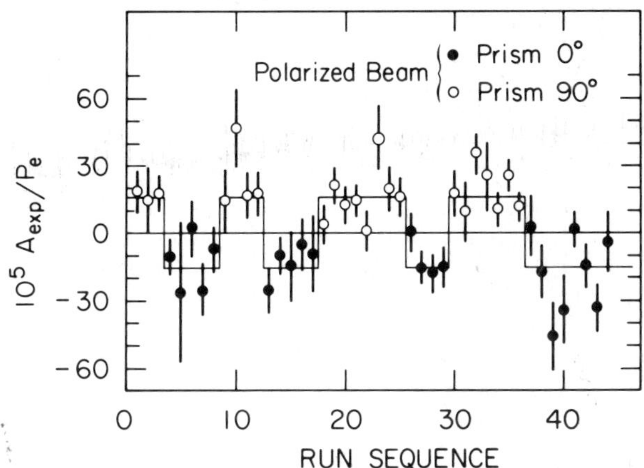

Figure 8: A sequence of runs in which the plane of linear polarization of the laser light was periodically alternated between 0 and 90 degrees.

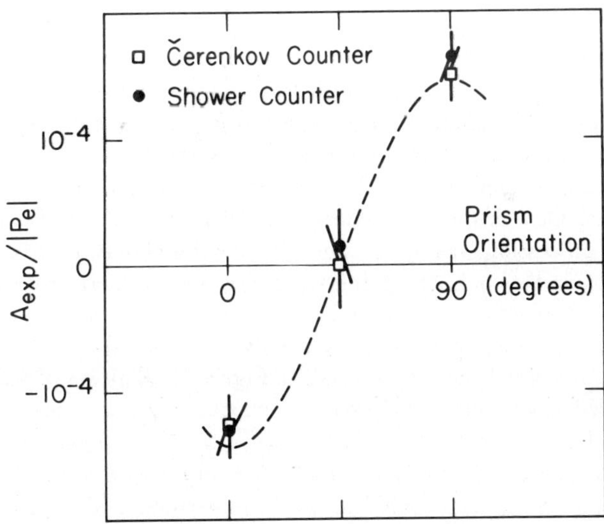

Figure 9: The experimental asymmetries for three values of the orientation of the linear polarizer. The two values of the asymmetry plotted at each setting of the angle came from two independent counters in the spectrometer.

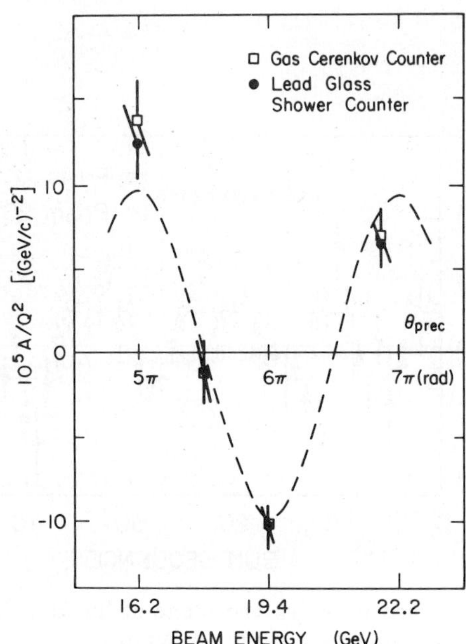

Figure 10: The expected $g-2$ precession of the asymmetries seen in the two independent counters in the spectrometer.

related, so the measurements were never combined. The lead-glass Cerenkov counter was chosen for the published data. With the completion of these runs, a final test was begun. That test, to look for the expected $g-2$ precession of the electron in the asymmetries, was part of the initial plan to serve as the ultimate check. The $g-2$ precession, discussed in Ref. 9, provided longitudinal spin orientation at the target every 3.237 GeV in the beam energy. At 19.42 GeV the precession was 6π. At 16.88 GeV, the precession was 5π, so the asymmetries should reverse sign. At 17.80 GeV, the precession was 5 $1/2\pi$, giving transverse spin and an expected null result. The 17.80 GeV point was an additional important check on systematic errors, since physics asymmetries must vanish at that energy.

The data at several energies is shown in Figure 10. This set of data demonstrated that the observed asymmetries followed the expected $g-2$ precession for an electron beam. No known systematic errors would have given such a curve. The $g-2$ data constituted the proof that we sought. The signals were at 10^{-4}, which were large relative to the designed capabilities of the experiment. The anticipated difficult measurements and the worry over null results had by now mostly vanished and the experiment proceeded in an orderly fashion.

Based on the strengths of the data taken in April 1978 which demonstrated

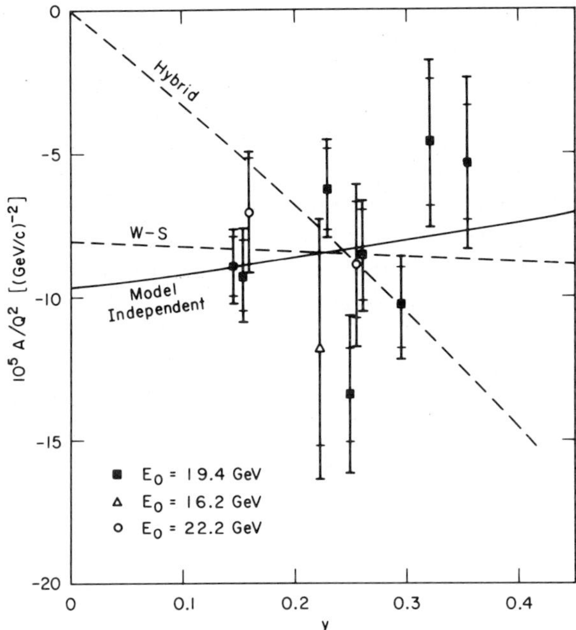

Figure 11: The y-dependence of the asymmetries ruled out the hybrid model. The model independent fit agreed with the Weinberg-Salam model.

rather convincingly parity non-conservation in electron scattering, the E122 experiment was scheduled for further data in the Fall of 1978. The need for further data was apparent. In order to separate different versions of the electroweak theory, a y scan was needed. The kinematic variable y is defined as $y = \nu/E_0$, where ν is the energy transferred to the nucleon in the scattering process, and E_0 is the incoming electron energy. In a parton model, y is closely related to an angular distribution. Measuring the y-dependence clearly separates different $SU(2) \times U(1)$ models. The reported null results in atomic bismuth experiments could be explained by assigning the right-handed electron e_R to a weak isospin doublet with an hypothesized heavy neutrino, while the model of Weinberg-Salam had specified that e_R be assigned to a singlet. The doublet assignment was called "mixed" or "hybrid", because right-handed quark components such as u_R, d_R, etc., were already shown to be consistent with singlet assignments from inelastic neutrino scattering. The efforts to explain null results in atomic parity non-conservation had led to a non-minimal $SU(2) \times U(1)$ model being proposed. The y-dependence for a "hybrid" model for deep inelastic electron scattering was quite different from that for the minimal version, as shown in Figure 11. The asymmetry vanished at $y = 0$ for the hybrid version. It was expected to be substantially non-zero for the Weinberg-Salam model. A measurement of the

y-dependence was needed to separate these. Figure 11 shows the data resulting from the Fall 1978 run. The Weinberg-Salam model was in good agreement, while the hybrid model was ruled out. These results thus convincingly confirmed the original Weinberg choice of the minimal $SU(2) \times U(1)$ assignment. The E122 experiment had made a set of statements of significance. They included: (i) weak-electromagnetic interference was seen in deep-inelastic scattering of electrons at a level of 10^{-4} at $Q^2 = 1(GeV/c)^2$. The weak force and the electromagnetic force were acting together in this process; (ii) the observation of interference between these forces implied that the weak neutral vector boson carried spin 1: (iii) A neutral current coupling of the electron was being observed; (iv) Within the context of the Weinberg-Salam model, the assignment of the electron to a left-handed doublet and a right-handed singlet was confirmed; and (v) The value of the mixing parameter $sin^2\Theta_W = 0.224 \pm 0.020$ (at the tree level) was determined and found to be in good agreement with the same parameter measured in neutrino scattering.[24,25]

The E122 experiment ended with these runs and data. Systematic errors were becoming a limiting factor, and the long run times involved for further improvements argued against continuing.

Parity Non-Conservation Studies at Bates

In a recently reported result, the techniques similar to those used in E122 were substantially improved and extended to measure parity non-conserving asymmetries in electron-carbon elastic scattering at the Bates Lab at MIT at somewhat lower energies.[26] The purpose in looking for parity non-conservation from nuclei is to isolate particular parts of the hadronic weak current. In particular, the ^{12}C nucleus is an isoscalar state. This process singles out the weak vector isoscalar part of the weak current. This coupling, $\tilde{\gamma}$, is related to the parity non-conservation asymmetry A by

$$A = \frac{e}{2}\tilde{\gamma}G_f Q^2(\sqrt{2}\pi\alpha)^{-1}$$

and was expected to be rather small, 10^{-6}, at Bates kinematics where $E_{beam} = 250$ MeV, and $q = 150$ MeV/c. The experiment was sensitive to potential new effects, such as extra gauge bosons possibly missed in previous experiments. The experiment developed a thorough and clean set of monitors to measure any systematic effects. Control of the systematic errors in A were less than 0.04 ppm, and the experiment reported a parity non-conservation asymmetry of 0.6 ppm, fully 150 times more sensitive than the E122 experiment. The coupling $\tilde{\gamma} = 0.136$ was in agreement with the Standard Model value 0.155. The superb control of systematic errors promises that polarized electron beams can become an excellent probe to study weak effects to the nuclei. Such experiments are an important piece of future experimental plans at electron facilities such as BATES, CEBAF, and in Europe.

Conclusion

In this brief tour of experiments, I have looked at polarized beam experiments at SLAC, CERN, and BATES which are part of Vernon's experimental career. This tour covers a lengthy period, from 1970 to the present. The experiments described required extensive developments in technology to support them. The long stretches of time between experiments were not times of inactivity. They represented major innovations in matters of targets, sources, detectors, and beam monitoring. I wanted to discuss these aspects of the work as well, in particular to give credit to the many physicists who participated in developing the experiments. The task of acknowledging the many contributions is overwhelming, so I can only leave it to the journal articles referenced in this report and elsewhere. I counted approximately 160 physicist names in these publications, representing approximately 29 institutions around the world. The fact that 160 physicists voted with their feet to join these efforts is a tribute to Vernon's vision, his leadership, and most importantly, his many successes in fundamental physics with beams of polarized electrons and muons.

References

1. R. P. Feynman, *Phys. Rev. Lett.* **23**, 1415 (1969).
2. J. D. Bjorken, *Phys. Rev.* **179**, 1547 (1969); *Phys. Rev.* **D1**, 1376 (1970).
3. M. Breidenbach et al., *Phys. Rev. Lett* **23**, 935 (1969).
4. R. P. Feynman, "Photon-Hadron Interactions", *Frontiers in Physics Lecture Notes Series*, (W.A. Benjamin, 1972).
5. J. Kuti and V. F. Weisskopf, *Phys. Rev.* **D4**, 3418 (1971).
6. M. J. Alguard et al., *Nucl. Instr. Meth.* **163**, 29 (1979).
7. M. J. Alguard et al., *Phys. Rev. Lett.* **37**, 1261 (1976).
8. M. J. Alguard et al., *Phys. Rev. Lett.* **37**, 1258 (1976).
9. P. S. Cooper et al., *Phys. Rev. Lett.* **42**, 1386 (1979).
10. J. D. Bjorken, *Phys. Rev.* **148**, 1467(1966).
11. J. Ellis and R. Jaffe, *Phys. Rev.* **D9**, 1444 (1974).
12. G. Baum et al., *Phys. Rev. Lett.* **51**, 1135 (1983).
13. J. Ashman et al., *Phys. Lett.* **B206**, 364 (1988).
14. J. D. Bjorken, *Phys. Rev.* **D1**, 1376 (1970).
15. I. B. Zel'dovich, *JETP* **33**, 1531 (1957).
16. I. B. Zel'dovich, *JETP* **36**, 964 (1959).
17. S. L. Adler, *Phys. Rev.* **143**, 1144 (1966).
18. W. B. Atwood et al., *Phys. Rev.* **D18**, 2223 (1978).
19. E. L. Garwin et al., *Helv. Phys. Acta* **47**, 393 (1974).
20. E. L. Garwin et al., *SLAC-PUB-1576* (1975).
21. L. L. Lewis et al., *Phys. Rev. Lett.* **39**, 352 (1977).
22. P. E. G. Baird et al., *Phys. Rev. Lett.* **39** 356 (1977).
23. L. M. Barkov and M. S. Zolotorev, *Phys. Lett.* **85B**, 308 (1979).

24. C. Y. Prescott et al., *Phys. Lett.* **77B**, 347 (1978).
25. C. Y. Prescott et al., *Phys. Lett.* **84B**, 524 (1979).
26. P. A. Souder et al., *Phys. Rev. Lett.* **65**, 694 (1990).
27. R. Carlitz and J. Kaur, *Phys. Rev. Lett.* **38**, 673 (1977); J. Kaur, *Nucl. Phys.* **B128**, 219 (1977).

It takes One to Know One

V. L. Telegdi
CERN, Geneva, Switzerland
and
California Institute of Technology, Pasadena, CA 91125

Introduction

Under the above title, I am expected to talk about a field which Vernon initiated and in which I have worked myself quite some time ago, namely on muonium. This will not be an easy task, since — as I discovered recently from a compilation kindly put at my disposal by the Yale Physics Department— Vernon has co-authored more papers on muonium than I ever wrote in all fields of physics put together!

Let me start out with some personal reminiscences. I first met Vernon at the University of Pennsylvania, in the very early fifties, when I attended there a conference on Photonuclear Reactions, a subject of great interest to me at that time. My first interaction with him led to something that may surprise you; a joint paper[1] (actually an abstract)! It concerned the g-factor of a muon bound in the atomic ground state. Even neglecting the anomaly, this is not equal to 2, but to $2(1-\alpha^2/3)$, a fact established by Breit in 1930, but according to a phone conversation with Vernon no longer remembered at Columbia University (in 1958). We submitted an abstract to the Washington meeting of the APS; it fell upon me to deliver it, because Vernon could not attend for family reasons (his son had broken a leg). Subsequently Vernon, in collaboration with some Los Alamos friends, published detailed calculations that allowed for the modified Coulomb potential in muonic atoms.

Contrary to what you may believe, Columbia and Chicago did not enter into a competition about muonium resonances right away (as to the discovery, see below). Upon returning from CERN to Chicago in 1960, I decided to build a muon channel for our synchrocyclotron. By 1962 it was operational and provided a greatly increased muon flux. I suggested to Vernon in 1967 that he continue his muonium work at Chicago instead of at Nevis; he declined. It was one of those moments in his career when rational considerations (more flux) were overpowered by emotional feelings (loyalty to Columbia). I would like to stress the fact that I had not proposed that Vernon collaborate with us, but only that he use our facilities. Perhaps the fact that we worked for several years independently on muonium was, all in all, good for physics; in any case, I would not have the pleasant task to talk here today had Vernon accepted my proposal!

The Discovery of Muonium

Everybody knows that Vernon and his collaborators discovered muonium. What is less well known, is how I had thought of this atom, how we tried to discover it in Chicago, and how (and why) we failed miserably.

As a junior faculty member at the University of Chicago, I was asked (1955) to sit on the Ph.D. exam committee of Sal Krasnen, one of S.K. Allison's students. This candidate, like many other Allison students, had worked on charge-exchange processes involving low-energy protons. My question to him was: "If you shot positive muons rather than protons through matter, what would you observe?" Then and there I coined the name "muonium" for the atom that could be formed when a muon (like a proton) moved through matter at the right (Bohr) velocity.

In the fall of 1956, when preparing the $\pi - \mu - e$ parity violation experiment in nuclear emulsions, muonium was uppermost in our minds. We realized that muonium formation was a potential depolarization mechanism, and that $^3S_1$ muonium, once formed, would precess in a given magnetic field 100 times faster than an unbound μ^+. Since the emulsions were to be exposed to pions at a point where the fringe field of the cyclotron was of the order of 10 G, Jerry Friedman and I spent lots of time on shielding them down to the mG level (this effort, although it delayed us, was not entirely wasted, since other groups, who did <u>not</u> take these precautions, failed to see parity violation in the "same" experiment!)

After the violation of parity conservation had been established (Jan. 1957), we relatively soon tried to actually demonstrate muonium formation. Having been originally trained as a chemical engineer, I realized that muonium would behave chemically as "hydrogen in statu nascendi", i.e. interact violently with most compounds. It was hence obvious that a noble gas at high density would be a suitable target. We built an apparatus very similar to the one used by Vernon and his collaborators in their actual discovery (Fig. 1), except that the two counters defining a muon stop within the gas were <u>inside</u> the pressure vessel (a trick that I still consider to be useful). That vessel we filled with argon, to be more precise with welding argon readily available from a bottle left on the floor of the experimental area. This turned out to be our downfall—we had forgotten that two microseconds are a long time on the scale of molecular collisions, so that the small traces of O_2 present in technical argon had ample time to "destroy" (by chemical reactions or spin exchange) any spin-polarized muonium! Vernon and his collaborators at Nevis succeeded (1960) where we had failed—simply by using specpure argon which they futhermore kept clean by continuous purification. I do not know who deserves the credit for this essential step, but I am sure that Klaus Ziock, who had lots of previous experience with vacuum technology, had lots to do with it. Figure 2 shows the signal observed.

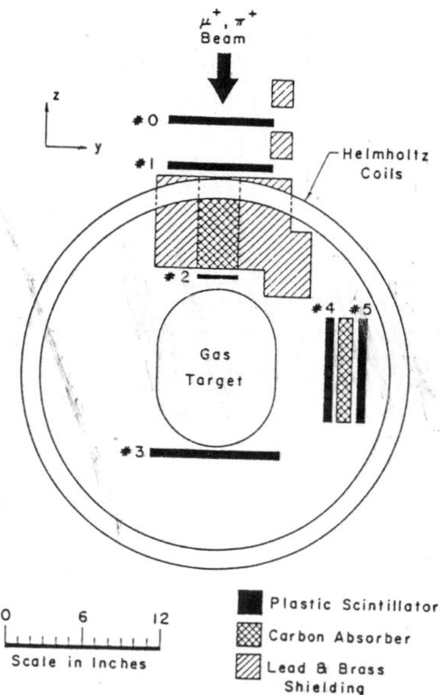

Figure 1: Apparatus used by Hughes et al., (1960) in the discovery of muonium.

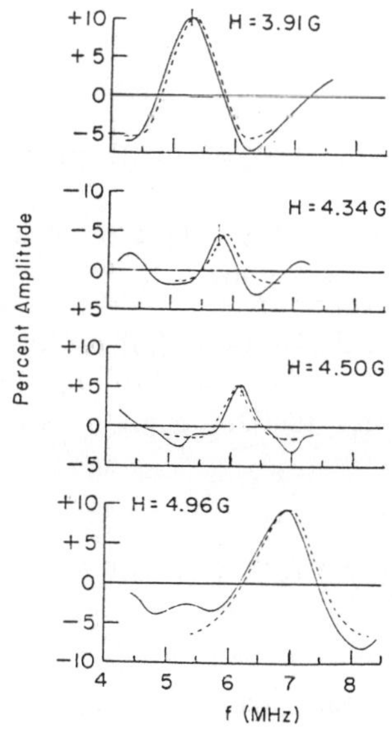

Figure 2: Signal observed with apparatus of Fig. 1 (Fourier analysis of the precession frequency).

We also failed in another, altogether different approach to muonium formation. Quartz is a rather inert chemical compound, and we knew that positronium is abundantly formed in crystalline quartz. Putting (perhaps incorrectly) two and two together, we tried to look for muonium precession in crystalline quartz. We saw no effect; this time the cause of our failure was more subtle—we worked at room temperature. Subsequently some Soviet workers observed a clear signal when their quartz was cooled to liquid nitrogen temperature.

The Road Towards High Precision

Anticipating what I shall say next, I present (Figure 3) the chronology of the developments. Next, before going into any experimental details, I would like to raise (and answer) the question **"What is muonium good for?"**

The h.f. "contact interaction" energy-first derived by Fermi is

$$\Delta E_F = \frac{32}{3}\frac{\mu_\mu \mu_e}{a_0^3}. \tag{1}$$

1956	We invented $\mu^+ e^-$, misnamed it *muonium*
1957-1958	We failed to produce it !
1960	Yale/Nevis discovered it...
1964	They got $\Delta\nu$ to 27 ppm (but...)
1969-1973	We had a couple of ideas, got $\Delta\nu$ to 0.5 ppm
1977-1982	Yale/Heidelberg/LAMPF threw us into **the dustbin of history** ($\Delta\nu$ to 0.04 ppm)

SIC TRANSIT GLORIA MUNDI!

Figure 3: Chronology

Neglecting a_e and reduced mass effects we rewrite this as

$$\Delta E_F = \frac{32}{3}\left(\frac{\mu_\mu}{\mu_B}\right)\frac{\mu_B^2}{a_0^3} \tag{2}$$

where

$$\mu_B = \text{Bohr magneton} = \alpha/2 \text{ a.u.}$$
$$[\mathrm{E}] = 2\,\mathrm{hc}\,R_\infty \tag{3}$$

Hence

$$\Delta\nu_F = \frac{16}{3}\alpha^2 c R_\infty\left(\frac{\mu_\mu}{\mu_B}\right) = \frac{16}{3}\alpha^2 c R_\infty\left(\frac{\mu_\mu}{\mu_p}\right)\left(\frac{\mu_p}{\mu_B}\right). \tag{4}$$

Allowing for the reduced mass effect and a_e, we have

$$\Delta\nu_F = \frac{16}{3}\underline{\alpha^2 c R_\infty\left(\frac{\mu_\mu}{\mu_p}\right)}\left(\frac{\mu_p}{\mu_B}\right)\left(1+\frac{m}{m_\mu}\right)^{-3}(1+a_e) \tag{5}$$

and allowing for the Breit (1930) correction

$$\Delta\nu = \Delta\nu_F(1+3\alpha^2/2). \tag{6}$$

Further, allowing for QED corrections other than a_e,

$$\begin{aligned} \Delta\nu &= \Delta\nu_F\left(1+3\alpha^2/2\right)\{QED\} \\ \{QED\} &= \{\text{Recoil} + \text{bound state}\} = 1+\delta_1 \\ \delta_1 &= [-179.2\,ppm + 23.7\,ppm]. \end{aligned} \tag{7}$$

From this we conclude: (a) assuming QED to hold, we can extract from $\Delta\nu_{exp}$ at best one of the quantities underscored in Eqn. 5; (b) conversely, we can check QED at best to the accuracy to which these quantities—essentially μ_μ/μ_p— are known. Assuming e.g. the latter to the 0.15 ppm, one can check QED to 0.1%, which is not very exciting compared to what is known about (g-2) or the Lamb shift, two purely QED quantities.

There remains one consolation award for the muonium (M) enthusiast: to use $\Delta\nu_{exp}$ to determine the spin-dependent polarizability δ_2 of the proton (P).

For this purpose we take the ratio

$$\left(\frac{\Delta\nu(M)}{\Delta\nu(H)}\right)_{exp} = \left(\frac{\mu_\mu}{\mu_p}\right)_{exp}\frac{(1+m/m_p)^3}{(1+m/m_\mu)^3}f, \tag{8}$$

where H stands for hydrogen, and

$$f = [1+\delta_1(M)]/[1+\delta_1(H)+\delta_2(P)]. \tag{9}$$

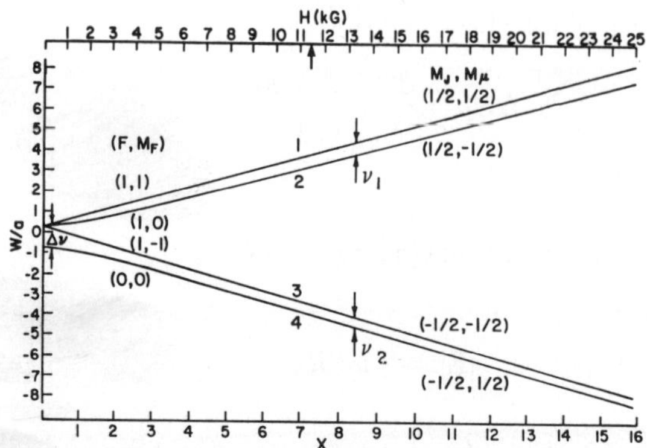

Figure 4: Breit-Rabi diagram of muonium. The arrows along the x and H scales indicate the "magic" field value.

With the most recent results of the Heidelberg-Los Alamos-Yale group we calculate

$$\delta_2(P) = (0.6 \pm 0.5)ppm, \tag{10}$$

which is consistent with zero as well as the available theoretical estimates.

Now that we know why one should study muonium, we can—before discussing some of the experiments—summarize what one needs for such studies:

(1) An intense, highly polarized, possibly low momentum μ^+ beam;
(2) a "wall-less" pressure vessel;
(3) a homogeneous magnet—or a highly field-free region;
(4) RF transition(s) with narrow line width(s).

The experiments fall into two categories, viz. RF transitions in high and in low (possibly zero, see point 3 above) magnetic fields. They both have their advantages and their shortcomings: the Zeeman levels being fully resolved, the high field experiments allow several RF transitions, and hence the determination of both $\Delta\nu$ and μ_μ/μ_p, whereas the low field measurements yield $\Delta\nu$ only.

In the high-field experiments one induces one or the other (occasionally both!) of the transitions indicated with ν_1 and ν_2 in the classic Breit-Rabi diagram reproduced in Figure 4. The transitions are detected by the change in the muon polarization best visualized in the Paschen-Back limit) which in turn manifests itself through the change in the observed forward-backward asymmetry. This method, pioneered by Vernon and his coworkers, is highly reminiscent of the "quenching" used in the Lamb shift and positronium experiments.

The first precise result for $\Delta\nu$ (claimed accuracy 27 ppm) was obtained by the Yale-Nevis collaboration. Their apparatus is shown in Figure 5, and the observed ν_1 resonance (obtained by sweeping the field) is reproduced in Figure 6. Their results

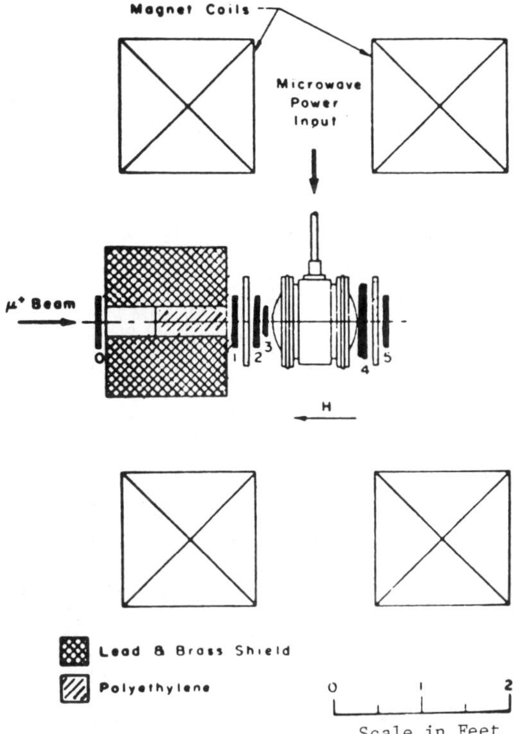

Figure 5: Apparatus used in the first RF experiment by Yale/Columbia

for α, derived by using an external value of μ_μ/μ_p (obtained by precession), lead to considerable controversy. On the one hand, it conflicted with the value of α derived from $\Delta\nu(H)$, on the other it seemed to confirm a value of α derived from the fine-structure splitting in H by Lamb and coworkers (a circumstance occasionally referred to as "Yale togetherness"!). A better agreement between the α values from $\Delta\nu(M)$ and $\Delta\nu(H)$ could be obtained by shifting, following Ruderman, the measured value of f_μ/f_p (i.e. μ_μ/μ_p) by 13 ppm to allow for differences in the diamagnetic shielding of protons and muons in the same sample (water). The discovery of the a.c. Josephson effect rapidly led to a new accurate determination of the fine structure constant, i.e. to α_{WQED} ("without QED"). Figure 7, taken from an exhaustive review[2] of the elementary constants by Taylor, Parker and Langenberg, shows the shockingly erratic evolution of α vs. time. Figure 8, taken from the same paper (but completed), illustrates the "controversy" attended to above.

In view of this situation, and of the fact that Vernon had declined our invitation to transfer his muonium research to the Chicago synchrocyclotron, it became imperative for us to "rethink muonium". Thanks to the Chicago muon channel, we

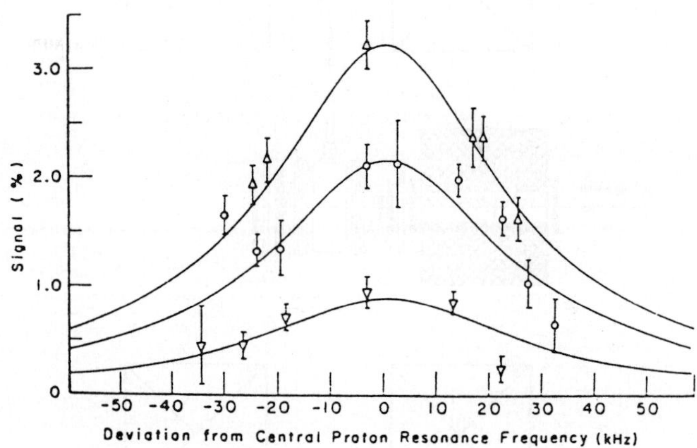

Figure 6: Resonance curves observed with the apparatus of Fig. 4. The curves of decreasing amplitudes correspond to argon pressures of 34, 24 and 25 atm. respectively.

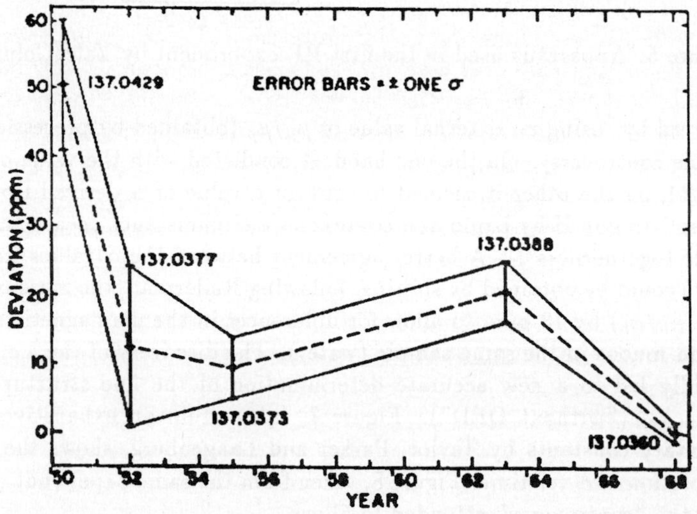

Figure 7: Recommended "best values" of α, 1950-1968. The latest entry is mostly based on the a.c. Josephson effect (i.e. α_{WQED}.)

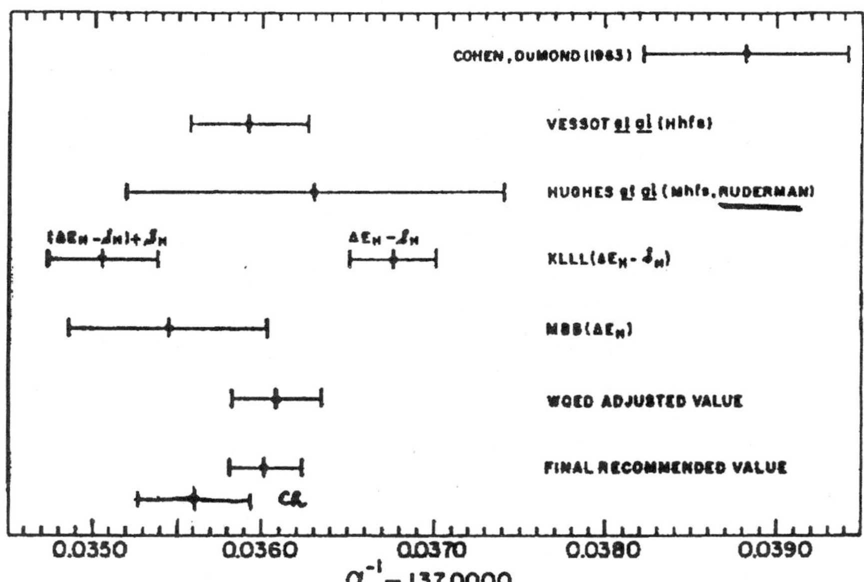

Figure 8: 1969 compilation of α-value presented by TPL, ΔE_H and S_H stand for fine structure interval and Lamb shift in H. The muonium value reported by the Chicago group has been added ("Ch").

had a sufficiently high μ^+ flux, we could work at much lower argon pressures than Vernon was forced to use; we proudly advertised that "our highest pressure point was at the lowest pressure ever used by Yale". We in fact suspected that the linear extrapolation of $\Delta\nu$ vs. density used by Yale was not justified—after all, the curve must turn flat once argon liquefies! To minimize the fraction of muons stopping in the entrance and exit windows of our pressure vessel, we mounted two proportional counters within the gas volume, which respectively in coincidence and anticoincidence with an external scintillator telescope defined a fiducial volume, namely that of the resonance cavity. There remained the problem of procuring a large magnet of sufficiently high uniformity—a task that greatly surpassed our financial and technical capabilities. This procurement problem was brilliantly solved (or, rather, altogether eliminated) thanks to a brilliant insight of my coworker Bob Swanson. He remarked that although, for the Breit-Rabi plot (Figure 4), ν_1 and ν_2 appear to vary monotonically with x (the field in natural units), there exists a "magic field" x_0 such that

$$\frac{\partial\nu_1}{\partial x}\Big|_{x_0} = \frac{\partial\nu_2}{\partial x}\Big|_{x_0} = 0 \tag{11}$$

i.e. where ν_1 and ν_2 are <u>extrema</u>, so that one has near x_0

$$\nu_i(x) = \nu_i(x_0) + \frac{1}{2}\frac{\partial^2\nu_i}{\partial x^2}(\Delta x)^2 + \cdots, \tag{12}$$

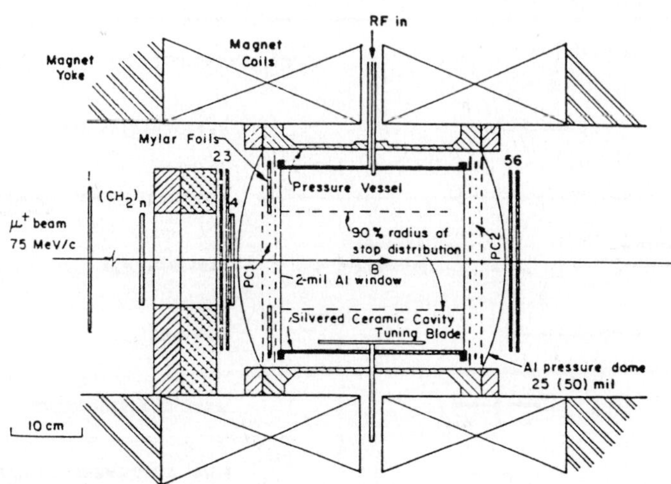

Figure 9: Setup for the "magic field" experiment at Chicago to measure $\nu_1(\nu_{12})$.

i.e. no linear dependence on field inhomogeneities $(\Delta x/x)$. For this idea to be practically useful, two conditions had to be met: (a) x_0 had to have a "reasonable" value, say <20 kG; (b) $\partial^2\nu_i/\partial x^2$ had to be small. But people with good ideas are mostly lucky—both conditions obtained. The magic field turned out to be about 11.3 kG, and the second derivative is so small that an inhomogeneity of 1% causes only a shift by 7 ppm (x_0 is indicated in Figure 4 by an arrow).

Strengthened by this knowledge, we could use a bubble chamber magnet lent by Roger Hildebrand for our high field experiments. Our set up is shown in Figure 9, while Figure 10 shows (a) a typical resonance curve for an observation interval of about 2 μ^+ lifetimes, and (b) the evolution of the signal vs. time, an amusing verification of the Rabi transition rate formula! From the experimental point of view it should be mentioned that the frequency, rather than the field, had to be scanned with this "magic" approach. Our "tunable, stable oscillator" consisted of 12 appropriately selected quartz crystals, inserted in succession by hand into a makeshift thermostat–a far cry from the deluxe synthesizer used in the East. In fact, the whole Chicago muonium work was generally done on a shoestring. Let me quote from one of our papers:

> This experiment was made possible through the generosity of many individuals and companies. We are indebted to Professor R. H. Hildebrand for loaning us his magnet, to Dr. E. L. Ginzton (Varian Associates) for a 4K3SJ klystron and several precision oscillators, to Mr. W. Hewlett (Hewlett-Packard Co.) for assorted microwave equipment, to Dr. G. Megla (Corning Glass Works), and Professor Ed Condon (University of Colorado) for the rf cavity, and to Dr. D. B. Sinclair (General Radio Corp.)

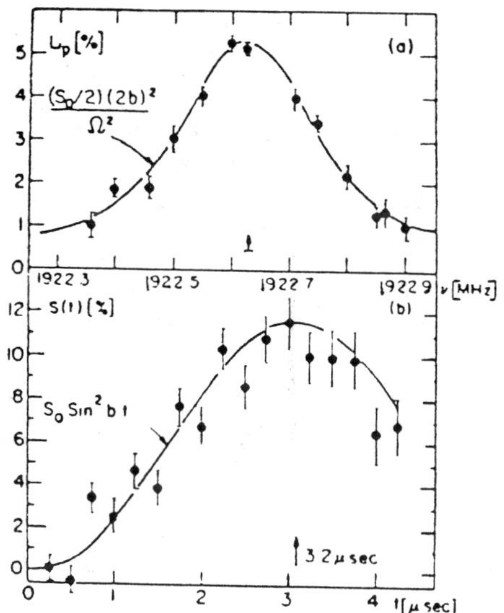

Figure 10: Results obtained with the setup shown in Figure 9: (a) Resonance curve: (b) Signal at the resonance frequency vs. the delay (electron) time after the μ^+ stop.

for a general purpose oscillator.

Our next step was to measure both ν_1 and ν_2 at the magic field, in order to determine both $\Delta\nu$ and μ_μ. To increase the efficiency, we decided to perform a "double resonance", i.e. to excite the cavity (whose geometry had been appropriately chosen) simultaneously with ν_1 and ν_2. The trick was to introduce the two frequencies

$$\nu_\pm \equiv \nu_1 \pm \nu_2 \tag{13}$$

and to scan them independently, i.e. keeping ν_- fixed while varying ν_+, or the converse. It is almost obvious that

$$\nu_+ = \Delta\nu, \tag{14}$$

yielding one of the two quantities sought.

To get the meaning of ν_-, one has to fiddle a little with the Breit-Rabi formula. One readily finds

$$\nu_-/\nu_+ = 2[G(1-G)]^{1/2}, \tag{15}$$

where

$$G \equiv g'_\mu(M)/[g'_\mu(M) - g_j(M)], \tag{16}$$

with $g'_\mu = g_\mu(m_e/m_\mu)$. The argument M serves to recall that bound state and diamagnetic corrections have to be allowed for.

These Chicago experiments yielded α to an accuracy of about 1 ppm, matching the best WQED value then available. We had the pressure shift of $\Delta\nu$ under complete control. We could clearly demonstrate that it contained a quadratic term, barely significant at our pressures but relevant at those used in the Yale/Nevis experiments.

All this success made us rather happy or—in retrospect—somewhat cocky. Thus, at the 1969 Intermediate Energy Physics Conference at Columbia(!) we presented a transparency which said:

Columbia Pictures Proudly Present
I WONDER WHERE MY ALPHA WENT
Starring
V. W. Hughes
IN CAMEO ROLES, ALSO
M. Ruderman W. Lamb, Jr. N. Ramsey

At the 1971 APS Meeting in Washington we had even more fun. I had been asked to present an Invited Paper, and proposed to give one on the Chicago muonium work. When the APS Bulletin arrived, I noted to my surprise that no other than Vernon was going to be my Chairman! I therefore went "well prepared" to the session. I started out by saying that muonium had been discovered by Hughes, that he had pioneered its precision measurements, that all subsequent work was based on his etc. etc. —Vernon was beaming. Then I continued: "Ladies and gentlemen, I must mention that there had been some priority questions in this field in the past. I am happy to say that by now they have all been resolved, as my first slide will show $\cdots$". That slide[1] reproduced a title page of the New York Times, with the headline "Irving confesses writing Hughes papers" (N. B. a bogus autobiography of the recluse millionaire Howard Hughes had been written by the Irving in question).

My group at Chicago next turned its attention to a direct measurement of $\Delta\nu$ by a zero-field transition, an experiment already done at Nevis. At our low argon pressures, the main remaining limitation was the line width. To overcome this radically, we introduced "Ramsey resonances", based entirely on Ramsey's separated oscillatory field method used in molecular beam resonances. There an atom (or molecule, or neutron) is subjected to the transition-inducing RF field in two distant regions, separated by a flight time T; the frequency resolution is of order 1/T. In its rest frame, the atom thus sees two successive RF pulses separated by T. Clearly, it was then enough to "hit" the muonium (at rest in the gas, so to speak) with two pulses separated by a T several times longer than τ, the muon life time. In the original Ramsey scheme, the atom sees a magnetic field B and its spin processes about it during the interval T. In our case, the muon spin simply oscillates back and forth between the two RF pulses, obviously with the frequency $\Delta\nu$! Figure 11 shows our "Ramsey resonance", actually obtained with its Silsbee variant to achieve a dispersion shape (a phase difference is introduced between pulses).

[1]Projected at the Conference.

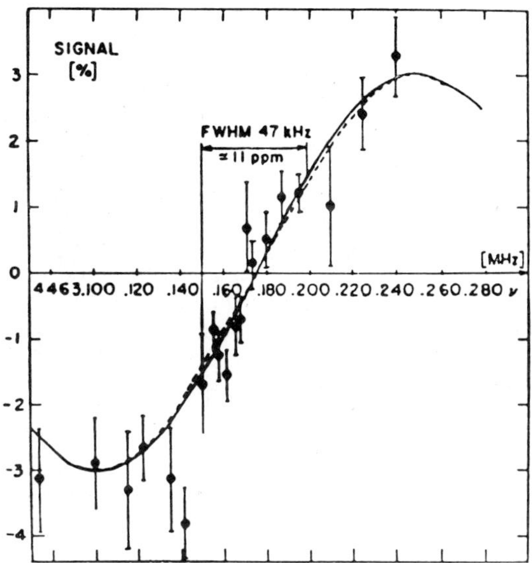

Figure 11: "Ramsey resonance" observed by the Chicago-San Diego Collaboration in Krypton.

This is almost the end of the story, as far as our own contributions to muonium research are concerned. In the summer of 1972 the Chicago synchrocyclotron was definitively shut down, in view of the fact that in the Fall of that year the rejuvenated Nevis cyclotron—to which we would have access— would start operating. That operation being delayed (actually to date, 1991, it is still not operating!!), we moved to the NASA synchrocyclotron for one last try: a "Ramsey resonance" in high field to get a better handle at g_μ'. History was repeating itself—just as the pressure shift of $\Delta\nu$ has been a limiting factor in the work of Vernon et al. at Nevis, now the pressure shift of quantity $g_j(M)$ was becoming our main bother.

Vernon's approach to physics seems to have been inspired by the following remarks of one of his eminent relatives

> "While it is never safe to affirm that the future of Physical Science has no marvels in store even more astonishing than those of the past, it seems probably that most of the grand underlying principles have been firmly established and that further advances are to be sought chiefly in the rigorous application of these principles to the phenomena which come under our notice.
>
> It is here that the science of measurement shows its importance—where quantitative results are more to be valued than qualitative work. An eminent physicist has remarked that the future truths of Physical Science are to be looked for in the sixth place of decimals.

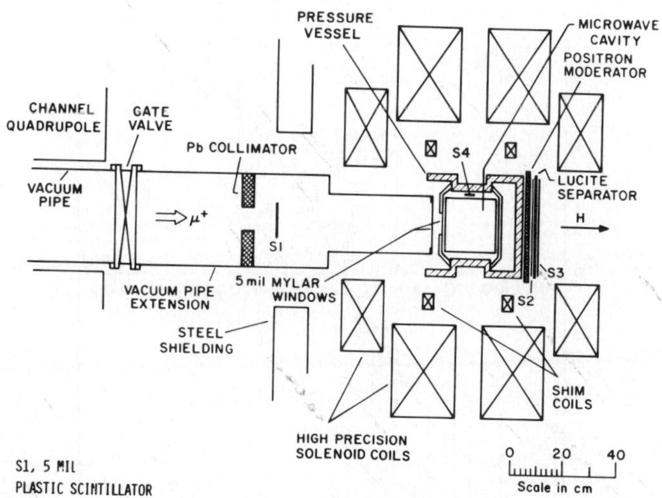

Figure 12: Apparatus used in the latest high field measurements by the Heidelberg-LAMPF-Yale collaboration.

> In order to make such work possible, the student and investigator must have at his disposal the methods and results of his predecessors, must know how to gauge them, and to apply them to his own work; and especially must be have at his command all the modern appliances and instruments of precision which constitute a well equipped laboratory,—without which results of real value can be obtained only at immense sacrifice of time and labor."

These remarks were written (in 1895) by A. A. Michelson, who was directly related to Vernon's first wife, Inge.

After 1973 the field of precision measurements on muonium became almost exclusively the fiefdom of the Heidelberg-Yale-LAMPF collaboration (I use the word fiefdom advertently, since the Heidelberg group is lead by a baron, namely Gisbert zu Putlitz). Figure 12 shows their high-field apparatus, and Figure 13 reproduces some of their fine resonance curves vs. H (note $\nu_{12} = \nu_1$, $\nu_{34} = \nu_2$). Finally, Figure 14 summarizes the evolution of the claimed accuracies in $\Delta\nu$ over a period of 30 years.

The field is far from exhausted. Vernon and his allies are pursuing the excited states of muonium. I am sure of two things: (a) at Vernon's 80th birthday party precision results on these states will be reported here, and (b) I shall not be asked to speak.

In closing, let me borrow a phrase from the IRS and wish you, Vernon, many happy returns!

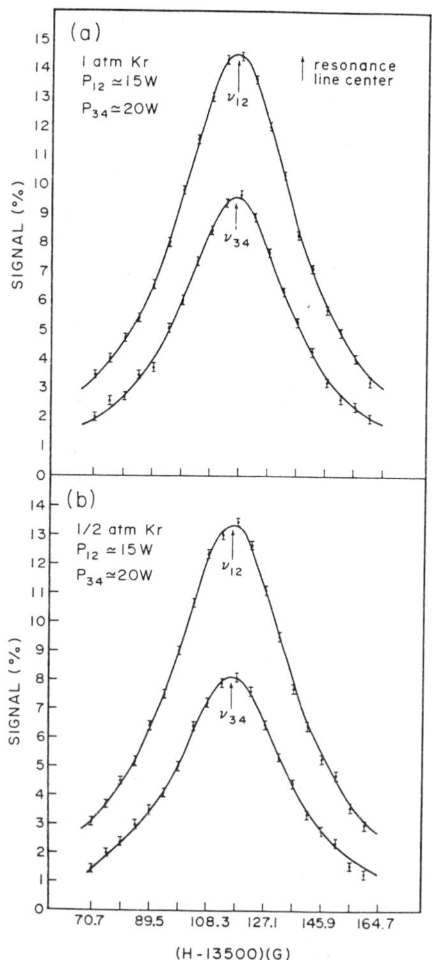

Figure 13: Resonance curves for ν_1 $(=\nu_{12})$ and $\nu_2(=\nu_{34})$ obtained with the apparatus of Fig. 12. Running time for each pair of curves 2 hours!

Postscript

This is the transcript of a talk and not a review article. For that reason, only very rarely do I quote references. The interested reader is referred to the article "Muonium" prepared by V. W. Hughes and G. zu Putlitz for Kinoshita's book on QED published by World Scientific (which should be available before the present text gets printed).

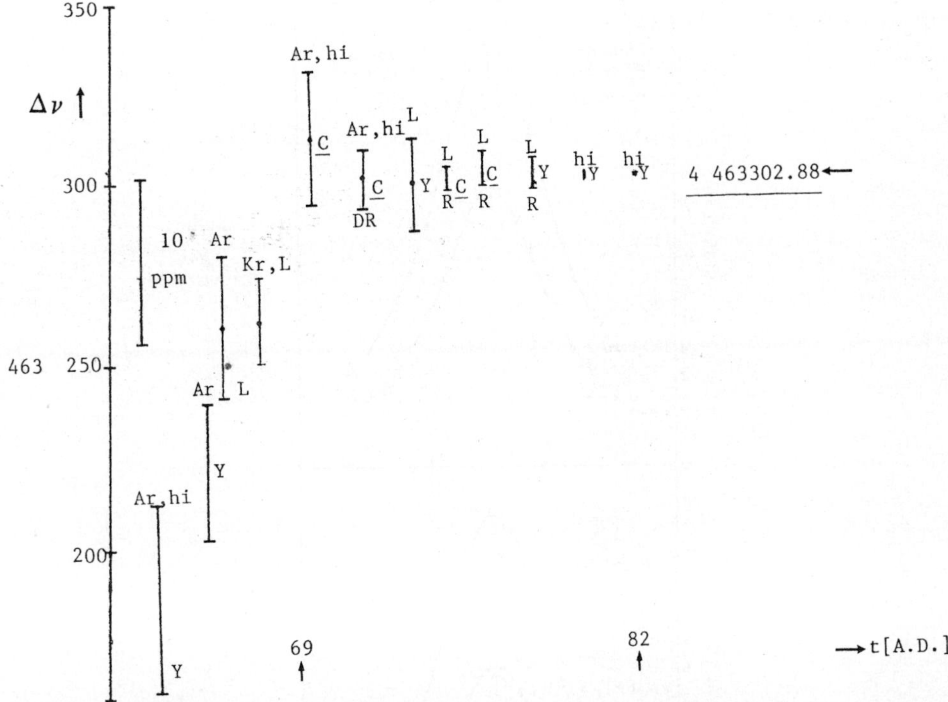

Figure 14: Accuracy in $\Delta\nu_{exp}$ vs. time hi=high, L=low field; DR=double resonance, R=Ramsey resonance, Y=Yale (et al.), C=Chicago (et al.).

References

1. V. W. Hughes and V. L. Telegdi, *Bull. Am. Phys. Soc.* **3**, 224 (1958).
2. B. N. Taylor, W. H. Parker and D. N. Langenberg, *Rev. Mod. Phys.* **41**, 375 (1969).

Muon g-2 and Muonium Hyperfine Structure: Theory Update

T. Kinoshita
Newman Laboratory of Nuclear Studies,
Cornell University, Ithaca, NY 14853

This paper is dedicated to Vernon Hughes in celebration of his seventieth birthday. I wish to convey my deep appreciation for his crucial contributions to high precision measurements in atomic and particle physics, and look forward to many more years of productive research.

Introduction

Today I should like to discuss two separate topics: the muon anomalous magnetic moment and the hyperfine structure of the muonium ground state. Of course, they are not entirely unrelated: both are concerned with learning more about the property of the muon and its interactions. Although theories of these topics have been reviewed recently,[1,2] it would not be useless to update them from time to time in preparation for the new experiments on the muon g-2 at the Brookhaven National Laboratory and the muonium hyperfine structure at the Los Alamos National Laboratory, both being pursued vigorously under Vernon's leadership.

One of the best ways to test the validity of QED is to compare the fine structure constant α obtained from theory and experiment of some selected processes. I shall examine the muonium hyperfine structure from this viewpoint and discuss how much its theory and measurement must be improved to be competitive with other high precision determinations of α.

Muon g-2

The best previous measurements of the muon anomalous magnetic moment a_μ are those carried out in the last CERN experiment.[3] The details of these experiments are discussed in a recent review article by Farley and Picasso.[4] They give the values

$$a_{\mu^-} = 1\ 165\ 936(12) \times 10^{-9},$$
$$a_{\mu^+} = 1\ 165\ 910(11) \times 10^{-9}. \tag{1}$$

Combining these results they obtain

$$a_\mu^{expt} = 1\ 165\ 923(8.5) \times 10^{-9}. \tag{2}$$

The theory of the muon's anomalous moment is formally very similar to that of the electron. However, the physics of the muon anomalous moment is quite different from that of the electron due to the fact that the internal momenta of the muon's structure scale as the muon mass rather than the electron mass. This makes the vacuum-polarization effect very important, and leads to logarithms of m_μ/m_e in the coefficients of the second and higher powers of α/π. For the same reason, the hadronic vacuum polarization and the weak interaction contribute significantly to a_μ. Thus, the theory of muon anomalous magnetic moment divides itself into three parts: pure QED contribution a_μ(QED), hadronic contribution a_μ(had), and weak interaction contribution a_μ(weak). The current status of these contributions will be discussed in the following subsections. Collecting all the theoretical results we find

$$a_\mu^{th} = 116\ 591\ 917(176) \times 10^{-11}, \tag{3}$$

as the best theoretical estimate. The difference between theory and experiment

$$a_\mu^{th} - a_\mu^{expt} = -3.8(8.7) \times 10^{-9} \tag{4}$$

is well within the experimental uncertainty quoted in (2).

The new Brookhaven muon g-2 experiment E821 aims at improving the experimental precision by about a factor of 20. Once this goal is achieved, however, it is the theoretical uncertainty in (3) that must be reduced substantially. This uncertainty comes predominantly from that of the hadronic contribution a_μ(had), which in turn results mainly from the rather poorly controlled systematic errors in the measurements of R, the ratio of total cross sections for hadron production and $\mu^+\mu^-$ production in e^+e^- collisions. The only way to reduce the error in a_μ(had) is to perform new measurements of R, particularly in the energy range from the two pion threshold to about 3 GeV, with strong emphasis on reduction of systematic errors. It is necessary to cut down the error of a_μ(had) to less than 30×10^{-11}.

When the theoretical error in the hadronic contribution is reduced to this size, the new g-2 experiment will enable us to test the validity of the standard model prediction for a_μ(weak), which is about 195×10^{-11}, to a precision of better than 30 percent. This in fact is the primary motivation for this experiment. Note that the weak interaction contribution arises from intermediate virtual states of one-loop order and hence is potentially sensitive to the physics beyond the range of current high energy accelerators.

It has recently been pointed out[5,1] that two-loop electroweak diagrams may contribute as much as 10×10^{-11} to a_μ. It is important to complete the two-loop calculation and remove this theoretical uncertainty before the experimental result becomes available.

In view of the recent measurements[6] of the mass and width of the Z^0 boson, which give a convincing confirmation of the standard model with three generations of lepton families whose neutrinos are (almost) massless, the calculated electroweak contribution to a_μ is very likely to be confirmed in the experiment E821. If this happens, a comparison of theory and experiment of a_μ may impose a very strict lower bound on the mass of not-yet-discovered particles in the energy range of SSC. In this sense, the importance of the Brookhaven muon g-2 experiment is further enhanced by recent results from LEP.

The QED Contribution to a_μ

This contribution can be written in the general form

$$a_\mu(\mathrm{QED}) = A_1 + A_2(m_\mu/m_e) + A_2(m_\mu/m_\tau) + A_3(m_\mu/m_e, m_\mu/m_\tau), \tag{5}$$

where m_e, m_μ, and m_τ are the masses of the electron, muon, and tauon, respectively.

The renormalizability of QED guarantees that the functions A_1, A_2, and A_3 can be expanded in power series in α/π with finite calculable coefficients:

$$A_i = A_i^{(2)}\left(\frac{\alpha}{\pi}\right) + A_i^{(4)}\left(\frac{\alpha}{\pi}\right)^2 + A_i^{(6)}\left(\frac{\alpha}{\pi}\right)^3 + ..., i = 1, 2, 3. \tag{6}$$

The value of A_1 has been evaluated to order α^4 in the calculation of the electron anomaly a_e :[7,8]

$$\begin{aligned} A_1^{(2)} &= 0.5, \\ A_1^{(4)} &= -0.328\ 478\ 965..., \\ A_1^{(6)} &= 1.176\ 13(42), \\ A_1^{(8)} &= -1.434(138). \end{aligned} \tag{7}$$

Here $A_1^{(6)}$ takes into account the recently completed analytic calculation of the diagrams containing a light-by-light scattering subdiagram.[8] Some parts of $A_1^{(6)}$ and all of $A_1^{(8)}$ are evaluated by the numerical integration routine VEGAS.[9] The uncertainties listed are also estimated by VEGAS.

If one chooses as α the most recent value based on the quantized Hall effect:[10]

$$\alpha^{-1} = 137.035\ 997\ 9(32), \tag{8}$$

one obtains from (6) an (7) the result

$$A_1 = 1\ 159\ 652\ 136.2(5.3)(4.1)(27.1) \times 10^{-12}, \tag{9}$$

where the first and second uncertainties come from the numerical uncertainties in $A_1^{(6)}$ and $A_1^{(8)}$, respectively, while the third reflects the uncertainty in α quoted in (8). Further improvement of $A_1^{(6)}$ and $A_1^{(8)}$ is in progress.

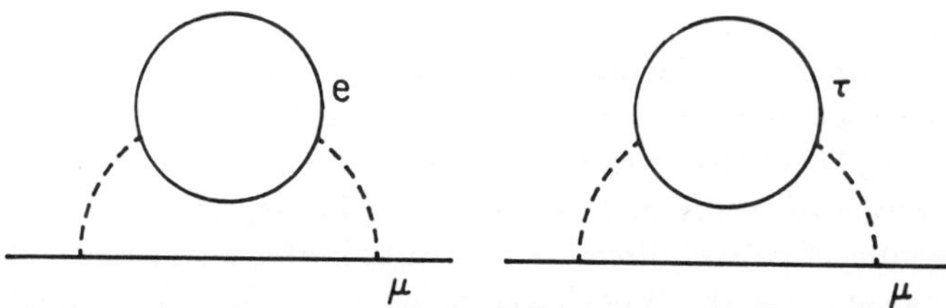

Figure 1: Fourth-order Feynman diagrams contributing to the muon anomalous magnetic moment. In this and following figures fermions are often assumed to propagate in a constant magnetic field. Thus no external magnetic field vertex is shown explicitly.

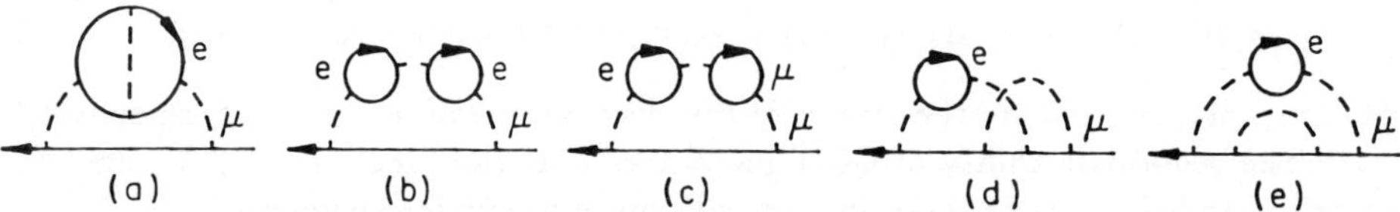

Figure 2: Sixth-order muon vertices obtained by inserting electron vacuum-polarization loops in the photon lines of second-and fourth-order muon vertices. The number of vertex diagrams represented by the diagrams (a), (b), (c), (d), and (e) are 3, 1, 2, 6, and 6, respectively.

As for A_2 and A_3, it is easy to see that $A_2^{(2)}=A_3^{(2)}= A_3^{(4)}=0$. In the fourth order, contributions to $A_2(m_\mu/m_e)$ and $A_2(m_\mu/m_\tau)$ come from the Feynman diagrams shown in Fig. 1. Both are known analytically.[11] Thus their numerical errors arise exclusively from the uncertainties in the measurements of the mass ratios m_e/m_μ and m_μ/m_τ. To the accuracy of interest these contributions are

$$A_2^{(4)}(m_\mu/m_e) = 1.094\ 258\ 28(5) \tag{10}$$

and

$$A_2^{(4)}(m_\mu/m_\tau) = 7.745(4) \times 10^{-5}. \tag{11}$$

The best mass values at present[12] are m_e=0.510 999 06(15) MeV (0.30ppm), m_μ = 105.658 389(34) MeV (0.32 ppm), and m_τ = 1784.1(+2.7/-3.6) MeV. Note, however, that I used in (10) the value 206.768 262(30) (0.15 ppm) for the ratio m_μ/m_e, which is more accurately known than m_e or m_μ separately.

In the sixth order there are 24 Feynman diagrams contributing to $A_2^{(6)}(m_\mu/m_e)$. These include 18 diagrams containing electron vacuum-polarization loops and 6 diagrams containing light-by-light scattering subdiagrams. Typical diagrams of both kinds are shown in Fig. 2 and Fig. 3, respectively.

Unfortunately these contributions are not known analytically at present. The most accurate value for the sum of contributions of diagrams of Fig. 2 is one obtained

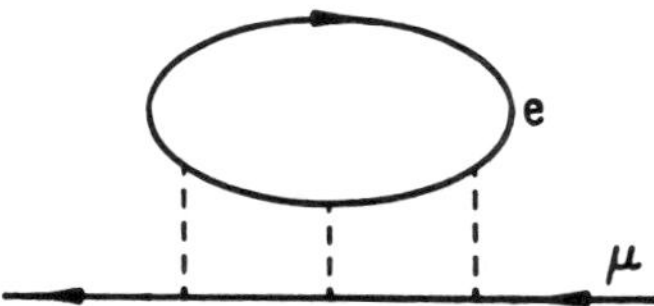

Figure 3: Sixth-order muon vertex obtained by insertion of an electron-loop light-by-light scattering subdiagram. This diagram represents six vertex diagrams in which the external magnetic field is inserted in electron lines in all possible ways.

by numerical integration:

$$A_2^{(6)}[\text{Fig.2}] = 1.920\ 50(46), \tag{12}$$

which includes a small improvement over the result quoted in Ref. 1. By far the most important numerically, however, is the contribution of six diagrams containing light-by-light scattering subdiagrams represented by Fig. 3:[1]

$$A_2^{(6)}[\text{Fig.3}] = 20.947\ 1(29). \tag{13}$$

From (12) and (13) one finds

$$A_2^{(6)}(m_\mu/m_e) = 22.867\ 6(30). \tag{14}$$

The sixth-order term containing the tauon vacuum-polarization contribution is small:[13]

$$A_2^{(6)} = (m_\mu/m_\tau) = 6.9 \times 10^{-5}. \tag{15}$$

Note that $\ln(m_\mu/m_e)$ and mass-independent terms of $A_2^{(6)}(m_\mu/m_e)$ can be determined algebraically by a renormalization group consideration.[14,15] However, these terms do not provide enough accuracy to check the result (12). The situation has been improved significantly by a recent calculation of Samuel and Li[13] who have evaluated all coefficients of terms up to $(m_e/m_\mu)^2$ analytically. Their result is

$$A_2^{(6)}[\text{Fig.2}] = 1.920\ 15 + O\left(\left(\frac{m_e}{m_\mu}\right)^3 \ln^2\left(\frac{m_\mu}{m_e}\right)\right). \tag{16}$$

This result confirms (12) within the uncertainties.

The eighth-order term $A_2^{(8)}(m_\mu/m_e)$ has contributions from 469 Feynman diagrams. As was reported previously[1], numerical evaluation of these integrals leads to

$$A_2^{(8)}(m_\mu/m_e) = 126.92(41), \tag{17}$$

where the dominant contribution comes from the diagrams of Fig. 4 with (l_1, l_2)=(e, e).

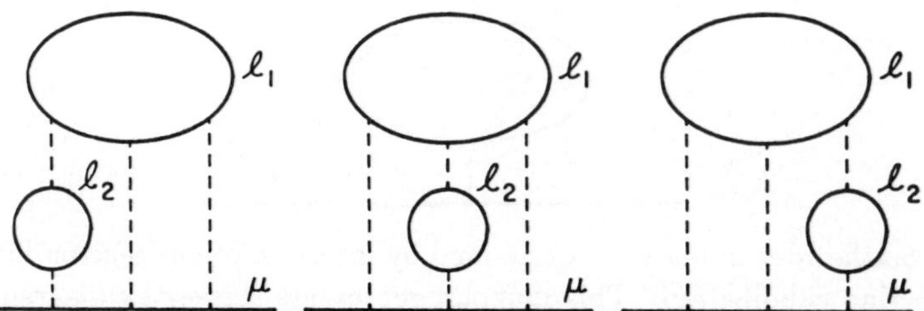

Figure 4: Muon vertex diagrams obtained by inserting a second-order electron vacuum polarization loop in the diagrams of Fig. 3, where $(l_1,\ l_2) = (e, e)$.

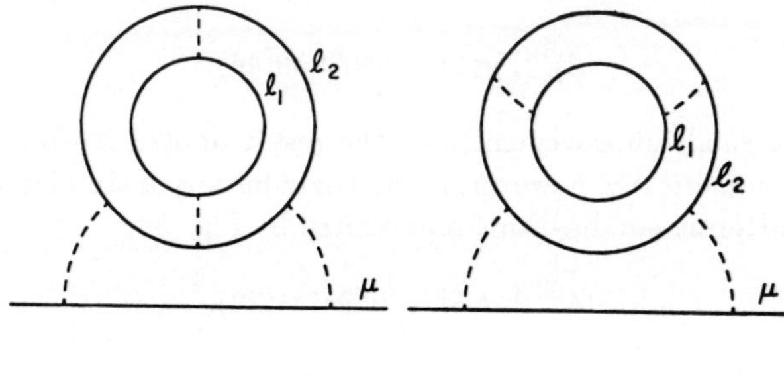

Figure 5: Muon vertex diagrams containing vacuum-polarization diagrams which have two electron loops. $(l_1,\ l_2) = (e, e)$.

For diagrams containing one-electron-loop vacuum-polarization subdiagrams, the leading $\ln(m_\mu/m_e)$ terms as well as mass-independent terms can be determined from renormalization group considerations.[14,16] Meanwhile, only logarithmic terms were known for diagrams whose vacuum-polarization subdiagram contains further electron-loop subdiagrams. Recently, however, it has been shown that mass-independent terms of such diagrams can also be determined by a renormalization group approach.[17] Applied to the diagrams of Fig. 5 (with (l_1, l_2)=(e, e)), this method yields

$$\begin{aligned} A_2^{(8)}[\text{Fig.5}(e,e)] &= \frac{1}{12}\ln^2\left(\frac{m_\mu}{m_e}\right) + \frac{1}{3}(\zeta(3)-2)\ln\frac{m_\mu}{m_e} \\ &+ \frac{1531}{1728} + \frac{5}{12}\zeta(2) - \frac{1025}{1152}\zeta(3) + ... \\ &= 1.452\ 570 + O\left(\frac{m_e}{m_\mu}\right), \end{aligned} \tag{18}$$

which is in good agreement with the numerically evaluated result:[1]

$$A_2^{(8)}[\text{Fig.5(e,e)}] = 1.441\ 6(18). \tag{19}$$

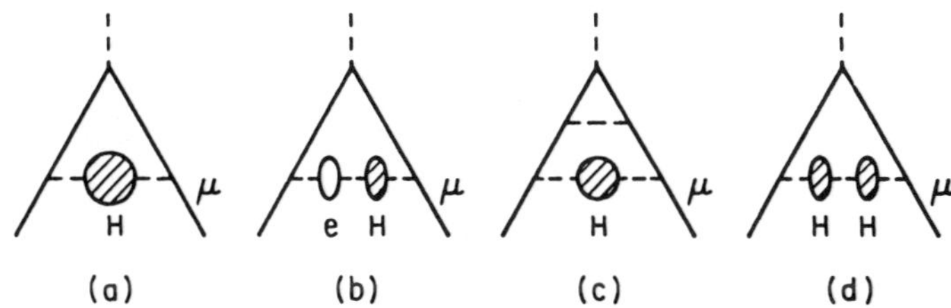

Figure 6: Various hadronic vacuum-polarization contributions to a_μ.

The tenth-order contribution to A_2 has also been estimated. The result is[1]

$$A_2^{(10)}(m_\mu/m_e) = 570(140). \tag{20}$$

The only nonnegligible sixth-order contribution to A_3 is[1]

$$A_3^{(6)}(m_\mu/m_e, m_\mu/m_\tau) = 5.24 \times 10^{-4}. \tag{21}$$

The leading contribution to the eighth-order term $A_3^{(8)}$ arises from a muon vertex that contains an electron light-by-light scattering subdiagram and a tauon vacuum-polarization loop and another in which the role of electron and tauon are interchanged. (See Fig. 4 with $(l_1, l_2) = (e, \tau), (\tau, e)$). Their values have been evaluated numerically. Their sum is given by[1]

$$A_3^{(8)}(m_\mu/m_e, m_\mu/m_\tau) = 0.079(3). \tag{22}$$

Collecting the results (9), (10), (11), (14), (15), (17), (20), (21), and (22), we obtain

$$a_\mu(\text{QED}) = 1\ 165\ 846\ 951(44)(28) \times 10^{-12}. \tag{23}$$

This result is in good agreement with the value quoted in Ref. 1. Note, however, that some of the contributing terms have undergone larger changes.

The Hadronic Contribution to a_μ

The sensitivity of the muon anomaly to the hadronic structure of the photon was first pointed out by Bouchiat and Michel[18] and Durand[19], who showed the dramatic enhancement effect of low-lying resonances such as the ρ resonance. There are several recent evaluations of the hadronic vacuum-polarization contribution to the second-order QED vertex diagram shown in Fig. 6(a):

$$
\begin{aligned}
a_\mu(\text{Fig.6(a)}) &= 7068(59)(164) \times 10^{-11}, && \text{(Ref.20)} \\
&= 7100(105)(49) \times 10^{-11}, && \text{(Ref.21)} \\
&= 684(11) \times 10^{-10}, && \text{(Ref.22)} \\
&= 7052(60)(46) \times 10^{-11}, && \text{(Ref.23)} \\
&= 7048(105)(46) \times 10^{-11}. && \text{(Ref.23)}
\end{aligned}
\tag{24}
$$

The apparently good agreement between these estimates is somewhat misleading. It is due to the fact that they are mostly dealing with the same set of experimental data on $\sigma_{total}(e^+e^- \to \text{hadrons})$.

The considerable difference in error estimates is partly due to different definitions of errors among different evaluations in (24) but primarily reflects different treatment of systematic errors, which are only poorly understood at best and very difficult to evaluate from the available information. The difficulty of dealing with the systematic errors was previously noted in footnote a to Table III of Ref. 20.

The first and second errors in the first two lines of (24) are statistical and systematic, respectively. The first errors in the last two lines of (24) are combinations of statistical and systematic errors with varying degrees of optimism, while the second errors are referred to as model errors. Although the evaluation of the last two lines of (24) relies on a more elaborate treatment of analytic structure of the pion charge form factor and utilizes some data not available earlier, the apparent improvement is mostly a result of rather optimistic estimates of systematic errors.

Very recently, another estimate of $a_\mu(\text{Fig.6}(a))$ has been made:[24]

$$a_\mu(\text{Fig.6(a)}) = 7244.7(66.3)(258.7) \times 10^{-11}. \tag{25}$$

This evaluation emphasizes the large systematic errors in the multi-hadron production cross section in the energy range of 1 to 3 GeV.

Clearly, a real progress in the evaluation of the hadronic contribution must wait for better measurements of R, the ratio of $\sigma_{total}(e^+ + e^- \to \text{hadrons})$ and $\sigma_{total}(e^+ + e^- \to \mu^+ + \mu^-)$. The most important contribution to $a_\mu(\text{Fig.6(a)})$ comes from the energy range below 1 GeV, which is dominated by the ρ resonance. The challenge in this region is to improve the already-well-measured R even further. New measurements of R at VEPP-2M at Novosibirsk are expected to contribute significantly to this improvement. However, the most promising method in reducing the error in this energy range will be along the line of the CERN NA7 experiment in which stationary e^- targets are bombarded with a high energy e^+ beam. A preliminary but very encouraging result[28] of this approach has been incorporated in (24). Further measurement along this line is urgently needed. A comparable precision may also be achievable using a uranium-liquid argon calorimeter as a detector.[29] This method might work for multi-pion production measurements, too. Use of asymmetric collider has also been explored.[30]

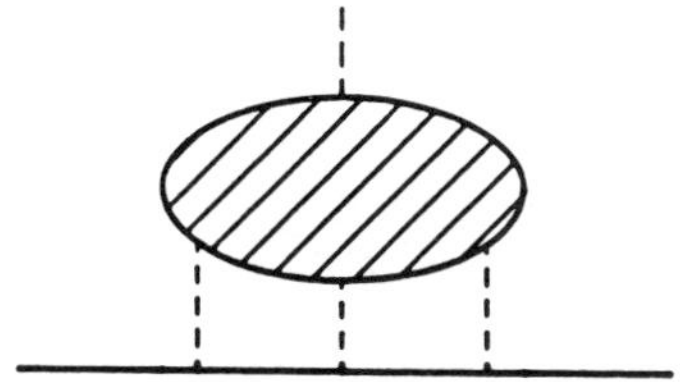

Figure 7: Hadronic light-by-light scattering contribution to a_μ.

The second major source of systematic error in a_μ(Fig.6(a)) is of a different nature. It is due to the relatively poor quality of experimental data in the 1 - 3 GeV range. (In fact, how to handle systematic errors in this region is the cause of difference between (24) and (25).) The large systematic error there reflects the difficulty of measuring exclusive multi-hadron production cross section. Computing R as a sum over exclusive channels results in systematic errors of about 20 percent of their contribution to a_μ. Here, what is needed is a brand new experiment in which $\sigma_{total}(e^+ + e^- \rightarrow \text{hadrons})$ is measured inclusively rather than summing up the contributions of exclusive channels. Currently, VEPP-2M is designed to cover energies up to 1.4 GeV. In view of the fact that there is no other suitable e^+e^- collider operating or being constructed in the relevant energy range, it is very important to extend the energy of VEPP-2M upwards as much as possible.

There are also small contributions from the higher-order hadronic terms represented by the diagrams of Figs. 6 (b), (c), and (d):[20]

$$a_\mu(\text{Figs.6}(b) - (d)) = -90(5) \times 10^{-11}. \tag{26}$$

Theoretically most difficult is the evaluation of the hadronic light-by-light scattering contribution shown in Fig. 7. No way has been found to relate this contribution to some observable hadronic amplitude or cross section. The best estimate thus far, obtained by saturating low-lying hadronic states with mesons, is[20]

$$a_\mu(\text{Fig.7}) = 49(5) \times 10^{-11}. \tag{27}$$

I emphasize that the prediction (27) is model-dependent. It is a challenge to theorists to make it less dependent on specific hadron model.

From (26), (27) and the first line of (24), we obtain

$$a_\mu(\text{had}) = 7\ 027(175) \times 10^{-11}, \tag{28}$$

as a tentative estimate of the hadronic contribution.

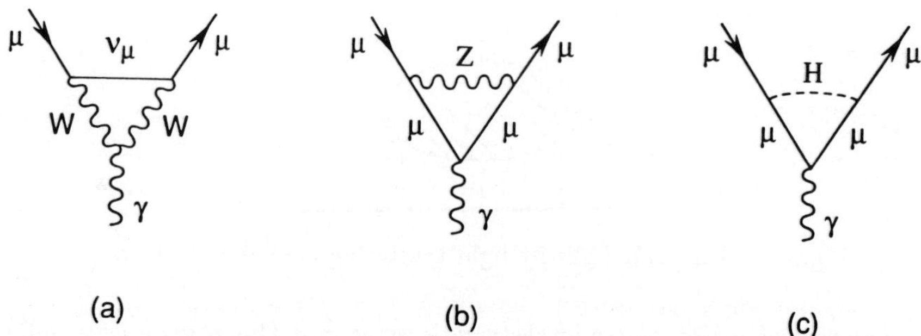

Figure 8: One loop weak corrections to a_μ in the standard model.

The Weak Interaction Contribution to a_μ

The one-loop weak corrections to a_μ in the standard model are graphically shown in Fig. 8. They consist of three parts

$$a_\mu^{weak}(\text{one}-\text{loop}) = a_\mu^W + a_\mu^Z + a_\mu^H \tag{29}$$

coming from diagrams involving W, Z, and Higgs bosons, respectively. Each contribution is separately finite and given by[25]

$$\begin{aligned} a_\mu^W &= \frac{10}{3}\frac{G_F m_\mu^2}{8\sqrt{2}\pi^2} + O\left(\frac{m_\mu^4}{m_W^4}\right), \\ a_\mu^Z &= -\left(\frac{5}{3} - \frac{(3-4\cos^2\theta_W)^2}{3}\right)\frac{G_F m_\mu^2}{8\sqrt{2}\pi^2} + O\left(\frac{m_\mu^4}{m_W^4}\right), \\ a_\mu^H &= 3F\left(\frac{m_H^2}{m_\mu^2}\right)\frac{G_F m_\mu^2}{8\sqrt{2}\pi^2} \end{aligned} \tag{30}$$

where $G_F = 1.166\ 37 \times 10^{-5}\text{GeV}^{-2}$, $\sin^2\theta_W \simeq 0.233$ and F is a known function. The current experimental information on the Higgs mass indicates that the a_μ^H is completely negligible.[26] Thus we obtain

$$a_\mu^{weak}(\text{one}-\text{loop}) = 195(1) \times 10^{-11}. \tag{31}$$

It has been recently pointed out, however, that the $\gamma\gamma Z$ electron triangle anomaly diagrams induce a relatively large two-loop contribution[5]

$$-\frac{3\alpha^2}{8\pi^2\sin^2\theta_W}\left(\frac{m_\mu^2}{m_W^2}\right)\ln\left(\frac{m_Z}{m_\mu}\right) \simeq -10 \times 10^{-11}$$

to the muon anomalous moment. This effect will be cancelled in part by analogous u and d quark triangle diagrams which remove the short distance anomaly. Summing over the three generations of leptons and quarks can bring the total contribution of the two-loop fermion triangle diagrams to $\sim 1\times10^{-10}$. There are in addition two-loop weak contributions of relative order $\alpha/\pi \sin^2\theta_W \simeq 0.01$ with respect to the one-loop effects in (30). They could potentially add up to $O(1\times10^{-10})$. Thus we assume that the uncertainty coming from the two-loop electroweak correction may be as large as $\pm\ 10 \times 10^{-11}$. The best estimate of the electroweak contribution to a_μ will thus be[1]

$$a_\mu^{weak} = 195(10) \times 10^{-11}. \tag{32}$$

A complete two-loop calculation should be carried out to eliminate this theoretical ambiguity.

"New Physics" Contributions to a_μ

One of the main purposes of the new muon g-2 experiment is to obtain strict constraints on "new physics" possibilities, such as extra gauge bosons, additional Higgs scalars, supersymmetry, and compositeness of leptons. Assuming that a_μ is measured to the precision of 40×10^{11}, one can obtain more stringent bounds than those available at present on "new physics" mass scales and couplings. For a review of these topics see Ref. 1. Additional work based on an extension of supersymmetric standard model can be found in Ref. 27.

Muonium Hyperfine Structure

The hyperfine splitting between the spin-0 and spin-1 levels of the muonium ground state is one of the most precisely measured quantities in atomic physics. The most recent measurement gives:[31]

$$\Delta\nu(\text{exp}) \;=\; 4\ 463\ 302.88\ (16)\text{kHz}. \tag{33}$$

For experimental details see the recent review article of Hughes and zu Putlitz.[32]

As is well known, the bulk of the hyperfine splitting is given by the Fermi formula

$$E_F = \frac{16}{3}(Z\alpha)^2 R_\infty \frac{\mu_\mu}{\mu_B}\left[1+\frac{m_e}{m_\mu}\right]^{-3}, \tag{34}$$

where Z is the charge of the muon in units of the electron charge, R_∞ is the Rydberg constant for infinite nuclear mass, μ_μ is the muon magnetic moment, μ_B is the Bohr magneton, and m_e and m_μ are the electron and muon masses, respectively. Of course Z is equal to one for the muon, but it is useful to keep it in the formula in order to distinguish contributions arising from binding effects ($Z\alpha$) and radiative corrections (α).

Many correction terms (of both α and $Z\alpha$ type) have been calculated since the work of Fermi. They have been reviewed recently by Sapirstein and Yennie.[2]

Conceptually, it is useful to separate the QED result into three types of contributions: non-recoil terms which include radiative corrections, pure recoil corrections, and radiative-recoil corrections. Following custom the contribution of the hadronic vacuum polarization is included in $\Delta\nu$(rad-recoil). Adding the weak interaction correction, the theoretical result may be written as

$$\Delta\nu(\text{theory}) = \Delta\nu(\text{rad}) + \Delta\nu(\text{recoil}) + \Delta\nu(\text{rad}-\text{recoil}) + \Delta\nu(\text{weak}). \tag{35}$$

The non-recoil terms obtained thus far can be written as

$$\begin{aligned}
\Delta\nu(\text{rad}) &= E_F(1+a_\mu)\Big(1+\frac{3}{2}(Z\alpha)^2 + a_e + \alpha\,(Z\alpha)(\ln^2 - \frac{5}{2}) \\
&- \frac{8\alpha(Z\alpha)^2}{3\pi}\ln(Z\alpha)(\ln(Z\alpha) - \ln 4 + \frac{281}{480}) + \frac{\alpha(Z\alpha)^2}{\pi}(15.38 \pm 0.29) \\
&+ \frac{\alpha^2(Z\alpha)}{\pi}\Big[-\frac{4}{3}\ln^2\frac{1+\sqrt{5}}{2} - \frac{20\sqrt{5}}{9}\ln\frac{1+\sqrt{5}}{2} \\
&+ \frac{608}{45}\ln 2 + \frac{\pi^2}{9} - \frac{38\pi}{15} + \frac{91639}{37800} - 0.310\;742\cdots + D_1'\Big]\Big).
\end{aligned} \tag{36}$$

Note that we use E_F as defined in (34), so that an explicit factor of $(1+a_\mu)$ appears, where a_μ represents the anomalous magnetic moment of the muon. The $\alpha(Z\alpha)$ and $\alpha(Z\alpha)^2$ radiative corrections are well known.[33] The $\alpha^2(Z\alpha)$ term is a recent evaluation[34] of the contribution of the diagrams a, b, c and d of Fig. 9. The D_1' term represents uncalculated radiative corrections involving two-virtual-photon exchange, as illustrated by the diagrams e and f of Fig. 9.

The presently known recoil corrections add up to

$$\begin{aligned}
\Delta\nu(\text{recoil}) &= E_F\Big(-\frac{3Z\alpha}{\pi}\frac{m_e m_\mu}{m_\mu^2 - m_e^2}\ln\frac{m_\mu}{m_e} \\
&+ \frac{\gamma^2}{m_e m_\mu}\Big[2\ln\frac{m_r}{2\gamma} - 6\ln 2 + \frac{65}{18}\Big]\Big),
\end{aligned} \tag{37}$$

where $\gamma \equiv Zm_r\alpha$. There is no new development concerning this term.[35]

The radiative-recoil contributions, which arise from both lepton lines and from vacuum polarization, are given by

$$\begin{aligned}
\Delta\nu(\text{rad}-\text{recoil}) &= E_F\frac{\alpha(Z\alpha)}{\pi^2}\frac{m_e}{m_\mu}\Big(-2\ln^2\frac{m_\mu}{m_e} + \frac{13}{12}\ln\frac{m_\mu}{m_e} \\
&+ \frac{21}{2}\zeta(3) + \zeta(2) + \frac{35}{9} + (1.91 \pm 0.26) \\
&+ \frac{\alpha}{\pi}\Big[-\frac{4}{3}\ln^3\frac{m_\mu}{m_e} + \frac{4}{3}\ln^2\frac{m_\mu}{m_e} + O\Big(\ln\frac{m_\mu}{m_e}\Big)\Big]\Big),
\end{aligned} \tag{38}$$

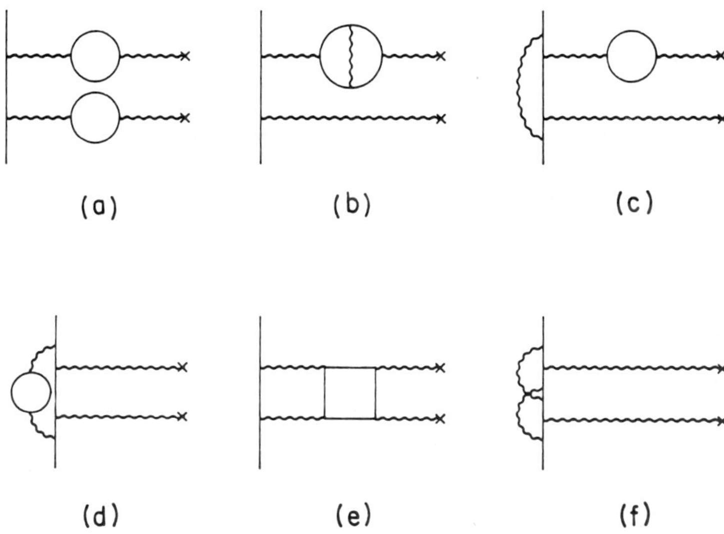

Figure 9: Representative diagrams for the $\alpha^2(Z\alpha)$ radiative corrections to the muonium hyperfine structure in which two virtual photons are exchanged between e^- and μ^+.

The importance of these terms was first pointed out by Caswell and Lepage[36], who evaluated the term proportional to $\ln^2(m_\mu/m_e)$. Other $\alpha(Z\alpha)$ terms have since been evaluated.[37] The hadronic vacuum polarization contribution is represented by the (1.91 ± 0.26) term. The $\ln^3$ and $\ln^2$ parts of the $\alpha^2(Z\alpha)$ term have been evaluated by Eides *et al.*[38]

Finally there is a small contribution due to the Z^0 exchange. The standard-model calculation gives[39]

$$\Delta\nu(\text{weak}) \simeq 0.065\text{kHz}. \tag{39}$$

If one uses the values of α^{-1} given in (8) and the values[12]

$$\begin{aligned} R_\infty &= 10\ 973\ 731.534\ (13)\text{m}^{-1}, \\ m_\mu &= 105.658\ 389(34)\text{MeV}, \\ \frac{m_\mu}{m_e} &= 206.768\ 262\ (30), \end{aligned} \tag{40}$$

the complete theoretical result, including the weak interaction contribution(39), is

$$\Delta\nu(\text{theory}) = 4\ 463\ 303.63\ (0.66)(0.21)(1.0)\ \text{kHz}, \tag{41}$$

where the first and second errors reflect the uncertainties in the measurements of m_μ and α^{-1} in (40) and (8) respectively, and the third is an order of magnitude estimate of the uncalculated radiative correction contribution D_1'.

The difference between theory and measurement is given by

$$\Delta\nu(\text{theory}) - \Delta\nu(\text{exp}) = +0.75\ (0.16)(0.66)(0.21)(1.0)\ \text{kHz} \tag{42}$$

where the first error comes from the $\Delta\nu$ measurement and the rest are carried over from (41). This shows that QED is valid down to the level of 0.28 ppm.

Another way to test the validity of QED is to calculate the fine structure constant α by comparison of $\Delta\nu$(theory) and $\Delta\nu$(exp). This leads to

$$\alpha^{-1}(\mu\ \text{hfs}) = 137.036\ 008(19)\ (0.14\text{ppm}). \tag{43}$$

As is seen from (33) and (41), the uncertainty in $\Delta\nu$(theory) is much larger than the current uncertainty in $\Delta\nu$(exp). The largest theoretical uncertainty comes from the uncalculated term D_1'. Its evaluation is a challenge to theorists. After that, uncalculated QED contribution will be at the level of one part in 10^8.

Another source of uncertainties in (41) is the muon mass. A more accurate measurement of the muon mass will become available before long.[40]

Muonium and High Precision Determination of α

At present the most precise test of internal consistency of QED is one that is obtained by comparing the muonium hfs and the anomalous magnetic moment of the electron. Unfortunately, such a test cannot go beyond 0.28 ppm because of the uncertainty in (41), in spite of the fact that the anomalous magnetic moment of the electron is known much more precisely both experimentally and theoretically. As of now, testing the validity of QED to a higher precision is feasible only by comparison of α's obtained from the electron anomalous magnetic moment and macroscopic quantum effects discovered in the realm of condensed matter physics, namely the quantized Hall effect and the ac Josephson effect. I shall review these high precision determinations of α briefly in order to indicate how far $\alpha(\mu$ hfs) must be improved before it becomes competitive with them.

Determination of α by the Electron Anomalous Magnetic Moment

Currently, the most accurate way to determine α is by the comparison of experiment and theory of the electron anomalous magnetic moment. The most recent experimental values of the magnetic moment anomaly of the electron and positron are[41]

$$\begin{aligned} a(e^-) &= 1\ 159\ 652\ 188.4(4.3) \times 10^{-12}, \\ a(e^+) &= 1\ 159\ 652\ 187.9(4.3) \times 10^{-12}, \end{aligned} \tag{44}$$

respectively.

The theoretical expression for the electron anomaly a_e may be written as

$$a_e = A_1 + \delta a_e, \tag{45}$$

where, to the precision of interest, A_1 is given by (9) and δa_e, which is the sum of the muon and tauon vacuum-polarization terms $(\alpha/\pi)^2 A_2^{(4)}(m_e/m_\mu), (\alpha/\pi)^2 A_2^{(4)}(m_e/m_\tau)$, the hadronic vacuum polarization term, and the weak interaction effect, has the value[42]

$$\delta a_e = 4.46 \times 10^{-12}. \tag{46}$$

By comparing (44) and (45), we obtain

$$\alpha^{-1}(a_e) = 137.035\ 992\ 25(94)\ (0.0069\ \text{ppm}), \tag{47}$$

where the experimental and theoretical errors contribute 0.0037 ppm and 0.0058 ppm, respectively. Work is in progress to reduce the theoretical error to the level of the experimental error.

Determination of α by the Quantized Hall Effect

When the electrons participating in a Hall effect measurement are confined to a very thin, almost two-dimensional, layer with a strong perpendicular magnetic field and the effect is observed at very low temperatures, there is a tendency for the Hall resistance to be quantized at certain particular values.[43] That is, as some external parameter such as the magnetic field or a gate voltage is varied (depending on the type of device), one observes certain 'plateaus' in the Hall resistance. The Hall resistance (R_H) is defined as the ratio of the voltage (V_H) across the sample to the current (I_H) flowing along the sample. With the use of the exactly defined constants μ_0, the permeability of the vacuum, and c, the speed of light, the quantized Hall resistance R_H determined from the plateau steps is related to the fine structure constant by

$$\alpha^{-1} = \frac{2R_H}{\mu_0 c}. \tag{48}$$

The latest measurement[10] of α based on (48) gives the value quoted in (8). Aside from uncertainties inherent in the measurement itself, this measurement requires precision reference voltage and current standards, as well as an assumption, or proof, that the theory is exact at the required level of interest. As a practical matter, one replaces the need for the precision voltage and current standards by a precision resistance standard. In turn, the resistance standard is calibrated by reference to a 'calculable capacitor'.[44] Any error in this calibration is reflected directly in α^{-1}.

The exactness of the theoretical expression is asserted by the work of Laughlin.[45] A discussion for non-specialists is given by Yennie.[46] Gauge invariance of condensed matter physics Hamiltonian is discussed by Kinoshita and Lepage.[47] Material independence of the quantized Hall effect has recently been established[48] to the level of

$\pm 3.5 \times 10^{-10}$.

Determination of α by the ac Josephson Effect

In the ac Josephson effect, a pair of electrons tunnels through a Josephson junction separating superconductors at two different potentials. Their energy change $2eV$ is related to the frequency ν associated with the transition through

$$h\nu = 2eV. \tag{49}$$

To determine α using the ac Josephson effect, the most precise way available at present is to combine four separate measurements as follows:

$$\alpha^{-2} = \frac{c}{4R_\infty \gamma'_p} \frac{\mu'_p}{\mu_B} \frac{2e}{h}. \tag{50}$$

The latest result based on (50) is[49]

$$\alpha^{-1}(\text{acJ}\&\gamma'_p) = 137.035\ 977\ 0\ (77) \qquad (0.056\ \text{ppm}). \tag{51}$$

The most critical of the four measurements are the ac Josephson effect, which gives a precise value for the combination $2e/h$, and the gyromagnetic ratio of the proton in water (called γ'_p). The frequency ν can of course be determined very precisely, and the voltage across the sample can be related to a voltage standard. The gyromagnetic ratio is the spin-flip frequency of the proton divided by the magnetic field. This frequency can be measured quite exactly, and the magnetic field is obtained by constructing a precision coil with a current known from a precision current standard. Two other important quantities are the Rydberg constant, R_∞, and the proton's magnetic moment (in water) in units of the Bohr magneton, μ'_p/μ_B.

Various theoretical corrections to the ac Josephson effect have been examined and evaluated in the literature.[50] In general these corrections seem to have a negligible effect upon $\alpha(\text{acJ}\&\gamma'_p)$ at least to a part in 10^8. Material independence of the ac Josephson effect has been established [51] to the level of $\pm 2 \times 10^{-16}$.

Determination of α by a Combination Method

It is also possible to determine α from the formula

$$\alpha^{-3} = \frac{R_H}{2\mu_0 R_\infty \gamma'_p} \frac{\mu'_p}{\mu_B} \frac{2e}{h}, \tag{52}$$

which is obtained by combining (48) and (50). Corresponding to (52) one finds[10]

$$\alpha^{-1}(\text{comb.}) = 137.035\ 984\ 0\ (51) \qquad (0.037\ \text{ppm}). \tag{53}$$

Measurements of (8) and (51) depend on a calculable capacitor and are thus sensitive to its uncertainty. The advantage of using (52) is that such a dependency cancels out in it and its error is mainly due to that of γ'_p. The relatively large uncertainty in (51) and (53) reflects the difficulty of constructing very high quality uniform magnetic field which is needed in the measurement of γ'_p. Thus it appears that the quantized Hall effect is more promising for better determination of α, in particular, if the uncertainty in the calibration of the calculable capacitor, which at present is 2.2×10^{-8}, can be reduced significantly.

Discussions

If one uses α of (47) determined from the electron anomalous magnetic moment, the muonium hyperfine splitting becomes

$$\Delta\nu(\text{theory}) = 4\ 463\ 304.00(0.66)(0.07)(1.0)\text{kHz}, \qquad (54)$$

and (42) is replaced by

$$\Delta\nu(\text{theory}) - \Delta\nu(\text{exp}) = +1.12(0.16)(0.66)(0.07)(1.0)\text{kHz}. \qquad (55)$$

If one assumes that QED is valid at least to the precision of 10^{-8}, one may predict that $D'_1 \simeq -2$. Clearly, complete calculation of D'_1 is a matter of high priority now. After that, accuracies of the numerically evaluated term on the second line of (36) and the hadronic vacuum-polarization contribution in (38) must also be improved.

Assuming that the new experiment being prepared at the Los Alamos National Laboratory achieves the precision of ± 0.03 kHz or better, and that theory manages to match this precision, it will be possible to determine $\alpha^{-1}(\mu \text{ hfs})$ to the precision of about 0.03 ppm, which is comparable to the present precision of α determined by the quantized Hall effect. Although this is not accurate enough to match the precision of α obtained from the electron g-2, it is still very important in the sense that it pushes the internal consistency test of QED by an order of magnitude in precision without relying on condensed matter physics. We will then be able to put predictions of condensed matter physics to a more rigorous test.

Acknowledgment

I thank E. Remiddi and M.A. Samuel for communicating their results prior to publication, D.R. Yennie for useful discussions, and Makiko Nio for confirming (39).This work is supported in part by the National Science Foundation. Part of the numerical work was conducted at the Cornell National Supercomputing Facility, which receives major funding from the U.S. National Science Foundation and the IBM Corporation, with additional support from New York State and members of the Corporate Research Institute.

References

1. T. Kinoshita and W. J. Marciano, in *Quantum Electrodynamics*, ed. by T. Kinoshita (World Scientific, Singapore, 1990), pp. 419-478.
2. J. R. Sapirstein and D. R. Yennie, in *Quantum Electrodynamics*, ed. by T. Kinoshita (World Scientific, Singapore, 1990), pp. 560-672.
3. J. Bailey et al., *Phys. Lett.* **B68**, 191 (1977).
4. F. J. M. Farley and E. Picasso, in *Quantum Electrodynamics*, ed. by T. Kinoshita (World Scientific, Singapore, 1990), pp. 479-559.
5. E. A. Kuraev, T. V. Kukhto, and A. Schiller, *Yad. Fiz.*, **51**, 1631 (1990) [*Sov. J. Nucl Phys.* **51**, 1031 (1990)].
6. G. S. Abrams et al., *Phys. Rev Lett.* **63**, 2173 (1989); L3 Collab., B. Adeva et al., *Phys. Lett.* **B231**, 509 (1989); ALEPH Collab., D. Decamp et al., *Phys. Lett.* **B231**, 519 (1989); OPAL Collab., M. Z. Akrawy et al., *Phys. Lett.* **B231**, 530 (1989); DELPHI Collab., P. Aarnio et al., *Phys. Lett.* **B231**, 539 (1989).
7. T. Kinoshita, in *Quantum Electrodynamics*, ed. by T. Kinoshita, (World Scientific, Singapore, 1990), pp. 218-321.
8. E. Remiddi and S. Laporta, private communication.
9. G. P. Lepage, *J. Comput. Phys.* **27**, 192 (1978).
10. M. E. Cage et al., *IEEE Trans. Instrum. Meas.* **38**, 284 (1989).
11. H. H. Elend, *Phys. Lett.* **20**, 682 (1966); **21**, 720 (1966).
12. E. R. Cohen and B. N. Taylor, *Rev. Mod. Phys.* **59**, 1121 (1987).
13. M. A. Samuel and G. Li, *Research Note* 249, Oklahoma State University, 1990.
14. T. Kinoshita, *Nuovo Cimento* **51B**, 140 (1967).
15. B. E. Lautrup and E. de Rafael, *Nuovo Cimento* **64A**, 322 (1969); B. E. Lautrup, *Phys. Lett.* **B32**, 627 (1970); S. J. Brodsky and T. Kinoshita, *Phys. Rev.* **D3**, 356 (1971); R. Barbieri and E. Remiddi, *Phys. Lett.* **B49**, 468 (1974).
16. B. E. Lautrup and E. de Rafael, *Nucl. Phys.* **B70**, 317 (1974).
17. T. Kinoshita, H. Kawai, and Y. Okamoto, *Phys. Lett.*, **B254**, 235 (1991); H. Kawai, T. Kinoshita, and Y. Okamoto, *Phys. Lett.* **B260**, 193 (1991); R. N. Faustov, A. L. Kataev, S. A. Larin, and V. V. Starshenko, *Phys. Lett.* **B254**, 241 (1991).
18. C. Bouchiat and L. Michel, *J. Phys. Radium* **22**, 121 (1961).
19. L. Durand, III., *Phys. Rev.* **128**, 441 (1962).
20. T. Kinoshita, B. Nizic, and Y. Okamoto, *Phys. Rev.* **D31**, 2108 (1985).
21. J. A. Casas, C. Lopez, and F. J. Yndurain, *Phys. Rev.* **D32**, 736 (1985).
22. L. M. Kurdadze et al., *Yad. Fiz.* **40**, 451 (1984) [*Sov. J. Nucl. Phys.* **40**, 286 (1984)].
23. L. Martinovic and S. Dubnicka, *Phys. Rev.* **D42**, 884 (1990).
24. F. Jegerlehner, private communication to Vernon Hughes.
25. R. Jackiw and S. Weinberg, *Phys. Rev.* **D5**, 2396 (1972); I. Bars and M. Yoshimura, *Phys. Rev. D* **6**, 374 (1972); K. Fujikawa, B. W. Lee, and A. I. Sanda, *Phys. Rev.* **D6**, 2923 (1972);W. A. Bardeen, R. Gastmans, and B. E. Lautrup, *Nucl. Phys.* **B46**, 319 (1972).
26. J. Gunion, H. Haber, G. Kane, and S. Dawson, *Higgs Hunter's Guide* (Addison-Wesley,

New York, 1990).
27. R. M. Francis, M. Frank, and C. S. Kalman, *Phys. Rev.* **D43**, 2369 (1991).
28. S. R. Amendolia et al., *Phys. Lett.* **B138**, 454 (1984).
29. W. Willis, private communication.
30. M. Greco, *Frascati Report* No. LNF-88/24(P), 1988(unpublished).
31. F. G. Mariam et al., *Phys. Rev. Lett.* **49**, 993 (1982); E. Klempt et al., *Phys. Rev.* **D25**, 652 (1982).
32. V. W. Hughes and G. zu Putlitz, in *Quantum Electrodynamics*, ed. by T. Kinoshita (World Scientific, Singapore, 1990), pp. 822-904.
33. N. Kroll and F. Pollock, *Phys. Rev.* **84**, 594 (1951); **86**, 876 (1952); R. Karplus, A. Klein, and J. Schwinger, *Phys. Rev.* **84**, 597 (1951); J. R. Sapirstein, E. A. Terray, and D. R. Yennie, *Phys. Rev.* **D29**, 2290 (1984); A. J. Layzer, *Nuovo Cimento* **33**, 1538 (1964); D. Zwanziger, *Nuovo Cimento* **34**, 77 (1964); S. J. Brodsky and G. W. Erickson, *Phys. Rev.* **148**, 26 (1966); J. R. Sapirstein, *Phys. Rev. Lett.* **51**, 985 (1983).
34. M. I. Eides, S. G. Karshenboim, V. A. Shelyuto, *Phys. Lett.* **B229**, 285 (1989); B249, 519 (1990).
35. R. Arnowitt, *Phys. Rev.* **92**, 1002 (1953); G. P. Lepage, *Phys. Rev.* **A16**, 863 (1977); G. T. Bodwin and D. R. Yennie, *Phys. Rep.* **43C**, 267 (1978); G. T. Bodwin, D. R. Yennie, and M. Gregorio, *Phys. Rev. Lett.* **48**, 1799 (1982); W. E. Caswell and G. P. Lepage, *Phys. Lett.* **B167**, 437 (1986).
36. W. E. Caswell and G. P. Lepage, *Phys. Rev. Lett.* **41**, 1092 (1978).
37. E. A. Terray and D. R. Yennie, *Phys. Rev. Lett.* **48**, 1803 (1982); J. R. Sapirstein, E. A. Terray, and D. R. Yennie, *Phys. Rev.* **D29**, 2290 (1984); M. I. Eides, S. G. Karshenboim, and V. A. Shelyuto, *Phys. Lett.* **B202**, 572 (1988); S. G. Karshenboim, V. A. Shelyuto, and M. I. Eides, *Zh. Eksp. Teor. Fiz.* **92**, 1188 (1987) [Eng. transl.:*Sov. Phys. JETP* **65**, 664 (1987)].
38. M. I. Eides, S. G. Karshenboim, and V. A. Shelyuto, *Phys. Lett.* **B216**, 405 (1989).
39. M. A. B. Bég and G. Feinberg, *Phys. Rev. Lett* **33**, 606 (1974); G. T. Bodwin and D. R. Yennie, *Phys. Rep.* **43C**, 267 (1978).
40. V. W. Hughes and G. zu Putlitz, *Comm. Nucl. Phys.* **12**, 259 (1984).
41. R. S. Van Dyck, Jr., P. B. Schwinberg, and H. G. Dehmelt, *Phys. Rev Lett.* **59**, 26 (1987); R. S, Van Dyck, Jr., in *Quantum Electrodynamics*, ed. by T. Kinoshita (World Scientific, Singapore, 1990), pp. 322-388.
42. See, for instance, T. Kinoshita in *New Frontiers in High Energy Physics*, ed. by B. Kursunoglu et al. (Plenum, 1978), pp. 127-143.
43. K. v. Klitzing, G. Dorda, and M. Pepper, *Phys. Rev. Lett.* **45**, 494 (1980).
44. A. M. Thompson and D. G. Lampard, *Nature* (London) **177**, 888 (1956).
45. R. Laughlin, *Phys. Rev* **B23**, 5632 (1981).
46. D. R. Yennie, *Rev. Mod. Phys.* **59**, 781 (1987).
47. T. Kinoshita and G. P. Lepage, in *Quantum Electrodynamics*, ed. by T. Kinoshita (World Scientific, Singapore, 1990), pp. 81-91.

48. A. Hartland et al., *Phys. Rev. Lett.* **66**, 969 (1991).
49. E. R. Williams et al., *IEEE Trans. Instrum. Meas.* **38**, 233 (1989).
50. R. H. Koch, D. J. Van Harlingen, and J. Clarke, *Phys. Rev. Lett.* **45**, 2132 (1980); A. O. Caldeira and A. J. Leggett, *Phys. Rev. Lett.* **46**, 211 (1981); A. Widom and T. D. Clark, *Phys. Rev. Lett.* **48**, 1572 (1982); V. Ambegaokar, U. Eckern, and G. Schon, *Phys. Rev. Lett.* **48**, 1745 (1982); M. R. Arai, Cornell University Ph. D. thesis, 1983.
51. J.-S. Tsai, A. K. Jain, and J. E. Lukens, *Phys. Rev. Lett.* **51**, 316 (1983).

The CERN Muon (g-2) Experiments: A Brief Survey

E. Picasso
CERN, Geneva, Switzerland
and
Scuola Normale Superiore Pisa, Italy,

Introduction

In this paper, dedicated to Vernon Hughes on the occasion of his 70th birthday, I only give the flavour of the various problems encountered in the three experiments performed at CERN from 1958 to 1976. All these arguments are much more fully dealt with in the final reports of the different experiments.[1,a,b,c] For a recent review article see F.J.M. Farley and E. Picasso.[2]

The g-factor of the muon is a dimensionless number which relates its magnetic dipole moment to its intrinsic angular momentum. Classically the dipole moments can arise from either charges or currents. For example, the circulating current, due to an orbiting particle with an electric charge e and mass m, has associated to it a magnetic dipole moment $\vec{\mu}_L$ given by

$$\vec{\mu}_L = \frac{e}{2mc}\vec{L} \tag{1}$$

where $\vec{L}$ is the orbital angular momentum. Alternatively, the electric dipole moment possessed by certain polar molecules is due to the relative displacement of the centre of positive and negtive charge distributions. Thus we have examples of a magnetic dipole moment and electric dipole moment both having their origins in the electric charge. It is of interest to note that all electromagnetic phenomena are explained in terms of electric charges and their current; there is no place, as yet, for magnetic charges. In particular the intrinsic magnetic dipole moments of all particles can be considered, in the classical picture, to be made up of circulating electric currents and not of distributed magnetic charges. This is just one aspect of the basic asymmetry between electricity and magnetism, which is apparent in Maxwell's equations.

The argument first proposed by Dirac,[3] that the existence of magnetic charges leads naturally to the quantitization of both magnetic and electric charges, still stands as a challenge to physicists, both theoretical and experimental, to find a proper place for the magnetic monopole in the electromagnetic theory and establish its physical reality.

For a particle with both magnetic and electric dipole moments, the electromagnetic interaction Hamiltonian contains a part

$$H = -\vec{\mu}_m \cdot \vec{B} - \vec{\mu}_e \cdot \vec{E} \tag{2}$$

where $\vec{B}$ and $\vec{E}$ are the magnetic and electric strengths and $\vec{\mu}_m$ and $\vec{\mu}_e$ are the magnetic and electric dipole moment operators.

Following the general form of eq.(1) and treating the electic dipole moment analogously to the magnetic dipole moment one can write [2]

$$\vec{\mu}_m = g\frac{e}{2mc} \cdot \frac{\hbar\vec{\sigma}}{2} \tag{3}$$

$$\vec{\mu}_e = \eta\frac{e}{2mc} \cdot \frac{(\hbar\vec{\sigma})}{2} \tag{4}$$

where the components of $\vec{\sigma}$ are the three Pauli spin matrices and for the negative muon we have to insert the change $e = - \mid e \mid$. Introducing the muon Bohr magneton $\mu_0 = \frac{e\hbar}{2mc}$, there equations can be simplified to

$$\vec{\mu}_m = g\mu_0\frac{\vec{\sigma}}{2}$$

$$\vec{\mu}_e = \eta\mu_0\frac{\vec{\sigma}}{2}$$

The expectation value of the electric dipole moment $\vec{\mu}_e$ must be zero for a particle described by a state of well-defined parity. The polar molecules that we have referred to above, are in a mixture of degenerate states with opposite parities, and so are not covered by the symmetry condition.

The g-factor represents a fundamental property related to electromagnetism; if the particle participates in any other interaction which endow it with an internal structure, then the value of its g-factor will reflect this departure from the point-like nature implied by the Dirac equation. Even in the absence of an intrinsic structure, the quantum nature of the electromagnetic interaction itself also modifies the g-factor. This modification is quite small and has become conventional to define a magnetic moment anomaly"a" such that

$$g = 2(1 + a) \tag{5}$$

The motivation of the muon (g-2) experiment is to verify that the quantum theory of electromagnetism correctly predicts the value of the magnetic moment anomaly. Due to the muon "energy scale", which is m_μ/m_e times that of the electron, a determination of the g-factor of the muon test quantum electrodynamics at shorter distances. The enhanced energy scale of the muons also means that it is most sensitive to vacuum polarization into other particles, and the strong interaction and possible weak interaction effects are brought into an otherwise pure electrodynamics system.

Quantum electrodynamics (QED) can be considered as a clearly-defined mathematical procedure whereby any process involving the interaction of the photon and charged lepton fields may be calculated to any order of approximation. It is not without its controversial points. The existence of infinites within its structure remains difficult to accept, in fact, their presence precludes any real understanding of the fundamental constants such as charge and mass. The increasing order of the approximation involves the calculation of more and more successive interactions between the photon and lepton fields.[4] All these terms depend upon the renormalization procedure in which the sum of the bare lepton mass and its (infinite) radiative correction is put equal to the observed rest mass of the particle, while the sum of the bare lepton charge and its (infinite) radiative correction is put equal to the observed electronic charge.

The target of the experimentalist is, therefore, to check the theory of smaller and smaller distances, searching for any evidence of the structure of leptons or of some new interaction which may explain the difference in mass between the electron and the muon. Experimentalists in doing so are measuring the effects of the higher order corrections, testing, therefore the validity of the renormalization procedure. These measurements need to be performed with extreme precision, and the atomic physics experiments, together with the measurements of the lepton magnetic moments, have provided the arena for such a precision.

Survey of the Theory

If the muon obeys the simple Dirac equation for a particle of its mass (206 times heavier than an electron), then $g = 2$ exactly; but this is modified by the quantum fluctuations in the electromagnetic field around the muon, as specified by the rules of quantum electrodynamics, making g larger by 1 part in 800. The quantum fluctuations require further correction for the very rare fluctuation, which include virtual pion states and strongly interacting vector mesons.

The gyromagnetic ratio is increased from its primitive value of 2, arising from the Dirac equation, to $g = 2(1 + a_\mu)$, where a_μ is the muon anomaly. In the QED theory, the lepton g-factor may be expressed as a power series in α/π:

$$a_\mu^{th} = A(\alpha/\pi) + B(\alpha/\pi)^2 + C(\alpha/\pi)^3 + \cdots \tag{6}$$

The calculation[4] of the coefficients in eq. (6) have given the following results:

$$\begin{aligned} A &= 0.5 \\ B &= 0.765\ 858\ 6 \\ C &= 24.067(3) \\ D &= 124.8(65) \\ E &= 570(140) \end{aligned} \tag{7}$$

Using[5]

$$\alpha^{-1} = 137.035\ 989\ 5(61) \quad (0.045 \text{ ppm}) \tag{8}$$

the result is

$$a_\mu^{\text{QED}} = 116\ 584\ 730\ (5)(5) \times 10^{11} \tag{9}$$

where the first error comes from the uncertainty in the coefficients C,D,E and the second reflects the error in α. Combining in quadrature the overall uncertainty in the QED value is (0.06 ppm).

To compare the experiment with the theory we have to examine the contributions to the anomaly which come from the strong and weak interactions. Strongly interacting particles do not couple directly to the muon, but if they are charged, they couple to the photon. Thus they can appear in the vacuum polarization loop with a pion pair replacing the e^+e^- pair. Because of the high mass of the pion, one would initially expect such amplitudes to be small, but there are strong resonances in the $\pi^+\pi^-$ system that enhance the effect. Only a vector resonance can contribute, because it alone can transform directly into the virtual photon which must have $J^{PC} = 1^{--}$ (one unit of angular momentum, negative parity, and negative charge conjugation); this process bears a resemblance to electron-positron annihilation into hadrons. This link is exploited by means of dispersion relations. Under the assumption that the annihilation is dominated by the single photon process, we may write this hadronic contribution to the anomaly as[6]

$$a_\mu(\text{hadronic}) = \frac{1}{4\pi^3}\int_{4m_\pi^2}^{\infty} \text{ds}\ \sigma_{e^+e^-\to\text{hadrons}}(\text{s})\text{K}(\text{s}) \tag{10}$$

where s is the total e^+e^- center-of-mass energy squared. The function K(s) is a purely QED quantity arising from the combination of the two lepton propagators and the propagator of a virtual photon with mass $\sqrt{s}$ at the lepton vertex; in the limit $s \gg m^2$ the K(s) has the value $1/3(m^2/s)$. With the additional assumption of the electron-muon universality, these formulae can be applied to the muon anomaly, and in fact the asymptotic dependence of the function K(s) on the square of the lepton mass indicates that the hadronic contribution to the muon moment will be some 10^5 times larger than the hadron part of the electron moment. This means that the electron moment is essentially a pure QED quantity.

The most recent determinations of the contribution to the muon anomaly are[4]

$$a_\mu(\text{hadron}) = 69(2) \times 10^{-9} \tag{11}$$

The contributions to the muon anomaly due to the weak interaction can be calculated unambiguously within the framework of the renormalizable spontaneously broken gauge theories. Calculations based on the Glashow-Salam-Weinberg theory[7] predict the weak contribution to be

$$a_\mu(\text{weak}) = 195(1) \times 10^{-11} \tag{12}$$

or

$$(1.7 \pm 0.01) \text{ ppm in } a_\mu$$

This contribution is a very small effect, at present masked by the uncertainty in the strong interaction contribution. So in order to utilize fully a higher-precision experimental value a_μ to determine the electroweak contribution, it is necessary to improve our knowledge of a_μ (hadron), implying a better measurement of the ratio

$$R(s) = \frac{\sigma(e^+e^- \to \text{hadrons})}{\sigma(e^+e^- \to \mu^+\mu^-)} \tag{13}$$

particularly in the low-energy region with invariant mass $< 1\ GeV/c^2$. Combining all the above values the overall theoretical predicition is [4]

$$a_\mu(\text{theory}) = 116\ 591\ 8(2) \times 10^{-9} \tag{14}$$

It is worth while to emphasize that over a period of many years the muon (g-2) has provided a very good constraint on theoretical constructs. Theories that attempt to generalize the Standard Model in one way or another all have their effect on the muon (g-2) value,[8] for example, the present precision of the muon (g-2) imposes a lower limit, of the order of 1 TeV, on the mass of the constituent particles of the muon.

Basic Principles of the (G-2) Precession Experiments

Many precise experiments on the electromagnetic properties of the muon have been made possible by the discovery[9] that it is fully longitudinally polarized at birth ($\pi \to \mu + \nu$) and gives an anisotropic distribution of electrons ($\mu \to e + \nu + \bar{\nu}$) which can be used to determine the mean polarization of an ensemble of muons. The polarizing and analysing mechanism thus provided, compensates for the low intensity of available muon beams and permits the measurement of many spin-dependent effects.

The principal features of the muon (g-2) experiment consist of a polarized source, a uniform magnetic field and a polarimeter (see Fig. 1).

In the uniform field region, assuming that the particle velocity is perpendicular to $\vec{B}$, the muon will follow a circular path with the cyclotron angular frequency

$$\omega_c = \omega_0/\gamma \tag{15}$$

with $\omega_0 = eB/m_\mu c$ and $\gamma = (1-\beta^2)^{-1/2}$. In the laboratory the spin $\vec{s}$ will also precess at a uniform angular frequency. This rate of precession is the Larmor frequency, which is invariant under Lorentz transformation from the muon rest frame, minus the Thomas precession frequency, which arises from the accelerated motion of the particle. This latter is given by

$$\omega_T = \frac{(\gamma - 1)}{\gamma}\omega_0 \tag{16}$$

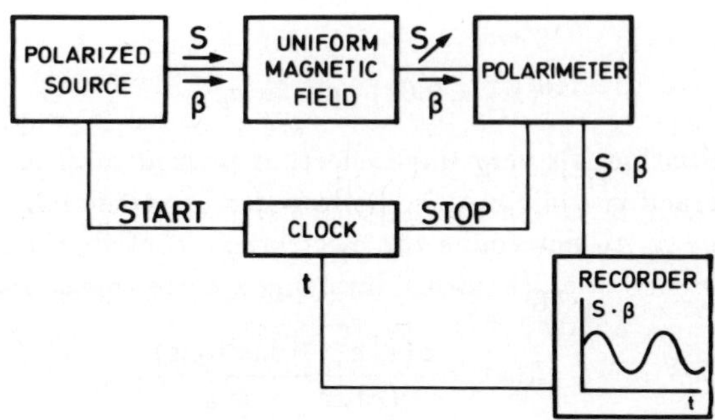

Figure 1: Block diagram of an ideal (g-2) experiment.

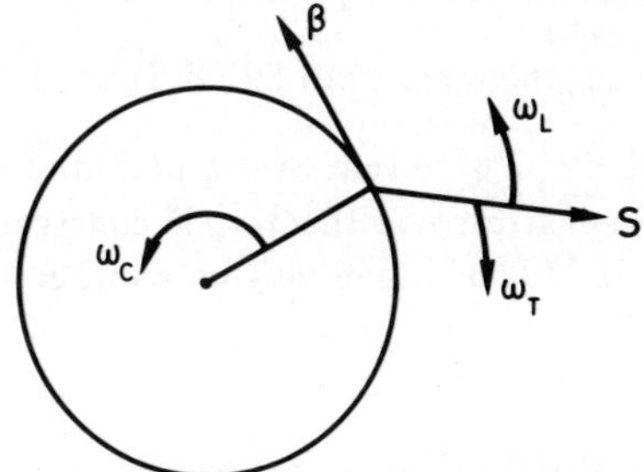

Figure 2: Relative motion of the spin and momentum in laboratory frame.

Thus the angular frequency of the spin is

$$\omega_s = \omega_L - \omega_T = \frac{1}{2} g\ \omega_0 - \frac{(\gamma - 1)}{\gamma}\omega_0 = (a_\mu + 1/\gamma)\omega_0 \tag{17}$$

The sense of these relations is illustrated in Fig. 2.

From this we can see that the anomaly a_μ causes the spin to rotate at a slightly higher frequency than the momentum vector, the difference in frequency being independent of γ and just equal to ω_a:

$$\omega_s - \omega_c = a_\mu\ \omega_0 = \omega_a \tag{18}$$

If the polarimeter measures the longitudinal polarization of the stored muon sample $(\vec{s} \cdot \vec{\beta})$ as a function of time, then a signal oscillating with the frequency ω_a will be obtained.

The experimental value of (g-2) has been determined by three progressively more precise measurements at CERN, the latest one achieving a precision of 7.3 ppm in the anomaly a_μ. All three of these experiments used muons from pion decay $\pi^+ \to \mu^+ + \nu_\mu$. Parity non-conservation in this process provides the initial polarization of the muons in the following way. In the pion rest frame the muons have a longitudinal polarization of 100%. Thus if the muons which are produced close to the forward direction are selected, then this yields a beam with high initial longitudinal polarization in the laboratory. Such muons are at the top of the allowed momentum range, and so by matching the momentum of the original pion beam with that of the selected muons, a sample of the latter can be obtained with over 95% initial polarization.

The analysis of the longitudinal polarization as a function of the time which the muon spends in the magnetic field is effected by making use of the asymmetry in the electron angular distribution, with respect to the muon spin direction, which arises from parity non-conservation in the muon decay process $\mu^+ \rightarrow e^+ + \nu_e + \bar{\nu}_\mu$. Access to this asymmtric distribution may be obtained directly by stopping the muons and thereby going to the muon rest frame, or by applying an energy cut on the detected electrons from decay in flight. The first CERN muon experiment used the former method, while the subsequent two experiments employed the detection of decays in flight. The process of applying an energy cut on these electrons is equivalent to selecting those which go forward in the muon rest frame. The expectation value of the longitudinal component of the muon spin oscillates with the (g-2) frequency and consequently the count rate of the selected decay electrons will be modulated with the frequency ω_a.

Discussion of Some Experimental Problems[2,10]

A few considerations are needed when we wish to describe a real (g-2) experiment. In such an experiment the quality of the storage region (trap) plays an important role because the uncertainty in a_μ is in part inversely proportional to the storage time t, and it is a great advantage to make this time as long as possible.

The first requirement for the quality of the trap is therefore that the particle be trapped or stored for a long period of time. The second requirement for the quality of the trap concerns the need to measure, with great precision, the average magnetic field seen by the particles. This quantity $\bar{B}$, must be known to a greater precision than that designed for a_μ itself. Any real beam of particles occupies a finite volume of phase space, and consequently the particles do not all follow the same orbits and the precise knowledge of $\bar{B}$ involves, therefore, a precise mapping of the magnetic field and a measurement of the distribution of particle orbits, or at least the calculation of this distribution in a reliable way. The extent to which the uncertainty in this distribution leads to an error in $\bar{B}$ depends on the uniformity of a magnetic field; if the field were perfectly uniform, we would only need to know that the particles were confined within the storage region, but not precisely where.

Another consequence of the finite extension of the phase space of the beam is that it will contain particles whose component of the velocity along the magnetic field lines is nonzero and therefore the motion of the particles is not restricted by a uniform field. In a precession experiment, the total path length travelled by the

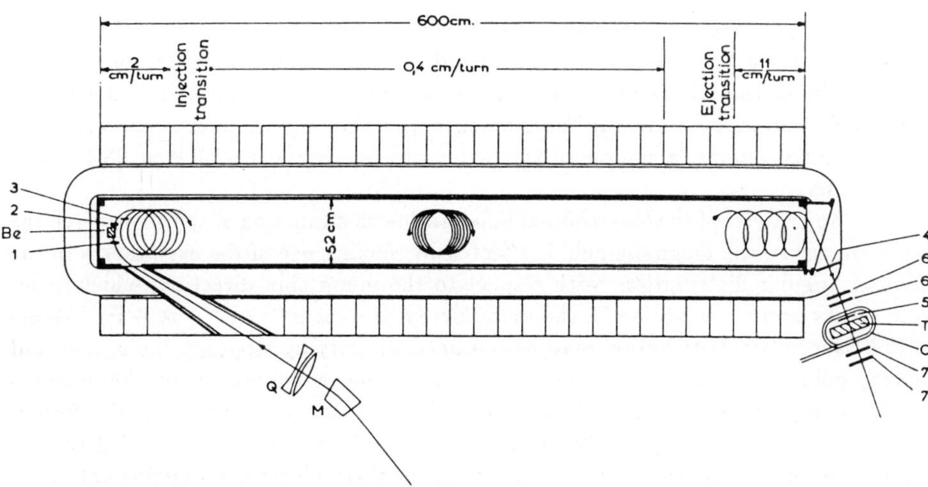

Figure 3: General view of the 6 m bending magnet used in the first CERN muon (g-2) experiment. The muons were deflected by small magnets M and focussed by the quadrupole pair Q before striking the beryllium moderator (Be). Injection muons were signalled by a coincidence between detectors 123, while ejected muons had the signature $466'5\bar{7}$. The decay electrons were separated into forward $[77'\bar{4}(\bar{6}\bar{6}')]$ and backward $[66'4,(\bar{7}\bar{7})']$ events and collected on 0.1 μs time bin as a fucntion of storage time (after reference 1a).

particles may reach several kilometers, and the confinement of these orbits within a reasonable volume implies some non-uniformity of the magnetic field itself, in the form of a gradient, or the application of an electric field. A magnetic gradient automatically makes it more difficult to determine $\bar{B}$, and thus there is a conflict between efficient trap, measuring the average magnetic field and ultimately the precision on a_μ.

The First Two (g-2) Experiments at CERN[1a,1b]

In the first (g-2) experiment a longitudinally polarized muon beam was formed by the forward decay of pions in flight inside the cyclotron. Scattering in a beryllium moderator reduced the muon momentum to 90 MeV/c and also injected the particle into the gap of a 6 m magnet (see Fig. 3). The 1.6T field of this magnet was so shaped that the muon orbits slowly drifted down to its length, making up to 1000 turns before reaching the ejection region. The final polarization was measured by stopping the muon in a non-depolarizing target, situated beyond the magnetic field

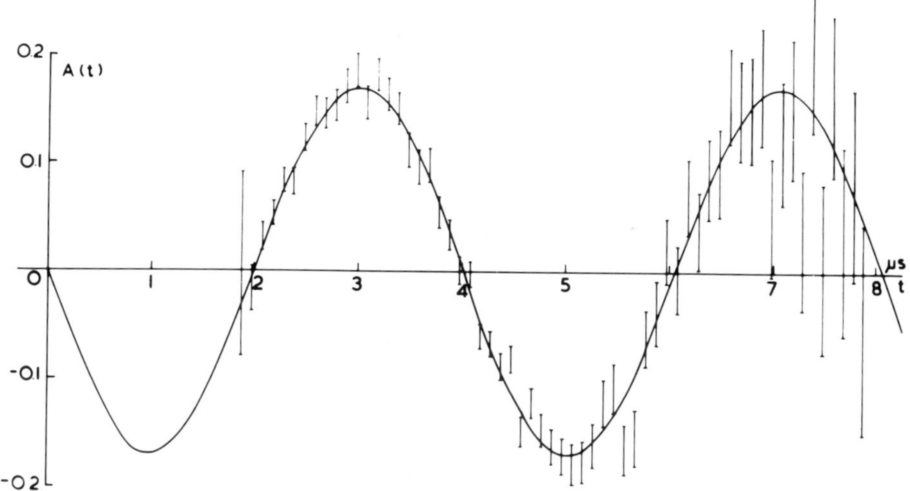

Figure 4: Precession curve due to the (g-2) factor of the muon obtained in the first CERN muon experiment (after reference 1a).

region, and then detecting the forward-backward asymmetry in the decay electron distribution (see Fig. 4). The first measurement of the anomalous part of the g-factor of free muons was obtained from the data containing about a million stopped muons, and yielded the result

$$a_\mu = (1\ 162 \pm 5) \times 10^{-6} \tag{19}$$

This number was of a similar precision to and in excellent agreement with the theoretical value at that time.

The second CERN muon experiment[1b] brought the considerable improvement of trapping muons of 1.27 GeV/c in a weak focussing storage ring ($n = -\frac{r}{B}\frac{\partial B}{\partial r} \sim 0.13$) having a diameter of 5m. The injection of polarized muons was accomplished by the forward decay of pions produced when a target inside the storage ring magnetic field was struck by a pulse of 10.5 GeV/c protons from the CERN Proton Synchrotron (see Fig. 5). The pions produced had momenta in a broad band, and so the initial polarization of the muon sample was diluted to about 26% by the presence of 1.27 GeV/c muons from pions of much higher momentum. The decay electrons were detected as they emerged on the inside of the ring and the (g-2) modulated count-rate was observed by applying an energy cut in the laboratory, as seen in Fig. 6. This figure shows the decay electron momentum space in the muon rest frame and represents a section through the axis of symmetry, which coincides with the direction of the muon velocity, β_μ. The parabolas illustrate the effect of a laboratory energy cut, E_t, in this space, the value of the energy cut being expressed in units $\frac{1}{2}m_\mu c^2 (\sim 53MeV)$. Thus the shaded region represents those decay electron momenta and directions which

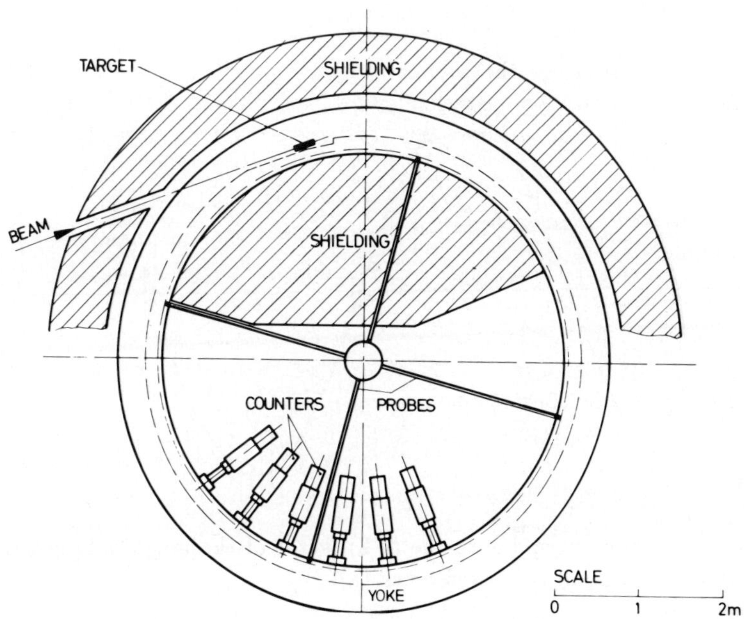

Figure 5: Plan view of the 5 m diameter storage ring magnet. The path of the proton beam and the muon path are indicated, as is that of the counter which detects the electrons from $\mu - e$ decay. The four probes of the NMR magnetometers are injected into specific locations on the muon orbit every 100 PS cycles (after reference 1b).

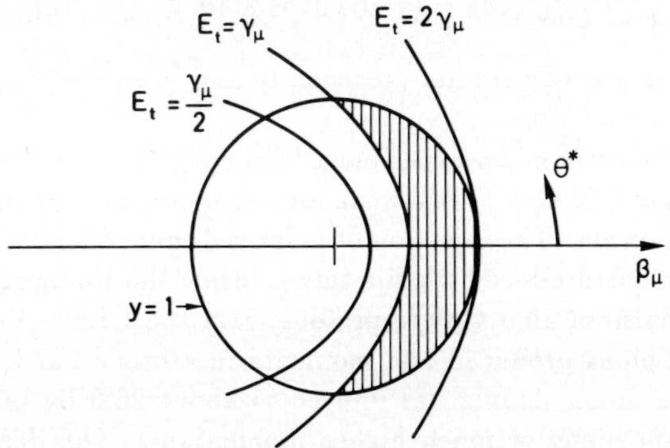

Figure 6: Decay electron momentum in the muon rest frame illustrating the paraboloidal boundaries equivalent to an energy threshold in the laboratory. The units of energy are $\frac{1}{2}m_\mu c^2$ and so $E_t = \gamma_\mu$ is equivalent to a cut at half the parent muon energy (after Combley and Picasso 1974, reference 10).

would be accepted for an energy cut equal to just one half the muon energy in the laboratory.

If now we envisage the asymmetric probability distribution rotating in the muon rest frame with angular frequency ω_a, we can see that the number of decay electrons selected in the shaded region will be modulated with the (g-2) frequency and have the form

$$N(t) = N_0 \exp(-t/\tau) \;\; [1 - A\cos(\omega_a t + \varphi)] \tag{20}$$

where τ is the muon lifetime and φ is the initial phase of the polarized muon sample.

The lifetime of the stored muons with momenta 1.27 GeV/c is increased over the value at rest, of 2.2 μs, by the relativistic time dilation factor $\gamma = 12$. The relative precession of the muon spin with respect to its momentum vector was followed, as a function of the time spent by the muon in the magnetic field, by energy selection of the decay electrons as discussed above. The stretching of the muon lifetime by time dilation enabled some 50 cycles of the (g-2) oscillations to be observed in the time spectrum of decay electron counts, (see Fig. 7). The frequency obtained from this data was then converted into a value for the anomaly by using the magnetic field strength which had been measured in terms of the proton magnetic resonance frequency in the field. In order to complete this conversion, the ratio λ of the muon precession frequency at rest to the proton magnetic resonance frequency, is also required.

It should be noted that in a weak focussing storage ring, the magnetic field is made non-uniform in order to confine the particle orbits vertically. There is a gradient in the field, and consequently any uncertainty in the position of the muon orbits lead to an uncertainty in the mean field seen by the stored sample. This is just an example of how the need to trap the particles conflicts with the aim of observing the spin motion in a precisely known magnetic field. The final error for this experiment contained a large contribution due to this source. The experimental value obtained for the anomaly was

$$a_\mu = (116\ 616\ 0 \pm 310) \times 10^{-9} \tag{21}$$

in good agreement with the theory.

The Third (g-2) Experiment[1c]

The third and most recent CERN muon (g-2) epxeriment very much evolved out of the experience gained with the second one. A source of major difficulty in that experiment had been the radial magnetic field gradient necessary to provide the vertical focussing; this had meant that the (g-2) frequency was a function of the mean radius of the muon orbits. This in turn had placed crucial dependence upon knowledge of the muon orbit distribution and was responsible for a major part of the experimental error. In response to this difficulty the new storage ring was designed to

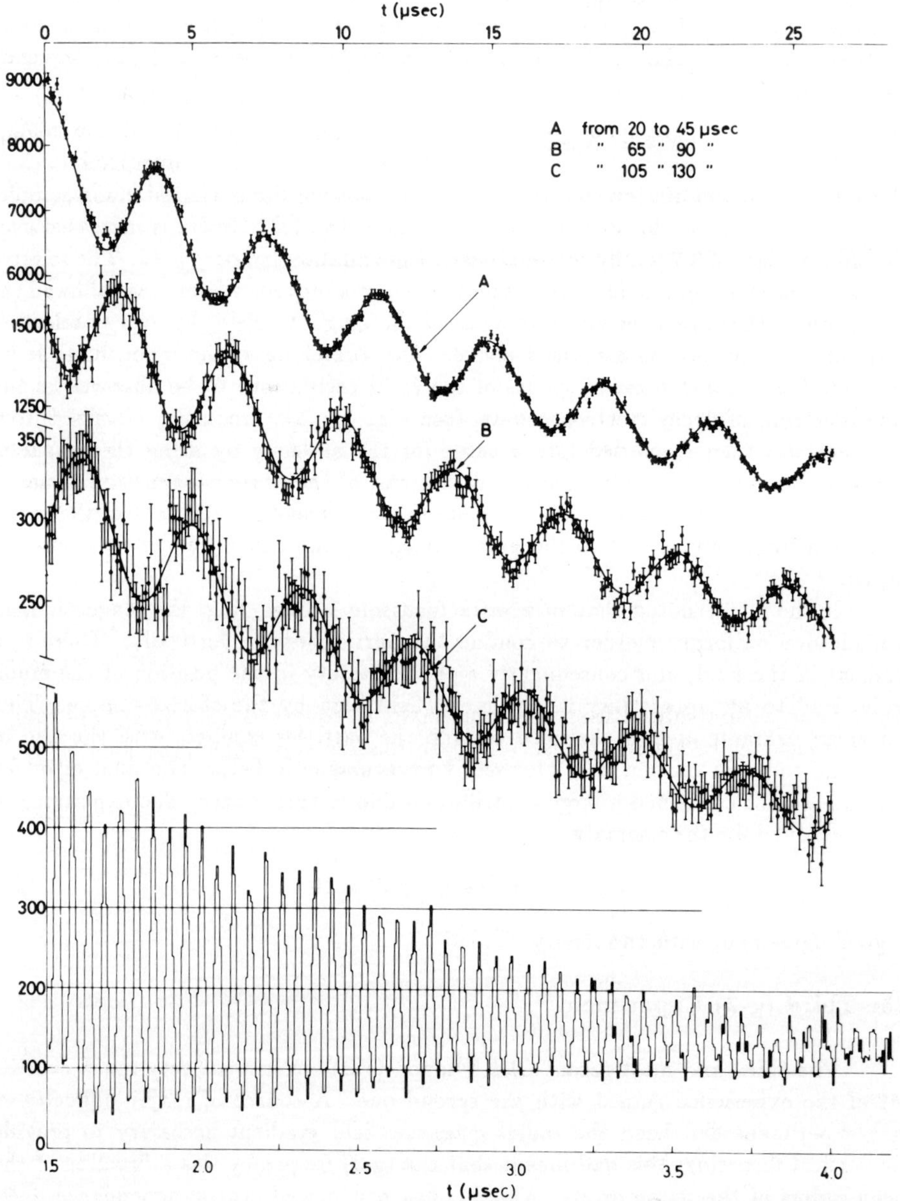

Figure 7: Decay electron count modulated by the (g-2) frequency as observed in the second CERN muon experiment. Decay electron count-rate at early time is modulated by the rotation frequency (after reference 1b)

provide a uniform magnetic field with vertical focussing of the muon orbits by electric quadrupoles.

A second difficulty in the previous rings had arisen from the method of injection. The large burst of particles coming from the target within the ring itself had upset the counting system and produced extensive background at an early time. As we have mentioned above, the method also produced low initial longitudinal polarization due to the wide range of pion momenta which contribute to the stored muon sample. In the new experiment a momentum selected pion beam was injected into the storage ring by means of a pulsed inflector; this gave the triple advantage of a much lower initial flash of unwanted particles, an increase in muon intensity due to the proper matching to the acceptance of the storage ring and a high initial polarization of the muon sample, ($\sim$ 95%).

We should discuss the effect that the chosen configuration of transverse fields ($\vec{\beta} \cdot \vec{B} = \vec{\beta} \cdot \vec{E} = 0$) has on the expression for the relative spin precession frequency. The forces that hold the muon in its orbit and give focussing for small deviations from equilibrium arise from what appears in the muon rest frame as an electric field, while the spin precessions arises from what appears there as a magnetic field. These two fields may be varied independently by applying suitable magnetic and electric fields in the laboratory frame. The equations of the motion in the configuration of transverse fields are [11]:

$$\frac{d\vec{\beta}}{dt} = \vec{\omega}_c' \times \vec{\beta} \tag{22}$$

$$\frac{d\vec{\sigma}}{dt} = \vec{\omega}_s' \times \vec{\sigma} \tag{23}$$

where $\vec{\omega}_c$ and $\vec{\omega}_s$ are shifted from the value given in equations (15) and (17) owing to the presence of the electric fields. The cyclotron and spin angular frequencies are now given by:

$$\vec{\omega}_c' = \frac{e}{mc}\left[\frac{\vec{B}}{\gamma} - (\frac{\gamma}{\gamma^2-1})\vec{\beta} \times \vec{E}\right] \tag{24}$$

$$\vec{\omega}_s' = \frac{e}{mc}\left[\left(a_\mu + \frac{1}{\gamma}\right)\vec{B} + \left(\frac{1-\gamma}{\gamma^2-1} - a_\mu\right)\vec{\beta} \times \vec{E}\right] \tag{25}$$

The precession of the spin relative to the velocity vector is, therefore:

$$\vec{\omega}_a' = \vec{\omega}_s' - \vec{\omega}_c' = \frac{e}{mc}\left[a_\mu\vec{B} + \left(\frac{1}{\gamma^2-1} - a_\mu\right)\vec{\beta} \times \vec{E}\right] \tag{26}$$

$$\vec{\omega}_a' = \vec{\omega}_a + \frac{e}{mc}\left(\frac{1}{\gamma^2-1} - a_\mu\right)\vec{\beta} \times \vec{E} \tag{27}$$

This equation emphasizes the point that the trapping potential shifts the observed frequency from the desired value $\vec{\omega}_a$. This shift is zero for the special choice of particle

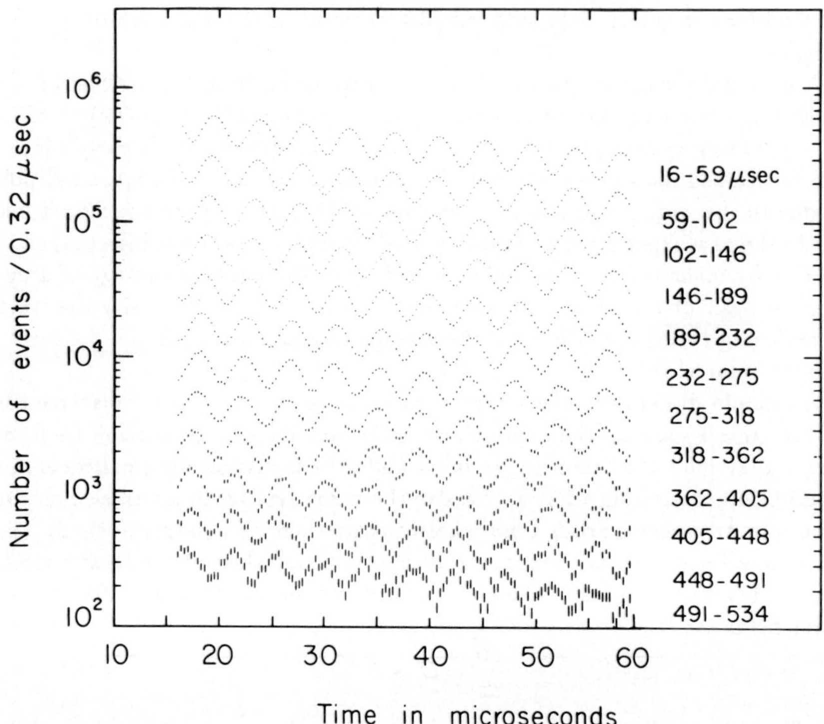

Figure 8: Decay electron count-rate from the last CERN muon (g-2) experiment. The distribution contains a total record of 1.4×10^8 electron counts (after reference 1c).

energy equal to:

$$\gamma = \left(1 + \frac{1}{a_\mu}\right)^{1/2} = 29.304 \tag{28}$$

For muons the actual momentum is 3.098 GeV/c which was the value chosen for the new muon storage ring, thereby fully exploiting this fortunate cancellation and reducing the corrections necessary for the effect of the electric field to 1.5 parts per million. Such corrections arose because not all the muons had exactly the so-called magic momentum.

The high relativistic γ-factor brought the additional advantage of a proportionally increased muon lifetime which enabled the (g-2) precession signal to be followed for over a hundred periods of the modulation (see Fig. 8).

A plan view of the storage ring is shown in Fig. 9. The uniform magnetic field of 1.4 T was provided by 40 bending magnets arranged so that their contiguous pole pieces formed a ring of 14 m diameter. The C-shaped magnets were open on the

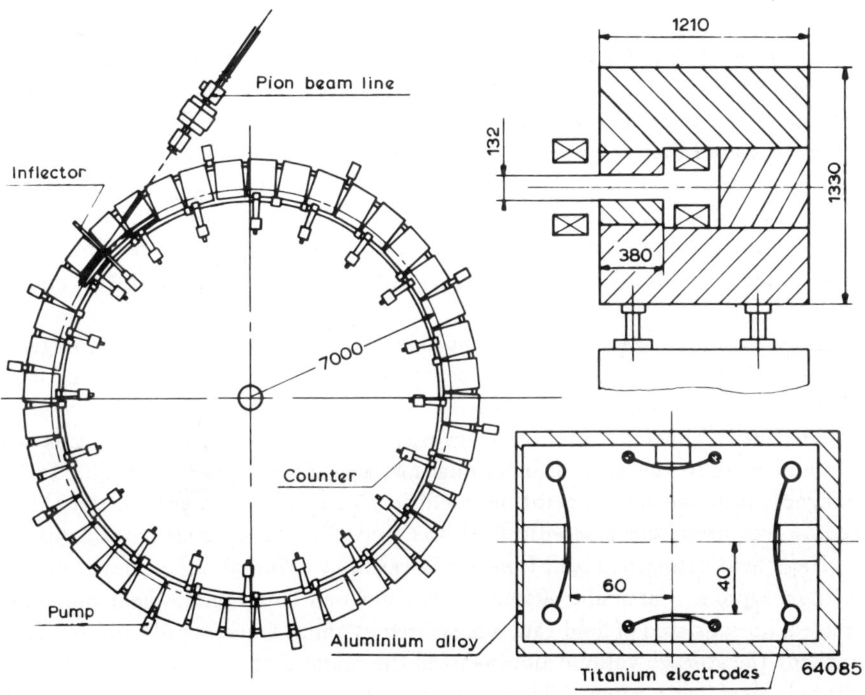

Figure 9: Plan view of the 7 m diameter muon storage ring, with cross sections of the magnets and electric quadrupoles. The open side of the C-magnet and the thin wall (3 mm) of the vacuum chamber face the center of the ring (after reference 1c)

inside to allow the decay electrons to emerge and strike the energy sensitive detectors of which there were 22 deployed around the ring.

The electrodes which provided the trapping potential for the vertical motion of the muons, were mounted inside the vacuum chamber. Each set of quadrupoles covered 36^0 in azimuth and there were eight such units. To accomodate the inflector to allow the pion orbits to enter the storage volume, a complete sector was left free of electrodes; in order to remove the first azimuthal harmonic in the focussing field and to reduce the closed orbit distortion, the sector directly opposite the inflector was also left without electrodes.[10]

At each cycle of the Proton Synchrotron a pion beam with a momentum spread of $\pm 0.75\%$ is injected into the volume where the muons are to be stored, by means of a pulsed inflector. The pions have slightly higher momentum than the 3.098 GeV/c for which the magnetic field of the ring is set. Consequently they tend to leave the ring on the outside, but before they do so, about a tenth of them decay. The overall trapping efficiency of stored muons produced per injected pion is about 10^{-4}. As pointed out above, the advantage of this method of producing the stored muon population is that the sample has high initial polarization in excess of 90% owing to the fact that the muons are selected from the top 1.5% of the available momentum range and hence maximum use is made of the complete polarization of the muons in the pion rest frame.

To maximize the stability and reproducibility of the field in the storage ring, each of the 40 magnets was stabilized separately by a control system which had an NMR probe and pick-up coil as its sensors. The signals from these devices were used to automatically control the current through additional compensating coils which were wound around the yoke of individual magnets close to each pole tip. To bring the magnets into the same operating condition for each run, a special switching-on procedure was used, the magnetic field was controlled to a precision of 1 ppm and the average field values deduced from maps taken at different times, never varied by more than 2 ppm. Particular attention was devoted to magnetic field mapping: a complete map consisted of field values measured at about a quarter of a million pointes throughout the storage volume and involved the removal of the vacuum chamber and quadrupole electrode system. This meant that the actual field seen by the circulating muons was slightly changed from the values obtained during large-scale mapping, and so these systematic effects due to the materials of the vacuum chamber and electrodes were carefully studies.[12] A mapping at some 400 points was made daily throughout the experimental runs. These measurements enabled any drifts in the mean field to be followed in the interval between the full field maps which were made before and after a sequence of runs.

In order to apply the small correction due to the slight inhomogeneity of the magnetic field and the effect of the electric field, it was necessary to know the distribution of the equilibrium orbits. The distribution of the muon orbits was determined from the rotation structure of the decay electron spectra just after injection. At an

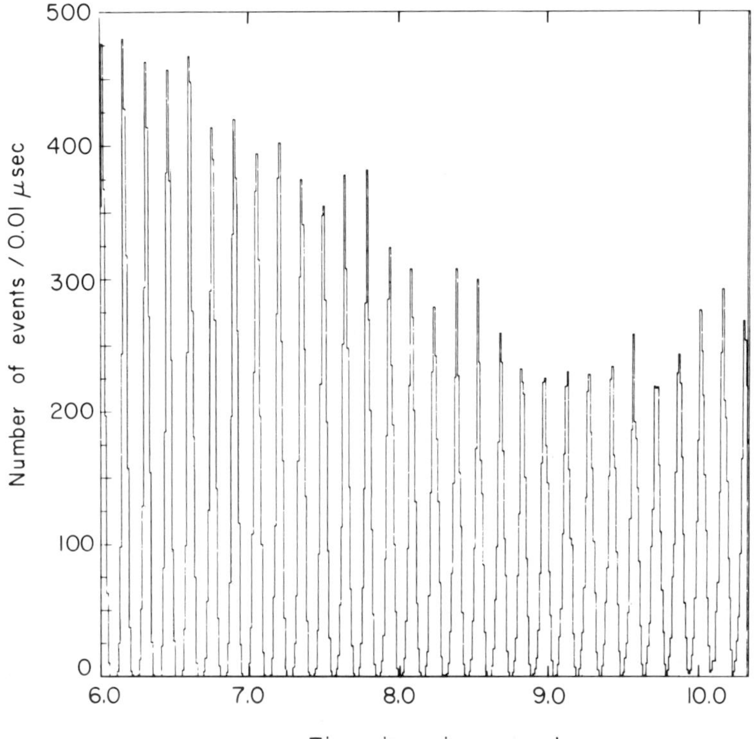

Figure 10: Decay electron count-rate at early time from the last CERN muon (g-2) experiment modulated at the rotational freqeuncy (after reference 1c).

early time the decay electron distribution is modulted with the cyclotron frequency and the determination of the rotational frequency allows to measure the mean radius of the muon orbit distribution, with a precision of 0.5 mm (see Fig. 10). The spread in radii about the center value (7000 mm) was measured to 0.1 mm. The parameters of the muon distribution were used to obtain the average value of the magnetic field and the small correction due to the electric field (see Fig. 11).

The measured decay electron count-rate is shown in Fig. 8, which contains the total sample of 1.4 x 10^8 record counts. The (g-2) modulation is clearly visible out to a storage time of 0.5 ms; a comparison of this figure with the previous results shown in Fig. 4 and 7 illustrates the considerable progress made throughout this sequence of three expriments. The experimental value for the angular frequency ω_a was obtained by fitting the data to a function of the basic form given in eq. (20). Nine separate runs were made over a period of two years in the values of the ratio $R = \omega_a/\bar{\omega}_p$ of the (g-2) frequency to the effective proton NMR frequency shown in Fig. 12 for these data sets, together with the average values for positive and negative muons which

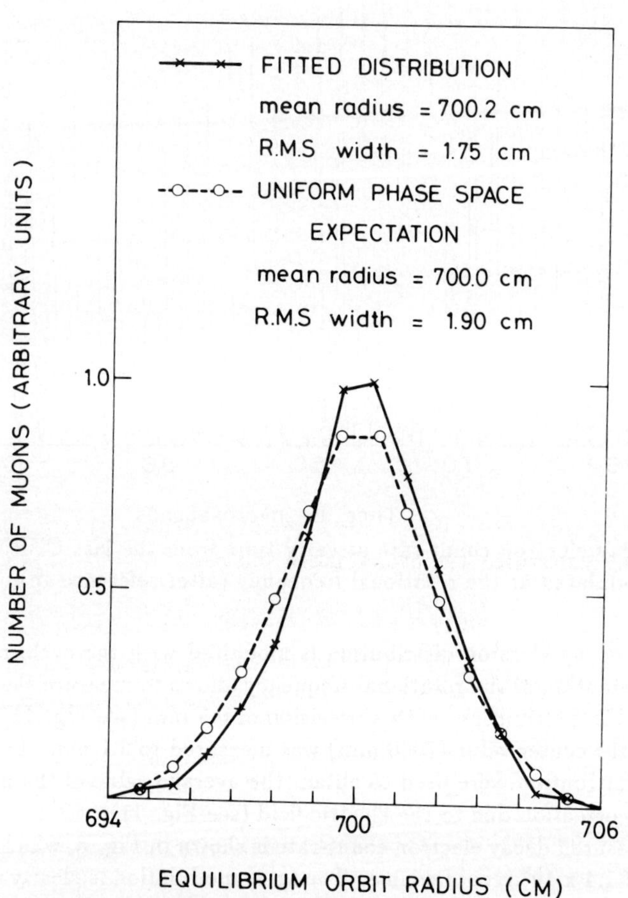

Figure 11: The muon distribution, in arbitary units, versus equilibrium orbit radius in cm. (after refernce 1c).

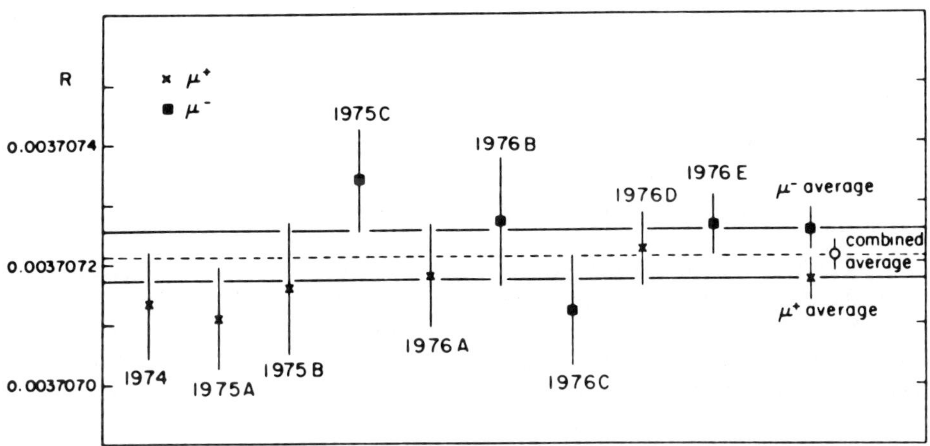

Figure 12: Values of the ratio $R = \omega_a/\bar{\omega}_p$ for the nine experimental periods of the third CERN muon (g-2) experiment together with the weighted averages (after reference 1c).

were:

$$R(\mu^+) = (37\ 071\ 73 \pm 36) \times 10^{-9} \tag{29}$$

$$R(\mu^-) = (37\ 072\ 56 \pm 37) \times 10^{-9} \tag{30}$$

In both cases the statistical error of 10 ppm and the systematic error of 1.5 ppm have been added in quadrature. The overall mean value is:

$$R(\mu) = (37\ 072\ 13 \pm 27) \times 10^{-9} \tag{31}$$

From the ratio R we can deduce the anomaly a_μ, knowing the ratio $\lambda = \omega_a/\bar{\omega}_p$, that is the muon precession freqeuncy relative to the proton precession frequency in the same field:

$$a_\mu = R(\lambda - R)^{-1} \tag{32}$$

and[13]

$$\lambda = 3.183\ 345\ 47(37) \tag{33}$$

The value of a_μ^+ and a_μ^- are:

$$a_\mu^+ = 1\ 165\ 910\ (11) \times 10^{-9}\ \ (10ppm) \tag{34}$$

$$a_\mu^- = 1\ 165\ 936\ (12) \times 10^{-9}\ \ (10ppm) \tag{35}$$

and for μ^+ and μ^- combined

$$a_\mu = 1\ 165\ 923\ (8.5) \times 10^{-9}\ \ (7ppm) \tag{36}$$

compared with $a_\mu^{th} = 1\ 165\ 918(2) \times 10^{-9}$ the result is

$$a_\mu^{exp} - a_\mu^{th} = (5 \pm 8.5) \times 10^{-9} \tag{37}$$

Plans for a new (G-2) Experiment

The last experiment carried out at CERN has verified the QED calculations up to the sixth order, the experimental uncertainty being equivalent to 1.2×10^{-6} in A, 3.5×10^{-3} in B and 4.7% in C. The hadronic contribution to the anomaly is confirmed to an accuracy of 20%. The existence of hadronic vacuum polarization has thus been established at the leve of five standard deviations. Finally it has shown that there is not evidence for a special coupling of the muon.[2]

With the advent of renormalizable gauge theory unifying the weak and the electromagnetic interactions, the calcultions of the weak interaction contribution to the muon anomaly has become reliable. Kinoshita[14] has recently reviewed this subject. The main change since the last (g-2) experiment, performed more than fifteen years ago, has been the firm establishment of the unified electroweak gauge theory[7] which seems to fit all known facts about the electromagnetic and weak interactions. The discovery of the W and Z bosons as free particles[15] and the recent results obtained at LEP and SLC[16] have confirmed the major premise of the theory. All the present tests of the weak interaction, however, involve only first-order processes, that is the direct exchange of W and/or Z bosons. Higher order processes involving virtual loops of W and Z are an essential of the theory, but have not so far been detected experimentally. It is important to establish their presence.

V.W. Hughes has been considering (g-2) measurements with a superconducting storage ring (possibly at 5 T) since about 1970. In 1984 he organized a meeting at Bookhaven to discuss whether a new (g-2) measurement could be initiated, and to work out the general parameters of the experiment. Further discussions led to a formal proposal.[17] The starting point was a muon storage ring working at the "magic energy", 3.1 GeV, which would, as before, allow a uniform magnetic field to be used with vertical focussing by means of electric quadrupoles. An improvement in acuracy by a factor of 20 was taken as the target, implying a 400-fold increase in counting statistics.

A limiting factor in the CERN experiment was a large perturbating pulse in the decay electron detectors, coinciding with the injection of pions into the ring. This was caused by a strong flash of light in the glass of the photomultipliers, light pipes and scintillators. Light continues to be produced at a lower level for many microseconds after injection, as epithermal neutrons are gradually captured and gamma rays from (n, γ) reactions give rise to Cerenkov light. At high injection intensities it can paralyse the detector electronics for excessive periods, leading to counting and timing errors.

In the CERN experiment these were controlled by less than 1 ppm in a_μ, negligible compared with the final accuracy of 7 ppm. But in a new measurement of 0.3 ppm level, the much higher injected intensities could clearly give problems. An innovation which I strongly recommend is to inject muons instead of pions. This injection scheme will increase the number of stored muons, while at the same time reducing the initial flash. However, in order to inject muons one must have a full aperture kicker in the ring, to kick the muons onto a permanently stored orbit; this magnetic kick must be switched off within 140 ns, before the bunch of muons makes one turn. There is no difficulty in principle in designing such a kicker, but it must be free of iron or ferrite. Also, because the muons continue to pass through the kicker at every turn, $20 \mu s$ after injection the residual field must be small and accurately measured (at a level of 0.1 ppm in the average field). It seems to me that the muon injection is a logical progression if I consider that in the first muon storage ring a proton beam was injected, whereas in the second muon storage ring it was a pion beam. This injection scheme, as pointed out in ref. 2 is not without problems, but at least must be studied as an alterntive to the pion injection scheme. I should like to call the attention to a second important problem of the new (g-2) experiment: A very accurate knowledge of the mean magnetic field $\bar{B}$ seen by the muons is needed. If the field of the storage ring is strictly uniform this is not a problem, but if the field varies by some parts per million over the volume of the storage ring, it is necessary to know the average field weighted over the muon distribution and corrected for the fact that the muons on the inside of the ring contribute more decay electron counts than those on the outside. It will be important at the new level of accuracy to measure directly the actual muon distribution, so that the average field can be calculated with greater confidence than in the previous experiments, where the assumption was made that the muons in the ring populated the available phase space uniformly.[2]

I am convinced that a new determination of a_μ is fully justified. The experimental value of a_e, is now known[18] 2000 times more accurately than a_μ. But new couplings common to the electron and muon usually imply a greater perturbation to a_μ by a factor $(m_\mu/m_e)^2 \sim 42,000$. Therfore the muon measurement, in general, still provides a more sensitive limit to exotic speculations.

After about twenty years from the conceptual design of the last (g-2) experiment, it is time to perform a new measurement, and I am deeply convinced that in the new muon (g-2) experiment at Brookhaven technical improvements have been introduced. But the achievement of reaching 0.3 ppm in a_μ still remains a great experimental challenge, if you consider that the muon lifetime at rest in only 2.2 μs.

It is rewarding for me and for my colleagues that Vernon and his collaborators have decided to follow the same method, the so-called "magic energy value" that we have discovered about twenty years ago.

It is a pleasure to devote this brief history of the (g-2) experiment to Vernon who is one of the major actors of the fourth muon (g-2) experiment. It was an enjoyable period of my life as a physicist to participate in these beautiful experiments and in particular to collaborate for more than sixteen years with my friends, F.J.M. Farley and J. Bailey.

References

1a. G. Charpak, F. J. M. Farley, R. L. Garwin, T. Müller, J. C. Sens and A. Zichichi, *Nuovo Cimento* **37**, 1241, (1965).

1b. J. Bailey, W. Bartl, G. vonBochmann, R. C. A. Brown, F. J. M. Farley, M. Giesch, H. Jöstlein, S. van der Meer, E. Picasso and R. .W. Williams, *Nuovo Cimento* **A9**, 369 (1972).

1c. J. Bailey, K. Borer, F. Combley, H. Drumm, C. Eck, F. J. M. Farley, J. H. Field, W. Flegel, P. M. Hattersley, F. Krienen, F. Lange, G. Lebée, E. McMillan, G. Petrucci, E. Picasso, O. Rúnolfsson, W. von Ruden, R. W. Williams and S. Wojcicki, *Nucl. Phys.* **B150**, 1 (1979).

2. F. J. M. Farley, and E. Picasso, in *Quantum Electrodynamics.* ed by T. Kinoshita (World Scientific, Singapore, 1990), pp.479-559

3. P. A. M. Dirac, *Proc. Roy. Soc.* **A133**, 60 (1931); *Phys. Rev.* **74**, 817 (1984).

4. T. Kinoshita, B. Nižic and Y. Okamoto, *Phys. Rev. Lett.* **52**, 717 (1984).

5. E. R. Cohen and B. N. Taylor, *Rev. Mod. Phys.* **59**, 1121 (1987).

6. C. Bouchiat and L. Michel, *J. Phys. Radium* **22**, 121 (1960); L. Durand, *Phys. Rev.* **128**, 44 (1962); M. Gourdin and E. de Rafael, *Nucl. Phys.* **B10**, 667 (1969); T. Kinoshita and W. B. Lindqvist, *Phys. Rev. Lett.* **47**, 1573 (1981); *Phys. Rev.*, **D27**, 853 (1983); T. Kinoshita, B. Nižic and Y. Okamoto,*Phys. Rev.* **D31**, 2108 (1985); J. A. Casas, C. López and F. J. Yndurâin, *Phys. Rev.* **D32**, 736 (1985); L. M. Barkov, A. G. Chilingarov, S. I. Eidelman, B. I. Khazin, M. Yu. Lelchuk, V. S. Okhapkin, E. V. Pakhtusova, S. I. Redin, N. M. Ryskulov, Yu. M. Shatunov, A. I. Shekhtman, B. A. Schwartz, V. A. Sidorov, A. N. Skrinsky, V. P. Smakhtin and E. P. Solodor, *Nucl. Phys.* **B256**, 365 (1985).

7. S. Weinberg, *Phys. Rev. Lett.* **19**, 1264 (1967); A. Salam, *Proc. Nobel Symposium Aspenasgarden*, 1968, ed. N. Svartholm (Almquist and Wiksell, Stockholm, 1968) 367; R. Jackiw and S. Weinberg, *Phys. Rev.* **D5**, 2473 (1972); G. Altarelli, N. Cabibbo and L. Maiani, *Phys. Lett.*, **B40**, 415 (1972); I. Bars and Y. Yoshimura, *Phys. Rev.* **D6**, 374 (1972); K. Fujikawa, B. W. Lee and A. I. Sanda, *Phys. Rev.* **D6**, 2923 (1972); W. A. Bardeen, R. Gastmans and B. E. Lautrop, *Nucl.Phys.* **B46**, 319 (1972); W. Marciano, *Proc. 11th International Symp. on Lepton and Photon Interactions at High Energies*, Ithaca, 1983, ed. D. G. Cassel and D. L. Kreinick (Cornell Univ. Press, Ithaca, 1983) 80; V. W. Hughes, *Phys. Scr.* **T22**, 111 (1988).

8. M. E. Peskin, *Proc. 10th International Symp. on Lepton and Photon Interactions at High Energies*, Bonn, 1981, (University, Bonn, 1981) 880. L. Lyons, *Part. Nucl. Phys.* **10**, 227 (1983); E. J. Eichten, K. D. Lane and M. E. Peskin, *Phys. Rev. Lett.* **50**, 811 (1983); M. Suzuki, *Phys. Lett* **153B**, 289 (1985).

9. T. D. Lee and C. N. Yang, *Phys. Rev.* **104**, 254 (1956); C. S. Wu, E. Ambler,

R. W. Hayward, D. D. Hoppies and R. P. Hudson, *Phys. Rev.* **105**, 1413 (1957); R. L. Garwin, L. Lederman and W. Weinrich, *Phys. Rev.* **105**, 1915 (1957); J. I. Friedman and V. L. Telegdi, *Phys. Rev.* **105**, 1681 (1957).

10. F. Combley and E. Picasso: *Phys. Reports* **14**, 1 (1974).
11. V. Bargmann, L. Michel and V. L. Telegdi, *Phys. Rev. Lett.* **2**, 435 (1959).
12. K. Borer, *Nucl. Instr, Methods* **143**, 203 (1977); K. Borer and F. Lange, *Nucl. Instr, Methods* **143**, 219 (1977).
13. E. Klempt, R. Schulze, H. Wolf, M. Camani, F. N. Gygax, W. Ruegg, A. Schenk and H. Schilling, *Phys. Rev.* **D25**, 652 (1982); F. G. Mariam, W. Beer, P. R. Bolton, P. O. Egan, C. J. Gardner, V. W. Hughes, D. C. Lu, P. A. Souder, H. Orth, J. Vetter, V. Moser and G. zu Putlitz, *Phys. Rev. Lett.* **49**, 993 (1982); E. R. Cohen and B. N. Taylor, *Rev. Mod. Phys.* **59**, 1121 (1987).
14. T. Kinoshita and W. J. Marciano, in *Quantum Electrodynamicsries*, ed. by T. Kinoshita (World Scientific, Singapore, 1990), pp. 419-478.
15. C. Rubbia, *Rev. Mod. Phys.* **57**, 669 (1985), (Nobel Lecture 1984).
16. F. Dydak, CERN-PPE/91-14 to appear in the Proceedings of High Energy Conference 90 held in Singapore.
17. C. Heisey et al.: "A New Precise Measurement of the Muon (g-2) Level at the Level of 0.35ppm" Brookhaven AGS Proposal 821, Sept. 1985, Revised Sept. 1986 (also see Design Report for AGS 821, March 1989); V. W. Hughes, *Phys. Scr.* **T22**, 111 (1988); and *Proc. 8th Int. Cong. on High Energy Spin Physics*, Minneapolis, 1988 (Am. Inst. Phys, New York 1989) 326; V. W. Hughes and T. Kinoshita, *Comments Nucl. Part. Phys.* **14**, 341 (1985).
18. R. S. Van Dyck Jr., P. B. Schwinberg and H. G. Demelt, *Atomic Physics 9*, ed. by R.S. Van Dyck, Jr. and E.N. Fortson (World Scientific, Singapore, 1984), pp. 53-74.

Vernon W. Hughes-Congratulations to the Scientist and Teacher of an International Community

G. zu Putlitz
Ruprecht-Karls - Universität Heidelberg
Heidelberg, Germany

Mr. Chairman, Dear Colleagues, Ladies and Gentlemen.

Dear Vernon!

An outstanding assembly of very distinguished scientists has come together here tonight to celebrate the forthcoming 70th birthday of Professor Vernon Hughes.

You, Vernon, have been a teacher and colleague to many of us over decades. Not only few have shared the privilege to work with you. For all of us you were a leader and a symbol how science should be done. Today we would like to express to you our admiration to your outstanding achievements as a scientist and as a teacher of a truly international community. More than fifty years of your life have been devoted to physics and the sciences at large. More than three hundred papers document impressively your success in research and your contributions for the understanding of nature. Numerous students and scholars have grown under your guidance to intellectual independence, to competence in their own field, and to creativity. The Yale Physics Departmentœ[B has made the most decisive progress in its recent history under your chairmanship. By doing most interesting modern research yourself you have set the preconditions for attracting the very best professors to Yale. National Laboratories all over this country and in the international scene requested your advice and participation. Earlier than others you recognized the importance of major instruments and facilities to be provided by these laboratories for academic research. You pleaded for joint action rather than jealous opposition between national laboratories and universities. Students, post-docs, and junior faculty from foreign countries were an integral part of the scientific community around you. Your laboratory served as a homestead for many-like myself in 1967 and thereafter. You created an open environment for all of us and shared with us the joy of physics. You understood that a major field in physics can survive only if it presents its contributions to the field at large in a series of international conferences. You created the International Conference on Atomic Physics, first held in 1968 in New York under your chairmanship and since then biannually with enduring success. Prizes, honorary lectures and honorary degrees have been bestowed on you. You have lectured in many countries spreading the progress in physics all over the world.

What a Successful Career!

Your contributions to physics have been outlined today in much more detail by most competent speakers, some of them being an integral part of your own history. For this reason I will try to summarize the highlights of your research rather briefly and not even comprehensively at all.

Your talent was quickly recognized when you entered the Physics Department of Columbia University right after the war. At that time Rabi was setting out to create a school of physicist so innovative, productive and unique that the influence of his aera and the impact of his school of physics extends until today and - I am certain - a generation beyond. In an atmosphere unparalleled until today - Norman Ramsey has given us a smell of this time in his talk this morning - you learned the basic laws of physics, how to ask the right questions, how to solve theoretical problems, and how to design and build up experiments, how to get the right results. There was a unique style in which the Rabi school carried out physics, a style oriented towards the most fundamental questions in our field. But beyond that there was Rabi's personality, full of imagination and phantasy, sometimes also of irony and wit. You experienced the influence of this great teacher, a man also concerned about the tremendous impact modern research had opened up with respect to applications. I believe it was this influence which among other things compelled you to work as an advisor and consultant to your government over many decades to ensure that they are recognizing their responsibilities.

Studying your long list of publications I have found many "First's". An electric beam resonance apparatus has been realized by you and your co-worker Grabner, where in 1950 you saw a radiofrequency - two quantum transition for the first time. Everybody knows how important multiquantum transitions are nowadays: in NMR several hundred quantum transitions have been achieved, in molecular physics infrared multiple quantum transitions lead to selective excitation, in laser spectroscopy the optical two quantum transition developed into a most general method for high resolution spectroscopy. Your scientific life was focussed on the most fundamental atoms in order to probe the validity of the laws of the electromagnetic interaction. You studied most rigorously the ^{3}He and ^{4}He triplet excited system with even increasing precision for more than 35 years, you took up positronium spectroscopy in 1955 and most important, discovered the muonium atom in 1960.

The fundamental exotic atoms remained Vernons companions for many decades!

Muonium even today is studied by you and your collaborators simultaneously at three major accelerator centers in the United States and Europe. Collaborations extending from Novosibirsk to Tokyo all around the globe work here together. The muonium hyperfine structure measurements have been pushed to a precision where tiny contributions of strong and weak interactions should be seen. Atoms so rare that they could hardly be seen initially are produced so copiously in these days that Lamb shift and Rydberg measurements may compete with those in the hydrogen atom eventually.

Vernon was intrigued by leptons, the funny particles with charge and magnetism but without spatial extension - particles not understood really until today.

Hence it is no surprise that in the beginning of your eighths decenium in life you are still the major driving force behind an experiment to measure the g-factor of the muon to a precision where spurious contributions of known and unknown interactions could be unvealed eventually. Your success is based on your capability to follow up a problem over many years, to do the experiment and the theory most rigorously, to achieve a precision much beyond what has been anticipated originally. Of course, this is well in the tradition of Rabi's famous school as we heard earlier today in numerous examples. You have set limits quite early on possible charge inequalities of electrons and protons, you were the first to demonstrate (in 1960) that spectroscopic lines of decaying systems can be narrowed below the average natural line width, a method later applied to optical double resonance experiments in Heidelberg (in 1964) and then brought to perfection by V.L.Telegdi in his competitive effort in muonium. Indeed, this competition had a very revitalizing effect on our own work in the sense "if there is no competition, there is no value".

Vernon is a man of precision

Of course, what it means to work as a student for a professor devoted to precision work can be anticipated only by those who suffer 'blood, sweat and tears' to obtain the ultimate number. But all the possible complaints by students and their spouses have to be seen in perspective also to Vernons own work habit, another important ingredient of his personal success. He is all the time the earliest in the morning and still works or observes the counters of our experiments when our eyes fell close at night. Really, this chapter in my speech about Vernons work habits should be rather given by somebody observing him all the time like Pat Fleming, his long time secretary. If you enter Vernon's office you realize immediately that he adores

New York. Reprints are arranged in such a way there that you are reminded of the skyscrapers constituting Manhattan's skyline. There is the 'Trump Tower' containing all the papers on deep inelastic scattering of polarized electrons, protons and quarks, a field Vernon pursues most actively with muons right now at CERN. The more ancient 'Algonquin' building piled up on a small chair is devoted to publications on muonium and positronium. The 'Chrysler Building' on the other hand contains theses of different kinds. A pile on the floor as large as the 'Rockefeller Center' contains material for the next funding request. Embedded in this skyscrapers landscape sits Vernon operating one, two or sometimes even three telephones at the same time. Mike Lubell in New York is told to finish his manuscript on polarized electron sources, people in Heidelberg report on their latest muonium experiment at Rutherford and all their problems getting sufficient beam time there, and a student reports from Los Alamos that Vernon's car broke down again. But really, Vernon handles all these problems at the same time with sovereignty, sometimes just summarizing them under friction losses in his truly worldwide international enterprise.

But let me come back to physics for one more moment. "If you want to discover something new you have to do something new" observes Lichtenberg, the great German physicist and philosopher of the 18th century in Göttingen. In order to do something new you have to have adequate tools. Knowing this, Vernon devoted a considerable part of his efforts to the promotion and development of tools, particularly accelerators and dedicated beams as well as targets being polarized or not, and withstanding the unfriendly word of high energy radiation. Considerable technical expertise was developed at Yale with respect to polarized electron sources as well as polarized proton targets, driving technology to the edge of the possible. This had, among other things, a very productive influence on the technical infrastructure of the Yale Physics Department at the time when these sources and targets were under construction. And again, Vernon and his colleagues went out to accelerators like SLAC or the one of MIT to probe the proton spin structure, to observe a possible parity nonconservation in the scattering of polarized electrons, to use muons later for probing the quark structure of the hadrons, possibly solving the "spin crisis" and obtain deeper inside where the spin angular momentum is located in nuclei composed of quarks and gluons.

Dear Vernon!

Let me come to an end after this long day of speeches given by your colleagues in your honour. What can we learn from your success in scientific research and teaching in a university? I think most important is to move ahead to new fields and frontiers all the time as you did. All efforts should be devoted to the right questions of fundamental nature because only then one can motivate colleagues and students to join you in your efforts. And of course, you tried as hard as possible all the time and never gave up. Your research in physics has brought you great recognition personally, expressed through the many honours you have received. I am sure you will continue in the future to pursue your research with the same vigour as before. Many of us had and will have the privilege doing work with you. We are looking forward for an exciting time ahead of us.

Today many of your former students and younger colleagues came here to celebrate with you. If you look around at this very moment and see them altogether here you realize that you have created a truly international community of scientists. And I hope, you will be proud to see that they adopted what you taught to them, recognized the value of their education through you, became successful in their own professional career and now follow their own lifeline in science. There is no better description of what binds us together as V.L. Telegdi's saying of a very strong "Yale togetherness".

Today's celebration of your forthcoming 70th birthday falls together with your official retirement from your active duties of teaching at Yale, the university to which you have given 37 years of your professional life so far. All of us are very honoured to be invited to share this moment with you. All of us hope that you will continue in your interest in physics and do research as before. As T.D.Lee remarked at the end of our scientific session today: "reaching the midpoint of such an extraordinary life is really worth a Fest".

So let me raise my glass and toast with you, dear Miriam, and with all the others to our friend and colleague Vernon Hughes.

Vernon Hughes and Japanese Physicists with Spinning Particles

K. Kondo
Institute of Physics, University of Tsukuba
Ibaraki 305 Japan

In the course of Vernon's work in physics, many Japanese have enjoyed opportunities to work with him. This is my personal recollection of their commitment and collaboration.

The Elder Japanese Physicists Vernon Met – and Should Have Met

Once Vernon told me that when he was looking for a thesis adviser at the graduate school of Columbia University, he was interested in H. Yukawa. Soon after World War II, Yukawa had been invited there and at that time he was working on non-local field theory. I imagine Yukawa was drawing on the blackboard everyday a circle which represented an elementary domain in his mind, as we heard from his contemporary theorists. Yukawa published his first paper on this subject[1] but the second one with the same title never appeared, as W. Pauli is said to have predicted to Yukawa.

Vernon took Yukawa's course in non-local field theory but chose I.I. Rabi as his adviser. This anecdote about Vernon and Yukawa tells us that Vernon's inclination towards fundamental aspects in physics was born in the very young days of his career. Vernon came to Japan in 1974, when he had a chance to visit Yukawa at the Research Institute of Fundamental Physics at Kyoto.

Vernon had no chance to meet with S. Tomonaga. When Rabi visited Nishina's Institute in Tokyo after the war, Tomonaga presented his work on renormalization theory, and Rabi is said to have referred to similar works by Feynman and Schwinger in the United States. In addition to being an eminent theorist, Tomonaga was a good teacher of physics. He once wrote a book entitled *Spin Spins.*[2] Actually, it describes the history of quantum mechanics. Spin played a great role in the foundations of quantum mechanics through the phenomenon of space or direction quantization. When we review Vernon's work, we are always impressed to find in it his good taste in physics. We clearly see a string of spin physics spreading from the early days of quantum physics, through Rabi's Columbia school to Vernon's work. I regret that I did not manage for Vernon and Tomonaga to see each other.

Whether or not due to Tomonaga's influence, there has been great activity in experimental work in Japan related to spins of particles with the high energy machines in Japan–the electron synchrotron at INS and the proton synchrotron at KEK. Thus, there has been good motivation for Japanese physicists to work with Vernon.

Japanese at Yale in 1960's

Back in the early 1960's, when I was a graduate student at Tokyo, I came across a paper by G. Cocconi and E. Salpeter[3] on the limit to the anisotropy of inertial mass. I realized that NMR should provide a most sensitive test on this subject and gave a talk to my colleagues at K. Shimoda's laboratory. About a week later one of my friends informed me about Vernon's paper which appeared in *Phys. Rev. Letters.*[4] K. Shimoda, my thesis adviser, had worked on microwave spectroscopy and was a collaborator of C. Townes in the early days of the maser.[5] So, being a student of Shimoda, I felt familiar with Columbia and with Rabi's school, and I knew Vernon's name and his article on molecular beams co-authored with P. Kusch.[6] After this happened about the inertial mass anisotropy, I felt more intimate with Vernon. In the far east, students used to read articles and books through which they felt they got to know the authors personally. Later in my career, I found from my American colleagues that the situation is different in the US where *Physical Review* articles are not to be read but only to be written.

After I finished graduate school, I found a job working for T. Nishikawa and started working on photoproduction experiments with a newly built electron synchrotron at INS. In the spring of 1968, I had the chance to work with Vernon at Yale. Around the same time, S. Dhawan from Nevis, who has since been working with Vernon on most of his experiments, and G. zu Putlitz from Heidelberg came to New Haven. Vernon had several experimental projects at that time, as now, and I decided to join a group working on a BNL experiment on kaon scattering by a polarized proton target. J. Rothberg was leading the high energy side of the experiment, and G. Rebka and A. Etkin built the polarized target. S. Mori, who got a degree from Cornell, was another Japanese in the group. We searched for an exotic quark state, a fashionable subject in those days, without any success.[7]

On the muonium side of Vernon's projects, S. Ohnuma was working on the design of a proton linear accelerator, and later K. Tanabe from Tokyo joined the group to work on the LAMPF muon channel. After he went back to Japan he contributed to the design of a muon channel at KEK. With this channel many nice experiments have been carried out on the muon and on muonium by T. Yamazaki, K. Nagamine and collaborators from this country.[8]

Yale-Japanese Collaboration in 1970's

In the spring of 1973, a collaboration program between the Hughes group at Yale and a group from Japan started with the support of the Japan Society for the Promotion of Science. The subjects included a BNL experiment on kaon scattering by a polarized proton target. M. Zeller, M. Mishina, and I. Nakano were involved in this experiment.

Another experiment in the collaboration was E-80 at SLAC: Deep inelastic scattering of polarized electrons by a polarized proton target. This started as Bielefeld-Tokyo (later Tsukuba)-SLAC-Yale collaboration. The experiment was discussed in detail this morning by W. Raith. Yale physicists were R. Ehrlich, A. Etkin S. Dhawan and P. Souder on the high energy side, and M. Lubell and others on the polarized electron beam side. From Bielefeld, G. Baum and P. Schüler, and from Japan, H. Kobayakawa, N. Sasao, S. Miyashita, K. Takikawa and I joined the collaboration. SLAC physicists working on this experiment were W. Ash, D. Coward, D. Sherden and C. Sinclair from SFG (Spectrometer Facility Group) R. Miller (Accelerator Physics) and S. St. Lorant from Low Temperature Facility. Later on, Chinese physicists, Y. N. Guo and Z. L. Mao joined the experiment.

In the early stage of the experiment it took longer than anticipated to prepare the polarized beam and the polarized target. I heard a rumor that W. Panofsky, the SLAC director at that time, said "E-80 will never fly". But with Vernon's persistence and the struggle of his collaborators we obtained the first data, which demonstrated the quark spin effect as predicted by the parton model.[9] We then proposed a second experiment, E-130, with which we accumulated more data, covering a wider kinematical range.[10]

In the later stage of the experiment, Vernon was enthusiastic to extend the experiment to observe the neutron spin structure by using a polarized deuteron target. The development of the target was continued with the collaboration of K. Morimoto as a polarized deuteron target expert. Unfortunately, this did not materialize as a scattering experiment, but I am so happy to hear that a new experiment is being prepared on this subject at CERN.

Muon g-2 Experiment and Japanese Commitment

In 1979, a program of US–Japan collaboration in high energy physics began with govermental support from the two countries. Among others, T. Nishikawa and L. Lederman, the KEK and Fermilab directors at that time respectively, made a great effort for initiating and developing the program.

In 1985, the muon g-2 experiment at BNL was selected as a subject. The Japanese commitment so far is limited to technical preparations for the experiment. From KEK, H. Hirabayashi, S. Kurokawa, and A. Yamamoto have been working on the superconducting storage ring (see Figure 1) and Y. Mizumachi on the beam

Figure 1: A photograph of the coil winding system at Brookhaven National Laboratory for the 14 m diameter superconducting coils and of the laboratory in which the superconducting muon storage ring will be located.

monitoring. The work is still going on, and I hope some Japanese will work on the physics side of the experiment in the near future.

Vernon's work in experimental physics has been a beautiful unification of techniques in different fields of physics: atomic physics, microwave techniques, and high energy physics. He never jumped into the first stage experiments with the new accelerators. Old accelerators, however, were revived by Vernon's touch and his new ideas, and produced wonderful results.

On the occasion of Vernon's Fest, we would like to thank him for his efforts and hospitality extended over many years to physicists across the ocean. I hope that international collaboration, of which Vernon was a great initiator and promoter, will develop further and help unite scientists in different parts of this spinning globe.

References

1. H. Yukawa, *Phys. Rev.* **77**, 219 (1950).
2. S. Tomonage, *Spin wa Meguru*, Chuokoron-Sha, Tokyo (1974).
3. G. Cocconi and E. E. Salpeter, *Phys. Rev. Lett.* **4**, 176 (1960).

4. V. W. Hughes, H. G. Robinson and V. Beltran-Lopez, *Phys. Rev. Lett.* **4**, 342 (1960).
5. K. Shimoda, T. C. Wang and C. H. Townes, *Phys. Rev.* **102**, 1308 (1956).
6. P. Kusch and V. W. Hughes, *Encyclopedia of Physics* (Springer-Verlag, Berlin, 1959), Vol. **37**, Part 1.
7. G. A. Rebka, Jr., J. Rothberg, A. Etkin, P. Glodis, J. Greenberg, V. W. Hughes, K. Kondo, D. C. Lu, S. Mori and P. A. Thompson, *Phys. Rev. Lett.* **24**, 160 (1970).
8. R. Kadono, J. Imazato, T. Matsuzaki, K. Nishiyama, K. Nagamine, and T. Yamazaki, *Phys. Rev.* **B39**, 23 (1989); S. Chu, A. P. Mills, Jr., A. G. Yodh, K. Nagamine, Y. Miyake and T. Kuga, Phys. Rev. Lett. **60**, 101 (1988).
9. M. J. Alguard, W. W. Ash, G. Baum, J. E. Clendenin, P. S. Cooper, D. H. Coward, R. D. Ehrlich, A. Etkin, V. W. Hughes, H. Kobayakawa, K. Kondo, M. S. Lubell, R. H. Miller, D. A. Palmer, W. Raith, N. Sasao, K. P. Schüler, D. J. Sherden, C. K. Sinclair and P. A. Souder, *Phys. Rev. Lett.* **37**, 1261 (1976).
10. G. Baum, M. R. Bergström, P. R. Bolton, J. E. Clendenin, N. R. DeBotton, S. K. Dhawan, Y. -N. Guo, V. -R. Harsh, V. W. Hughes, K. Kondo, M. S. Lubell, Z. -L. Mao, R. H. Miller, S. Miyashita, K. Morimoto, U. F. Moser, I. Nakano, R. F. Oppenheim, D. A. Palmer, L. Panda, W. Raith, N. Sasao, K. P. Schüler, M. L. Seely, P. S. Souder, S. J. St. Lorant, K. Takikawa and M. Werlen, *Phys. Rev. Lett.* **51**, 1135 (1983).

Remarks

Vernon W. Hughes

The principal sentiment I want to express is my deepest gratitude to all of you, my friends, for coming and making this wonderful birthday party today. I am deeply moved and indeed overwhelmed by this affair. As some of you know, my real birthday is May 28, so I am not 70 but only in my late 60's. Today is my public birthday as declared by my good friend Mike Zeller, Chairman of our Physics Department.

I thank especially the speakers and session chairmen, who have worked toward and provided a most interesting and intellectually exciting day for us. Several of them in particular were kind enough to travel a considerable distance. Val Telegdi came from Geneva. Val's being here somehow makes the day for me, because muonium has been central in my scientific life and Val's own important work, deep understanding, and clever ideas in this field, as well as his lively criticism and competition, were always most stimulating. In some way the title of his talk, "It Takes One to Know One—Remembering Muonium" expresses our long-standing common interest in muonium. You can appreciate that it has become positively dull to work in the field since Val left it.

Bill Raith came from Bielefeld, Germany. Bill and I were very close and friendly collaborators in the early development of sources of polarized electrons at Yale and in achieving the first high energy polarized electron beam at SLAC. We also worked with Mike Lubell at Yale on the first measurement of spin-dependent asymmetries in very low energy atomic collisions of polarized electrons on polarized hydrogen atoms. It has always been a great pleasure and joy to work with Bill.

I would like to be able to say that Emilio Picasso came here from Pisa, or at least Geneva; actually he is visiting Chicago for a few months and came here from Chicago. I have been very interested in the muon g-2 measurement for close to three decades. Indeed, Francis Farley very kindly invited me to join him in the second CERN g-2 experiment, which he was leading and developing. I visited CERN many times during the last CERN experiment led by Emilio and learned an enormous amount from him and his colleagues. He has been most helpful and supportive of our new muon g-2 experiment at BNL. CERN did a fine prototype muon g-2 experiment; we believe we can improve the precision of the measurement by a factor of 20, using the same basic principle with very many differences in important details.

Charlie Prescott came here from SLAC. When in 1971 Charlie learned of our intention to develop a polarized electron source for SLAC, he proposed the idea of looking for parity non-conservation in a predominantly electromagnetic interaction by measuring the helicity dependence of the electron-nucleon deep inelastic scattering cross section, and we joined forces to do this experiment together. He described very well the background of the parity nonconservation experiment at SLAC. Charlie is a highly professional and dedicated physicist and a fair and honorable man. It was a great pleasure to work with him. Currently he is the mentor of my younger son, Emlyn, who has become a physicist.

By travelling from Cornell to Yale, Toichiro (Tom) Kinoshita made a well-known second-forbidden transition. Tom has had an enormous influence on our work. Without him there would certainly be no muon g-2 experiment at BNL. It is a great pleasure to share physics interests with him. Tom calculates difficult, important quantities of enormous complexity to very high accuracy. One can't avoid wondering whether he is giving the right answer. At CERN a few weeks ago I had lunch with Eduardo de Rafael and asked his opinion about this matter. De Rafael answered that firstly, Tom understands fundamental field theory so deeply that it is inconceivable that he would make any error of principle; secondly, he builds many checks into his calculations, has well-trained colleagues, and is himself so careful that it is very unlikely that any calculational errors could survive. Although I am, and indeed have always been, basically satisfied that Tom gives the correct answers, it would be reassuring if someone else had the courage and ability to do the muon g-2 calculations independently.

Norman Ramsey and Dan Kleppner came down from Cambridge/Boston which isn't so far away, fortunately, so I often get to visit them. (I have always enjoyed Van Vleck's comparison of Boston and New Haven. He said he preferred New Haven to Boston because, by train, Boston was 4-1/2 hours from the City, whereas New Haven was only 1-1/2 hours away.) Norman and I first met at Columbia in 1946, when I returned from war-work at MIT on radar for graduate studies at Columbia. I attended his seminar on molecular beams and worked a bit in his laboratory. Within six months or so, Norman went off to Harvard. I have followed his great work carefully and seen him often. Norman has been a sort of older brother to me in the field of physics.

Dan Kleppner was Norman's wonderful student and I admire greatly his basic, versatile, and creative work in many fields of atomic physics. He is the eloquent spokesman for small, human-scale physics.

T. D. Lee, Gary Feinberg, and Chien-Shiung Wu came up from the greatest of all places, Columbia and New York. T.D. was kind enough to remark that my relations with Columbia are perhaps as close as with Yale. (I also appreciate very much the expression this evening of Columbia's hospitality to me by Frank Sciulli, the Chairman of Columbia's Physics Department.) Emotionally, that is certainly true. Columbia was my early home in physics and I love to get down there. I always enjoy and learn so much from them. Through his own brilliant work in virtually all fields of modern theoretical physics, his deep understanding, important suggestions and strong personality, T.D. establishes in a major way the tone and spirit of physics at Columbia. T.D. is a great teacher and friend.

Gary's ideas have been very stimulating for much of our work—searching for muonium-to-antimuonium conversion, and the Bates parity nonconservation experiment on e^-_{pol}-carbon elastic scattering (suggested by Gary)—and we share a common interest in simple atoms.

Chien-Shiung and I have worked closely together on positronium and exotic atoms, and we edited a series of three volumes entitled *Muon Physics.* She really introduced me to nuclear physics in our positronium work in the 1950's and during that period I was privileged to hear about her great experiment which discovered parity nonconservation in the weak interactions. I have enjoyed and profited enormously from her deep knowledge of physics, her broad interest in science, her great spirit, and her warm friendship.

Columbia has been generous enough to have me as Visiting Associate Professor in the late 1950's, when I was looking for muonium, and as visiting I. I. Rabi Professor for a term in 1984, when we were developing the proposal for a new muon g-2 experiment at BNL. They now kindly call me Adjunct Professor and post my name (among others) on a door in Pupin Laboratory.

Gisbert zu Putlitz—or more properly, as Telegdi reminded us, Gisbert Freiherr zu Putlitz—came to this event from the city and University of Heidelberg to which New Haven and Yale are in many ways comparable. I was very fortunate that in 1967 Gisbert, as a young man, decided to come and work with us at Yale on muonium. He spent 1 1/2 years in residence at Yale and, since his return to Heidelberg, his group and our Yale group have collaborated closely—some 23 years until the present time—to develop our knowledge and understanding of the wonderful atom, muonium. Our long collaboration has been most profitable, productive, and enjoyable. It has been characterized by complete trust. The close friendship of Gisbert's family and mine has been and continues of great importance to me.

Kunitaka Kondo of the University of Tsukuba travelled from Tokyo, or perhaps from Fermilab. It is unfortunate that we invited him a bit too late to make after-dinner remarks, but I am delighted that he has contributed remarks for this volume. In the late 1960's Kuni came from the University of Tokyo to work with us and was in residence at Yale for about 2 years. A few years later, when he was established at the University of Tsukuba, Kuni brought several younger colleagues and students to join our SLAC experiment to study the spin-dependent structure function of the proton through deep inelastic scattering of high energy polarized electrons by a polarized proton target. Our collaboration on the SLAC experiment was a major, rewarding, and most pleasant one. It was, I believe, one of the early significant collaborations between United States and Japanese physicists, before the formal program of US-Japan Collaboration in High Energy Physics started. Largely through Kuni I have had the pleasure of knowing many fine Japanese physicists. Kuni was most helpful to me in developing our important collaboration with KEK and University of Tokyo physicists on the BNL muon g-2 experiment. He has been my host during many visits to Japan, including a most notable one when I spent a three week period as Visiting Professor of the Japan Society for the Promotion of Science, lecturing at various Japanese Universities. Kuni has been a close friend and a most valued colleague.

I regret that Willis Lamb was unable to attend. As a sophomore at Columbia College in 1939 I learned to use the micrometer from Willis and became a great admirer of his. Indeed Willis was my close colleague here at Yale for eleven years beginning in 1962. For me Willis provides an ultimate standard of taste and intelligence in physics.

Mrs. Helen Rabi seriously considered undertaking the trip from New York for this evening. I am honored by her consideration but thankful that she thought better of making what would have been an arduous trip for her.

Yale University has been most generous to host this symposium. Of course the Physics Department has been central. In particular, our Chairman, my fine young(er) colleague and friend Michael Zeller, initiated and has been primarily responsible for its organization. I strongly suspect that in the midst of all the work involved, he may have had second thoughts about the wisdom of its initiation. But he has refused to admit it and I am deeply grateful to him. Nancy Soper and Jeanette DeFranco worked hard to make this a successful and pleasant event and I thank them. Pat Fleming, my invaluable Administrative Associate, was uncustomarily in the background on this occasion but played an essential role in providing her suggestions and transmitting mine for this fine affair. Without Pat's efficient, enthusiastic, and dedicated help over the past twenty-three years I might not have accomplished enough to justify this party.

Being descended from Iowa stock, which included an evangelical Methodist minister, I should probably have some moral, wise thoughts to offer. However, I don't have anything very original to say. Surely the important driving force for research is one's own interest in the topic. One shouldn't be too influenced by other people's opinions, or by current fads, particularly theoretical fads or speculations. I have found, on several occasions, progress in research has been slowed at least partly due to the injection of criteria extraneous to the research itself, e.g. consideration of credit to individuals, a teaching objective or some political funding matter. Effective collaboration in small or large groups requires the closest possible communication among their members and maximum input from everyone. I have derived great satisfaction from the warm human relationships with physicists with whom I have worked or shared common interests.

On an occasion such as this , one naturally reflects on one's past. I have had wonderful opportunities. I had the good fortune to get the right start in my physics career at Columbia College. In my time there just before World War II there was no excessively protective concern for the student so I had the freedom to find out that physics was a very attractive subject, indeed much more so than law, which was my initial goal. After a year of graduate work at Cal Tech and four years of war-work on radar at MIT, I returned to Columbia to work with Professor Rabi. I chose to build a new molecular beam electric resonance apparatus by myself and to study molecular spectroscopy for my Ph.D. research. In retrospect, it was a fine exercise and we did observe the first two-photon transition; however, the subject was not a wise choice because I had also the opportunity to do the first measurement of the hyperfine structure interval in hydrogen with Professor Rabi and this experiment by Nafe, Nelson and Rabi reported the first discrepancy with quantum electrodynamics, which soon came to be understood as due to virtual radiative corrections to the electron spin magnetic moment. It was, of course, the greatest privilege to study with Professor Rabi. His taste in physics, his deep insight into how atoms work, his commitment to fundamental physics, his great common sense and wisdom and his brilliant articulateness were of enormous influence on all of us who had the opportunity to know and work with him. I have been fortunate to have an enduring friendship with Rabi and his family. In the post-war years Columbia was an almost unbelievably exciting and significant center of physics. The goal and spirit of its education was to become a physicist in a broad sense, not an experimental, theoretical, atomic or nuclear type. A full understanding of what, how and why one was doing something was expected. One had the great models of Rabi, Lamb, Townes, and Ramsey and the remembrance of Fermi. Nowadays, specialization has probably become more essential – at least for almost all normal mortals; still, I believe the ideal of a broad physicist is valid, valuable and wholesome.

Since 1954, that is for 37 years, Yale University has offered me a superb opportunity to pursue teaching and research in physics and also to contribute to the development of our Physics Department. Because of its distinction Yale attracts many wonderful students, both undergraduate and graduate—it is certainly the graduate students who do the major share of day-to-day research. Excellent postdoctoral fellows, young teaching and research faculty, and senior visiting physicists have come to work in our laboratories at Yale.

Our great Yale University is a relatively small community (President Brewster used to say that he knew personally, through at most one intermediary, all faculty and students at Yale.) Through my activities as Chairman of the Physics Department from 1961 to 1967 and as a member of many University committees, I have come to know our Presidents and a number of major administrative officers. President A. Whitney Griswold was very encouraging and supportive in the late 1950's and early 1960's before his premature death. President Kingman Brewster led a determined effort to strengthen the sciences and the graduate school at Yale and created an inspiring, progressive and friendly climate. Deans John Perry Miller and Georges May and Provost Charles Taylor were his strong administrative colleagues during a long and exciting tenure. I did not know President Bart Giamatti well but came to admire him as one of Yale's important presidents. I am sorry that President Benno Schmidt could not be here this evening, but I sincerely thank him for the encouraging and warm letter he wrote to me for today.

After the early days when I worked principally in atomic physics, Columbia's Nevis Labortory, our U.S. National Laboratories—including especially Brookhaven, Los Alamos, and SLAC—and, more recently, CERN have provided the centers for our research. They have offered unparalleled opportunities, and I thank especially the following leaders of these laboratories for their faith in and support of our work: Leon Lederman of Columbia; Maurice Goldhaber, Bob Palmer, Bob Adair (also of Yale) and Nick Samios of Brookhaven; Louis Rosen and Darragh Nagle of LAMPF at Los Alamos; Pief Panofsky and Dick Taylor of SLAC; and Carlo Rubbia, Pierre Darriulat and Friedrich Dydak of CERN.

Since 1962 I have been a trustee from Yale of Associated Universities, Inc., which manages Brookhaven National Laboratory and National Radioastronomy Observatory. These are both first class scientific organizations. It has been a privilege to serve on AUI with many dedicated colleagues and with its fine presidents—during my tenure Ted Reynolds, I. I. Rabi, Gerald Tape, and currently Robert Hughes. I am most honored that Bob and his wife Lou came to Yale for this day.

I have felt for a long time, as I believe many of us do, that a particular importance of our profession is its international aspect. We have a rather unique opportunity, largely independent of politics, to communicate and work and to become friends with physicists from almost any country which has activity in our field. Our collaborations are not only a great professional and personal pleasure but also an important contribution to international understanding and peace.

This fine occasion is a combination 70th birthday and retirement party. In accordance with the retirement policy of Yale, I am retiring as professor at age seventy. It is expected that in about two years our national policy will change so that mandatory retirement based on age will become illegal. I think, however, that some sort of change in status should occur at about my age and I am in no way unhappy about retiring as Professor. Yale University, the Physics Department, and Mike Zeller are being kind enough to offer me a new active status as Senior Research Scientist, in addition to Professor Emeritus. Since DOE seems happy to continue to support our research, I plan to remain active, free from class teaching obligations and committee meetings. I very much appreciate and value this opportunity for which I thank the Physics Department and Mike Zeller.

Since this isn't a memorial service, all my next of kin aren't here, most notably my two wonderful young grandsons, Ariel and Isaac, and my daughter-in-law, Miriam. But I am happy that my two sons, Gareth and Emlyn, are here and, of course, also my superb wife Miriam.

Well, let me conclude by, once again, offering my heartfelt thanks to all of you for coming here and making this such a warm and wonderful occasion.

Selected Publications of Vernon W. Hughes

Reprinted from THE PHYSICAL REVIEW, Vol. 79, No. 2, 314–322, July 15, 1950
Printed in U. S. A.

The Radiofrequency Spectrum of $Rb^{85}F$ and $Rb^{87}F$ by the Electric Resonance Method*

VERNON HUGHES** AND LUDWIG GRABNER
Columbia University, New York, New York

(Received February 24, 1950)

The radiofrequency spectrum of the two molecular species $Rb^{85}F$ and $Rb^{87}F$ was studied by the molecular beam electric resonance method. The electrical quadrupole interaction constant, eq_1Q_1/h, of Rb in RbF was determined for $Rb^{87}F$ in rotational states $J=1$ and $J=2$ for the first few vibrational states, and for $Rb^{85}F$ in rotational state $J=1$ for the first few vibrational states. The interaction constants are unusually large for an alkali halide molecule; for the zeroth vibrational state of $Rb^{85}F$, $eq_1Q_1/h=-70.31\pm0.10$ Mc/sec. and for the zeroth vibrational state of $Rb^{87}F$, $eq_1Q_1/h=-34.00\pm0.06$ Mc/sec. The absolute value of the interaction constant decreases about 1.1 percent from one vibrational state to the next higher one; the values of the interaction constants for rotational states $J=1$ and $J=2$ are the same within the limit of error. The ratio of the electric quadrupole moment of Rb^{85} to that of Rb^{87} is $+2.07\pm0.01$.

Results are reported on the spin-orbit coupling $c_2(\mathbf{I}_2\cdot\mathbf{J})$, between the spin of the fluorine nucleus and the molecular rotational angular momentum. For the zeroth vibrational state of $Rb^{85}F$, $|c_2/h|=11\pm3$ kc/sec. and for the zeroth vibrational state of $Rb^{87}F$, $|c_2/h|=14\pm4$ kc/sec. No variation in $|c_2/h|$, with vibrational state was observed.

Finally, an unpredicted line group was observed in the spectrum at one-half the frequency of one of the line groups in the $Rb^{85}F$ spectrum. Possible origins of this line group are discussed, including that of a double quantum transition.

The vibrational frequency, ω_0, of $Rb^{85}F$ is 340 ± 68 cm^{-1}.

I. INTRODUCTION

THE radiofrequency spectrum of the two isotopic species of RbF—$Rb^{87}F$ and $Rb^{85}F$—have been studied by the molecular beam electric resonance method.[1,2] This technique permits the study of the energy levels in a static electric field of polar diatomic molecules which are in the $^1\Sigma$ electronic state, in a low vibrational state, and in the rotational state $J=1$. Sometimes molecules in the rotational state $J=2$ can also be studied. The research reported in this paper is a continuation of the studies of the internal interactions of the alkali fluoride molecules.[2,3] The apparatus used was the one described in references 1 and 2, and further details on the experimental technique are given in reference 3.

FIG. 1. Spectrum observed for $Rb^{85}F$ and $Rb^{87}F$ under low electric field conditions (i.e., $\mu^2E^2/\hbar^2/2A\ll|eq_1Q_1|$). Change in beam intensity is expressed in cm of deflection on a galvanometer scale. All the lines were not taken under comparable conditions so line intensities cannot be compared indiscriminately. Fine structure of the lines is not indicated.

Under low field conditions the frequencies of the transitions induced are determined primarily by the internal molecular interactions. The largest of these is the electrical quadrupole interaction associated with the Rb nucleus, and it was determined for each molecular species for several vibrational states. The interaction constants for the RbF molecules are very much larger than those for CsF^2 or $K^{39}F^3$ or, indeed, than those for any other alkali halide molecule yet studied.

A splitting of the spectral lines of both $Rb^{87}F$ and $Rb^{85}F$ was observed. This is caused primarily by a spin-orbit interaction between the fluorine nucleus and the molecular rotational angular momentum. The magnetic dipole-dipole interaction between the two nuclei also contributes slightly to this splitting.

The appearance of an unpredicted line group at exactly one-half the frequency of a line group in the weak field spectrum of $Rb^{85}F$ is reported. Possible explanations for these lines are discussed, including the one that they are double quantum transitions in which two one-half frequency quanta provide the energy for the transitions. However, none of these explanations seems to account for all the observations.

II. EXPERIMENTAL DATA

The spectra for the two isotopic species $Rb^{87}F$ and $Rb^{85}F$, were observed simultaneously[4] under conditions of low electric field intensity, i.e., the interaction energy of the external electric field with the molecule was small compared to the electrical quadrupole interaction energy associated with the Rb nucleus. The spectrum (see Fig. 1) consists of some 26 lines occurring in the

* This research has been supported in part by the ONR.

** Submitted by Vernon Hughes in partial fulfillment of the requirements for the degree of Doctor of Philosophy in Physics at Columbia University.

[1] H. K. Hughes, Phys. Rev. 72, 614 (1947).
[2] J. W. Trischka, Phys. Rev. 74, 718 (1948).
[3] L. Grabner and V. Hughes (to be published).

[4] Simultaneous observation of the spectra of both isotopic species was possible because the A and B field voltages which refocus one molecular isotope refocus the other molecular isotope as well. All the data were taken using the $(1,\pm1)_A-(1,0)_B$ refocused beam as described in reference 3.

frequency range from about 3 Mc/sec. to 21 Mc/sec.; a search was made over the more extended range from 1 Mc/sec. to 45 Mc/sec. but no lines were found other than those shown. These lines are analyzed into line groups which are designated by Roman numerals with subscripts. The lines within a group correspond to the same transition for different vibrational states, and they are distinguished by the vibrational quantum numbers which are written directly above the lines. It will be shown that with the exception of the line group near 3 Mc/sec., the spectrum can be understood as due to transitions of molecules with rotational states $J=1$ and $J=2$.

III. THEORY

The following Hamiltonian is assumed for the diatomic molecule in a static electric field.[2]

$$\mathcal{H}=\frac{\hbar^2}{2A}\mathbf{J}^2-\boldsymbol{\mu}\cdot\mathbf{E}-\frac{eq_1Q_1}{2I_1(2I_1-1)(2J-1)(2J+3)}\times\{3(\mathbf{I}_1\cdot\mathbf{J})^2+\tfrac{3}{2}(\mathbf{I}_1\cdot\mathbf{J})-\mathbf{I}_1{}^2\mathbf{J}^2\}+c_2(\mathbf{I}_2\cdot\mathbf{J})+\frac{g_1g_2\mu_N{}^2}{r^3}\left\{\mathbf{I}_1\cdot\mathbf{I}_2-3\frac{(\mathbf{I}_1\cdot\mathbf{r})(\mathbf{I}_2\cdot\mathbf{r})}{r^2}\right\}. \quad (1)$$

The first term in which $\mathbf{J}$ is the rotational angular momentum and A is the moment of inertia is the rotational energy term. The second term gives the interaction of the external electric field $\mathbf{E}$ with the permanent electric dipole moment $\boldsymbol{\mu}$ of the molecule. The third term is the electrical quadrupole interaction between the electric quadrupole moment of the Rb nucleus, Q_1, and a second derivative of the electric potential at the position of the Rb nucleus, q_1, produced by those charges in the molecule which are exterior to the nucleus.[5] (The subscript $_1$ refers to the Rb nucleus.) $\mathbf{I}_1$ is the spin of the Rb nucleus. There is no electrical quadrupole interaction with the F nucleus because its spin is $\frac{1}{2}$. The fourth term is a cosine coupling between the spin of F designated by $\mathbf{I}_2$ and the rotational angular momentum of the molecule.[6,7] The last term is the magnetic dipole-dipole interaction between the Rb and F nuclei, in which $\mathbf{r}$ is the internuclear distance, g_1 and g_2 are the gyromagnetic ratios for the Rb and F nuclei respectively, and μ_N is one nuclear magneton.[8]

Several vibrational states and two rotational states are expected to be present in the refocused beam, so the observed spectrum will be a superposition of lines from molecules in these different rotational and vibrational states. Hence we shall observe the variation of the terms appearing in the Hamiltonian with vibrational and rotational state.

TABLE I. Quadrupole interactions of Rb^{85} in $Rb^{85}F$ and of Rb^{87} in $Rb^{87}F$.

	Line	Rotational state	Transition $F_1\to F_1'$	Vibrational state	Line frequency (center of line*) in Mc./sec.	Quadrupole interaction $-(eq_1Q_1/h)$ in Mc./sec.
$Rb^{85}F$	${}^0II_1$	1	$3/2\to7/2$	0	6.328	70.312
	${}^1II_1$	1	$0.09(eq_1Q_1/h)$	1	6.259	69.544
	${}^2II_1$	1		2	6.184	68.712
	${}^3II_1$	1		3	6.119	67.988
	${}^4II_1$	1		4	6.048	67.200
	${}^0II_2$	1	$5/2\to7/2$ $0.21(eq_1Q_1/h)$	0	14.800	70.476
	${}^0II_3$	1	$3/2\to5/2$	0	21.103	70.344
	${}^1II_3$	1	$0.3(eq_1Q_1/h)$	1	20.863	69.544
	${}^2II_3$	1		2	20.632	68.774
$Rb^{87}F$	${}^0I_1$	1	$1/2\to5/2$	0	6.800	34.000
	${}^1I_1$	1	$0.2(eq_1Q_1/h)$	1	6.725	33.626
	${}^2I_1$	1		2	6.640	33.200
	${}^0I_2$	1	$3/2\to5/2$	0	8.520	34.080
	${}^1I_2$	1	$0.25(eq_1Q_1/h)$	1	8.425	33.700
	${}^2I_2$	1		2	8.330	33.320
	${}^0I_3$	2	$1/2\to5/2$	0	14.570	33.996
	${}^1I_3$	2	$3/7(eq_1Q_1/h)$	1	14.418	33.642
	${}^2I_3$	2		2	14.250	33.250
	${}^0I_4$	1	$1/2\to3/2$	0	15.319	34.042
	${}^1I_4$	1	$0.45(eq_1Q_1/h)$	1	15.143	33.552
	${}^2I_4$	1		2	14.971	33.268

* See text for further discussion of how the "centers" of the lines are chosen.

In the application of the perturbation theory to the calculation of the eigenvalues of (1), it is important to know that $|eq_1Q_1|\gg|c_2|$ and $(g_1g_2\mu_N{}^2/r^3)$. Furthermore, the electric field intensity is such in our experiments that either:

$$\mu^2E^2/(\hbar^2/2A)\ll|c_2| \quad (\text{"very weak" field}) \quad (a)$$

or

$$|eq_1Q_1|\gg\mu^2E^2/(\hbar^2/2A)\gg|c_2| \quad (\text{"weak" field}). \quad (b)$$

In both cases there is a strong coupling of $\mathbf{I}_1$ and $\mathbf{J}$, which is the electrical quadrupole interaction associated with Rb; hence there is a well-defined $\mathbf{F}_1=\mathbf{I}_1+\mathbf{J}$ vector. We regard the zeroth order part of our Hamiltonian as:

$$\mathcal{H}_0=\frac{\hbar^2}{2A}\mathbf{J}^2-\frac{eq_1Q_1}{2I_1(2I_1-1)(2J-1)(2J+3)}\times\{3(\mathbf{I}_1\cdot\mathbf{J})^2+\tfrac{3}{2}(\mathbf{I}_1\cdot\mathbf{J})-\mathbf{I}_1{}^2\mathbf{J}^2\} \quad (2)$$

$\mathcal{H}_0$ has only diagonal matrix elements in a representation (I_1, J, F_1, m_{F_1}), in which m_{F_1} is the projection of F_1 in the field direction. These energy values depend upon F_1 but are degenerate in m_{F_1}. In case (a) $\mathbf{F}_1$ and $\mathbf{I}_2$ are coupled by the internal interactions involving $\mathbf{I}_2$ (4th and 5th terms in (1)) to form a total angular momentum $\mathbf{F}=\mathbf{F}_1+\mathbf{I}_2$. In a representation (I_1, J, F_1, I_2, F, M), in which M is the projection of $\mathbf{F}$ in the field direction, the first and third terms in the Hamiltonian (1) are diagonal. The diagonal matrix elements of the fourth

[5] J. Bardeen and C. H. Townes, Phys. Rev. **73**, 97 (1948). Note that in the initial report of this research in the abstracts of the June, 1949, American Physical Society meeting at Cambridge, q' rather than q_1 was used. ($q'=q_1/2e$ in which e is the proton charge.) Since our first report it has been agreed in this laboratory, in the interests of uniformity, to use q as defined by Bardeen and Townes.

[6] H. Foley, Phys. Rev. **72**, 504 (1947).

[7] G. C. Wick, Phys. Rev. **73**, 51 (1948).

[8] Kellogg, Rabi, and Ramsey, Phys. Rev. **72**, 1075 (1947).

and fifth terms of (1) depend on F but not on M. There are off-diagonal elements of the fourth and fifth terms in (1) which connect different F_1 states, but these are neglected in our computations. The $-\boldsymbol{\mu}\cdot\mathbf{E}$ interaction has no diagonal matrix elements in this representation, and its effect is computed by second-order perturbation theory. The degeneracy for different M values is partially removed by the field interaction, but states differing only in the sign of M still remain degenerate. In case (b) we use a representation (I_1, J, F_1, m_{F_1}, I_2, m_2) in which m_2 is the projection of $\mathbf{I}_2$ in the field direction. The first and third terms in (1) have only diagonal matrix elements and are considered first. The interaction with the field, $-\boldsymbol{\mu}\cdot\mathbf{E}$, is computed by second-order perturbation theory and finally the effects of the $c_2(\mathbf{I}_2\cdot\mathbf{J})$ interaction and of the nuclear dipole-dipole interaction are considered as smaller perturbations which couple $\mathbf{F}_1$ and $\mathbf{I}_2$. The field interaction is the same for states with $+m_{F_1}$ and $-m_{F_1}$ and the interactions involving $\mathbf{I}_2$ are the same for states with the same value of the product $m_{F_1}m_2$. Hence some degeneracy remains even after all interactions have been considered. The eigenvalues and eigenfunctions for the two cases are given in detail elsewhere.[9,10]

IV. ANALYSIS OF SPECTRUM

(A) Spectrum of $Rb^{85}F$

The lines assigned to the more abundant (73 percent) molecular species $Rb^{85}F$, in which the Rb nucleus has a spin of 5/2, are shown in Fig. 1. The three line groups, designated II_1, II_2, and II_3, are assigned to rotational state $J=1$. The uniqueness of this assignment rests on the facts that: (1) Only a few of the lower rotational states can be present in the refocused beam because only these are deflected sufficiently to get around the wire stop. Indeed only rotational states $J=1$ and $J=2$ are known to have been refocused as yet in electrical resonance experiments. (2) The spectral pattern is different for different rotational states.

The lines within a group are assigned to different vibrational states as indicated by numbers written directly above the lines. There is a fine structure for the lines which is not indicated in Fig. 1. This structure is due to the $-(\boldsymbol{\mu}\cdot\mathbf{E})$ interaction and to the fourth and fifth terms in the Hamiltonian (1), and will be discussed in the next paragraphs.

The positions of the lines are determined primarily by the change in the electrical quadrupole interaction-energy of Rb-term (3) in Hamiltonian (1). This change can be designated by the two F_1 values involved and in Table I the transition to which each line corresponds is indicated. The line frequency is taken at the center of the fine structure and on this basis the electrical quadrupole interaction constant is calculated. The reliability of the values shown for the interaction constants can only be evaluated after we have discussed the fine structure of the lines; however, it can be remarked that the equality of the quadrupole interaction constants evaluated from the different line groups is the proof that the identification of the observed lines with certain transitions is correct.

That the quadrupole interaction constant, eq_1Q_1/h, must be negative will now be shown with the aid of Fig. 2. It is to be remembered that transitions between states at low fields will be observed only if the two states involved correspond to states with different $|m_J|$ at high fields.[3] It is seen from Fig. 2, which is drawn for $eq_1Q_1/h<0$, that transitions from the levels $F_1=3/2$ to $F_1=7/2$ will be observable. If a diagram similar to Fig. 2 is drawn for the case in which $eq_1Q_1/h>0$, then it will be seen that a transition from a state with $F_1=3/2$ to a state with $F_1=7/2$ would be between two states which at high fields had $|m_J|=1$, and hence

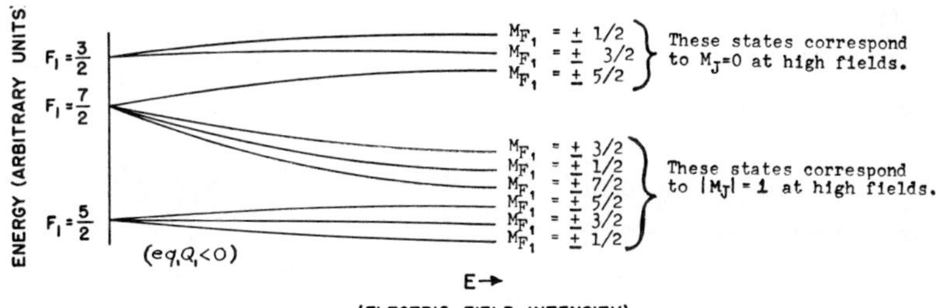

FIG. 2. A diagram indicating which low field states of $Rb^{85}F$ go into states with $M_J=0$ at high fields and which go into states with $|M_J|=1$. (M_J is the projection of $\mathbf{J}$ in the field direction—M_{F_1} is the projection of $\mathbf{F}_1$ in the field direction.) States with $J=1$ and $I_1=5/2$ are shown. An adiabatic change in the field parameter is assumed. The level spacings are not drawn to scale and there may be some crossing of levels with different MF_1 values, but these matters are not important to the discussion for which this figure is used. The correct ordering of the levels at high and low fields is shown. See reference 10 for the method of establishing the correspondence between low and high field states.

[9] Nierenberg, Rabi, and Slotnick, Phys. Rev. **73**, 1430 (1948).
[10] V. Hughes and L. Grabner (to be published).

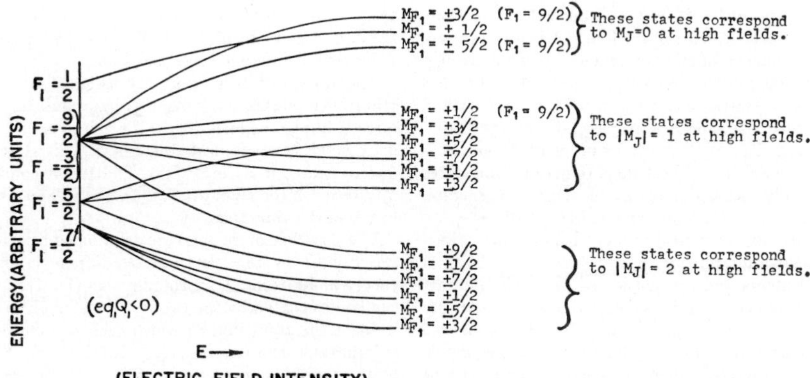

FIG. 3. A diagram indicating the correspondence between low and high field states for $Rb^{85}F$ ($I_1=5/2$, $J=2$). Other remarks apply as given in Fig. 2.

would not be observed. Since this transition was observed experimentally, eq_1Q_1/h must be negative.[11]

The number of line groups observed will now be considered in view of the selection rules and with the assumption of the observability criterion.[12] For the rotational state $J=1$ three transitions are allowed and all three were observed. It can be noted from Fig. 2 that the transition $F_1=5/2$ to $F_1=7/2$ is expected to be weak because most of the component lines are non-observable; it is indeed true that this line is of weak intensity. The non-appearance of any $J=2$ lines for $Rb^{85}F$ can be made plausible with the aid of Fig. 3. It is pointed out in reference 3 that because of the nature of the refocusing process if the observability criterion holds a transition at low fields will be observed only if it is between two states which at high fields correspond to $m_J=0$ and $|m_J|=2$. Of the seven lines allowed by the selection rule $\Delta F_1=\pm 1, \pm 2$ four will not be observed because they do not satisfy the above condition. Three transitions—($F_1=1/2$, to $F_1=5/2$), ($F_1=9/2$ to $F_1=7/2$), and $F_1=9/2$ to $F_1=5/2$)—satisfy the observability requirement but they would be expected to be weak lines for the following reasons:

(a) Only a few of the component lines for these F_1 transitions fulfil the observability requirement.

(b) The fraction of the total number of $J=2$ state molecules issuing from the oven that appear in the refocused beam is smaller than the analogous fraction for $J=1$ state molecules, because a larger fraction of $J=2$ state molecules impinge on the wire stop. Hence it is expected that these lines would be weak, and they were not observed at all.

Fine Structure

For a study of the fine structure we consider the line group II_1 (Fig. 4) which was taken under "very weak" field conditions. Five lines were observed corresponding to vibrational states $v=0$ to $v=4$; each line is a doublet with a separation between the two components of about 20 kc/sec. The interaction of the electric field with the molecule is less than 1 kc/sec., and hence is negligible.[13]

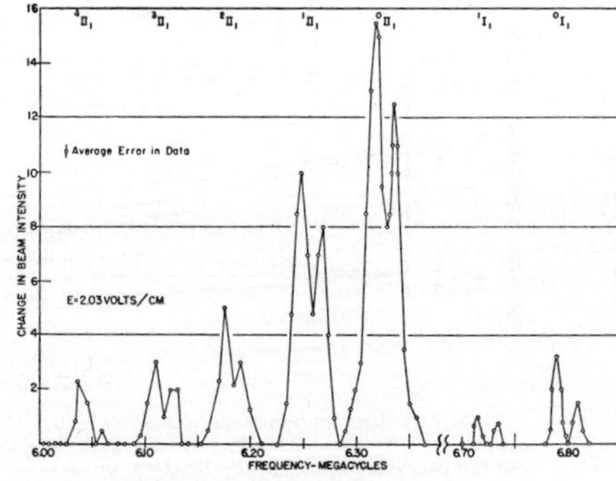

FIG. 4. Detailed spectrum of line groups II_1 and I_1 (see Table I). For line group I_1 radiofrequency field, $E_{rf}\sim 4$ volts/cm: for line group II_1, $E_{rf}\sim 8$ volts/cm. Change in beam intensity is expressed in cm of deflection on a galvanometer scale.

[11] The correspondence between states at low fields and at high fields which must be known in order to draw a figure such as figure 2 is established by the requirements that a state has the same m_{F_1} value at all fields and that states with the same m_{F_1} values cannot cross. See reference 10.

[12] See reference 10. The selection rule is $\Delta F_1=0, \pm 1, \pm 2$. The transitions with $\Delta F_1=0$ would involve frequencies of the order of the fine structure of our lines; such transitions were not studied. The observability criterion is that transitions will be observed only if they occur between states which can be adiabatically transformed into states at high fields with the proper $|m_J|$ values for the refocussing process. The existence of the refocused beam $(1, \pm 1)_A-(1, 0)_B$ indicates the occurrence of nonadiabatic transitions in the interfield, regions and hence it may be that the observability criterion is not strictly applicable. For further discussion see references 3 and 10.

[13] This interaction is calculated using a value for μ of 10.6 debye.

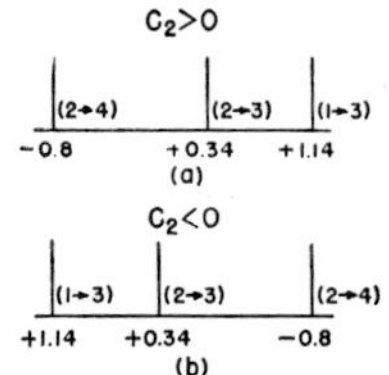

FIG. 5. Line splitting predicted from interaction $c_2(\mathbf{I}_2\cdot\mathbf{J})$ for transition $F_1=3/2$ to $F_1=7/2$ of $Rb^{85}F$ ($J=1$). The transition in F value is indicated in parentheses. Figure (a) applies when $c_2>0$ and figure (b) applies when $c_2<0$. No attempt is made to indicate relative line intensities. Zero value for the abscissa refers to the line position due to the quadrupole interaction alone.

TABLE II. Internal interactions of $Rb^{85}F$ and $Rb^{87}F$.

	Rotational state	Vibrational state	Quadrupole interaction $-eq_1Q_1/h$ in Mc./sec.	$\lvert c_2/h\rvert$ kc/sec.
	1	0	70.31±0.10	11±3
	1	1	69.54±0.10	13±3
$Rb^{85}F$	1	2	68.71±0.10	10±3
	1	3	67.99±0.10	10±3
	1	4	67.20±0.10	14±4
	1	0	34.00±0.06	14±4
	1	1	33.63±0.06	14±4
$Rb^{87}F$	1	2	33.20±0.06	—
	2	0	34.00±0.06	—
	2	1	33.62±0.06	—
	2	2	33.25±0.06	—

It is natural to assume that the doublet should arise from an interaction involving the fluorine nucleus. Two such interactions are given in the fourth and fifth terms of the Hamiltonian of Eq. (1). Using the known values of g_1 and g_2[14] and a calculated value of the internuclear distance,[15] it is found that the maximum splitting that the nuclear dipole-dipole interaction can produce is about 1 kc/sec. Hence the major portion of the splitting must be assigned to the spin-orbit interaction between the F nucleus and the molecular rotational angular momentum.

Since this line group was taken under "very weak" field conditions, an appropriate representation for the states involved is (I_1, J, F_1, I_2, F, M); the energy depends on M only because of the field interaction. The transition is $F_1=3/2$ to $F_1=7/2$ so the states involved are $(5/2, 1, 3/2, 1/2, 1, M)$, $(5/2, 1, 3/2, 1/2, 2, M)$ and $(5/2, 1, 7/2, 1/2, 3, M)$, $(5/2, 1, 7/2, 1/2, 4, M)$. The general selection rule allows $\Delta F=0, \pm1, \pm2$ but not[12] $\Delta F=\pm3$, so three components are allowed for the line of each vibrational state. Actually, the lines are split into two components with the larger intensity component appearing at the lower frequency.

The line splitting expected for the $c_2(\mathbf{I}_2\cdot\mathbf{J})$ interaction is shown in Fig. 5 for the cases of c_2 negative and positive. The positions of the three component lines were computed from the weak field formula given in reference 9. A half-width of about 10 kc/sec is predicted for our lines from the relationship $\Delta\nu\Delta t\sim1$ in which $\Delta\nu$ is the half width in c.p.s. and Δt is the time spent by the molecule in the C-field.[2] Thus in view of the magnitude of the observed line splitting (~20 kc/sec) we would not expect to resolve the two closer-lying component lines $(F=3\rightarrow F=1)$ and $(F=3\rightarrow F=2)$. One of the lines of the observed doublet must therefore be regarded as the superposition of the two line components $(F=3\rightarrow F=1)$ and $(F=3\rightarrow F=2)$.

To say which of the two observed lines corresponds to the transition $(F=4\rightarrow F=2)$ requires a theory of relative line intensities. It does not seem, however, that experimental conditions are sufficiently well known to allow such a theory. Thus non-adiabatic transitions are known to occur in the absence of an applied radiofrequency field as the molecule traverses the apparatus, and such transitions could affect relative line intensities. It is believed that these transitions occur in the interfield region between the A and C fields, but little is known of their exact character. Furthermore, the amplitude of the radiofrequency field at the position of the beam is adjusted so that the transition probability is believed to be rather high. But the r-f amplitude is not sufficiently well known so that relative transition probabilities can be calculated. Several simplifying assumptions have been tried, but none of them explained the relative intensity data for the $c_2(\mathbf{I}_2\cdot\mathbf{J})$ splitting in both the $Rb^{85}F$ and $Rb^{87}F$ spectra. A further discussion of relative line intensities is given in reference 10.

The inability to say which of the two observed lines corresponds to the transition ($F=2$ to $F=4$) means that only the absolute value of c_2/h can be determined. If we assume that either one of the observed lines appears midway between the predicted positions of the transitions $(F=3\rightarrow F=1)$ and $(F=3\rightarrow F=2)$—i.e. at $0.74\ c_2/h$—and that the other observed line appears at the position of the transition $(F=4\rightarrow F=2)$—i.e. at $-0.8c_2/h$—we easily compute that $|c_2/h|=(11\pm3)$ kc/sec. for vibrational state $v=0$. The results for the other vibrational states as well are given in Table II. The limiting error of ±3 kc/sec. is assigned because of

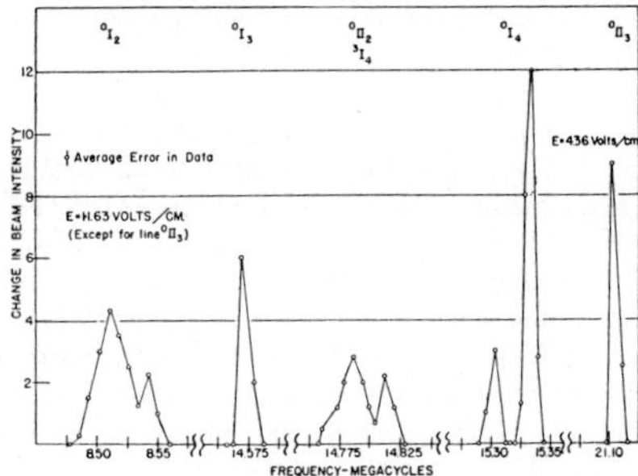

FIG. 6. Detailed spectrum of zeroth vibrational state lines for several line groups (see Table I). $E_{rf}\sim4$ volts/cm. Change in beam intensity is expressed in cm of deflection on a galvanometer scale.

[14] J. B. M. Kellogg and S. Millman, Rev. Mod. Phys. 18, 323 (1946).

[15] Schomaker and Stevenson, J. Am. Chem. Soc. 63, 37 (1941).

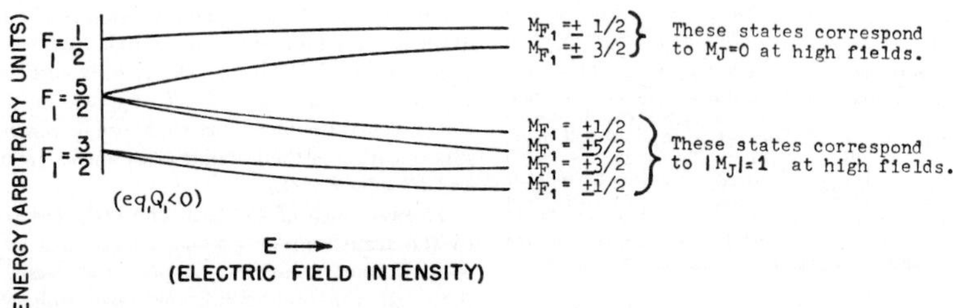

FIG. 7. A diagram indicating the correspondence between low and high field states for $Rb^{87}F$ ($I_1=3/2$, $J=1$). Other remarks apply as given in Fig. 2.

the uncertainty in the peak positions of the observed lines and because of the uncertainty as to whether the observed line which corresponds to the transitions ($F=3\rightarrow F=1$) and ($F=3\rightarrow F=2$) appears exactly at the midpoint between the predicted positions of these two transitions.

We can now discuss the accuracy with which the determination of the quadrupole interaction constant is made. Our understanding of the fine structure splitting of the $F_1=7/2$ to $F_1=3/2$ line allows us to specify the position at which this line would appear in the absence of the $c_2(\mathbf{I}_2\cdot\mathbf{J})$ interaction. These frequencies are given in Table I for the line group II_1 and are called the positions of the "centers" of the lines. Because of the experimental uncertainty in the determination of the peaks of the lines, these positions can only be given to ± 5 kc/sec. This implies that the quadrupole interaction constant can be given to about ± 0.1 percent. Another source of error arises from the theory used, which neglected off-diagonal matrix elements of the quadrupole operator connecting different J states.[16] The percentage error thus introduced is of the order of the quadrupole interaction energy divided by the energy difference between rotational states J and $J+2$ and for our case is about 0.1 percent. The correction of the theory has not been made because the error introduced is of the order of the error in the experimental data.

Little can be learned from the fine structure of the other $Rb^{85}F$ lines, because both lines were obtained under electric field conditions such that the $(\boldsymbol{\mu}\cdot\mathbf{E})$ term as well as the small internal interactions (terms (4) and (5) in Hamiltonian (1)) contribute significantly to the fine structure. The zeroth vibrational states of the two other line groups (II_2 and II_3) assigned to $Rb^{85}F$ are shown in Fig. 6. Line ${}^0II_3$ was obtained for a field condition intermediate between "very weak" and "weak." It is a superposition of some 10 unresolved components which are produced by a splitting due to the field interaction of 10 kc/sec. and by a splitting due to the $c_2(\mathbf{I}_2\cdot\mathbf{J})$ interaction of 8 kc/sec. Line ${}^0II_2$ was obtained under "weak" field conditions and is composed of 4 components with a total width of about 40 kc/sec. In the observed structure there also occurs the vibrational state $v=3$ for line group I_4 of $Rb^{87}F$. Line ${}^0II_2$ appears with low intensity and the higher vibrational states were not observed. The positions of the centers of the lines in line groups II_2 and II_3 (defined as the centers of the observed structures) provide confirmation for the values of the quadrupole interaction constants determined from line group II_1 (See Table I).

[16] See U. Fano, J. Research Nat. Bur. Stand. **40**, 215 (1948).

(B) Spectrum of $Rb^{87}F$

We consider those lines in the spectrum which are assigned to the less abundant (27 percent) molecular species $Rb^{87}F$ (see lines designated by I in Fig. 1). The nucleus Rb^{87} has a spin of 3/2; hence we expect the spectrum of $Rb^{87}F$ to be similar to that of $K^{39}F$ which is reported upon in detail in another paper.[3] Indeed, four line groups were observed under low electric field conditions as was also the case for $K^{39}F$. Three of these groups, designated I_1, I_2, and I_4, are assigned to rotational state $J=1$; and line group I_3 is assigned to rotational state $J=2$.

The transition in F_1 value to which each line corresponds is indicated in Table I. The quadrupole interaction constants are calculated by taking as the line positions the centers of the fine structure of the lines. The sign of the quadrupole interaction constant, eq_1Q_1, is determined to be negative by the existence of the line group $F_1=1/2$ to $F_1=5/2$, as explained in reference 3 (see Fig. 7). The argument is similar to that used earlier in this paper to prove that the quadrupole interaction constant for $Rb^{85}F$ is negative.

The number of line groups expected on the basis of

$C_2>0$

(1→3) (1→2) (0→2)

−0.67 +0.53 +1.2

(a)

$C_2<0$

(0→2) (1→2) (1→3)

+1.2 +0.53 −0.67

(b)

FREQUENCY (IN UNITS OF C_2/h)

FIG. 8. Line splitting predicted from interaction $c_2(\mathbf{I}_2\cdot\mathbf{J})$ for transition $F_1=1/2$ to $F_1=5/2$ of $Rb^{87}F$ ($J=1$). The transition in F value is indicated in parentheses. Figure (a) applies when $c_2>0$ and figure (b) applies when $c_2<0$. No attempt is made to indicate relative line intensities. Zero value for the abscissa refers to the line position due to the quadrupole interaction alone.

the observability criterion were observed. For the rotational state $J=1$ three transitions are allowed and all three were observed.[12] For the rotational state $J=2$ five transitions are allowed;[12] two of these are coincident with transitions for the $J=1$ state and a third is observed.[17] The other two allowed transitions for the $J=2$ state were not observed because they occur between states which experience the same deflection in the B field; i.e., have the same value of $|m_J|$ at high fields. This matter is discussed in detail in reference 3 for the exactly similar case of $K^{39}F$, and an analogous discussion has been given earlier in this paper for the isotopic species $Rb^{85}F$.

Fine Structure

The fine structure of $Rb^{87}F$ is similar to that of $Rb^{85}F$. Line group I_1, shown in Fig. 4, was taken under "very weak" field conditions for which the effect of the field term is negligible. Two lines were observed corresponding to vibrational states $v=0$ and $v=1$.[18] The doublet character of the lines is due primarily to the interaction $c_2(\mathbf{I}_2\cdot\mathbf{J})$. The contribution to the splitting caused by the nuclear spin-spin interaction is less than 1 kc/sec.

The line splitting expected for the $c_2(\mathbf{I}_2\cdot\mathbf{J})$ interaction is shown in Fig. 8 for the cases of c_2 negative and positive. The two component lines $(F=2\rightarrow F=0)$ and $(F=2\rightarrow F=1)$ of Fig. 8 will be unresolved, so the doublet which is observed is expected.

As in the analogous case for $Rb^{85}F$ it is not possible to say which of the two observed lines corresponds to the transition $(F=3\rightarrow F=1)$ because of the lack of understanding of line intensities. Thus only the absolute value and not the sign of c_2/h can be determined. By a comparison of the observed doublet with the line pattern of Fig. 8 it is found by reasoning similar to that used for $Rb^{85}F$ that $|c_2/h|=(14\pm4)$ kc/sec. for both the $v=0$ and $v=1$ vibrational states. This value is the same within experimental error as the value of $|c_2/h|$ obtained for $Rb^{85}F$. This agreement is expected since by its nature the $c_2(\mathbf{I}_2\cdot\mathbf{J})$ interaction depends only on the fluorine nucleus and the molecular electronic configuration, and so should be substantially the same for the two isotopic species.

As for the case of $Rb^{85}F$ we use this knowledge of c_2 to specify the positions at which the lines $F_1=5/2$ to $F_1=1/2$ would appear in the absence of the $c_2(\mathbf{I}_2\cdot\mathbf{J})$ interaction. These positions are given as the "centers" of the lines in line group I_1 in Table I, and from them are computed the quadrupole interaction constants given in Table II. The accuracy with which the quadrupole interaction constants can be stated is limited in exactly the same way as for $Rb^{85}F$.

The zeroth vibrational state lines of the other $Rb^{87}F$ line groups are shown in Fig. 6 and the transitions to which they correspond are indicated in Fig. 1. They were obtained under "weak" field conditions so that the $-(\boldsymbol{\mu}\cdot\mathbf{E})$ interaction contributes significantly to the fine structure. It is predicted that line ${}^0I_2$ will be split into five components by the field interaction and the $c_2(\mathbf{I}_2\cdot\mathbf{J})$ interaction, with a separation of about 60 kc/sec. between the outermost components. This is in agreement with the observed line width, but the resolution was not sufficient to distinguish all this structure. For line ${}^0I_4$ a splitting due to the field interaction into two components of about the observed separation is predicted. Each of these components will in principle be further split by the interactions with the F nucleus; the magnitude of this splitting is below the resolution of the experiment. Finally, the internal structure of line ${}^0I_3$, which is assigned to rotational state $J=2$, caused by the field interaction and by the interactions with the F nucleus is less than the resolution width.

The positions of the centers of lines I_2, I_3, and I_4 (defined as the centers of the observed structures) confirm the values of the quadrupole constants determined from line I_1 (see Table I).

V. THE EXTRA LINE

An extra line group shown in Fig. 9 was observed in the neighborhood of 3 Mc/sec. This line group appears in all detail at one half the frequency of line group II_1 which was shown in Fig. 4. This extra line group is not predicted from the Hamiltonian (1) for either molecular species for rotational states $J=0$ to $J=3$, and no higher rotational states are deflected sufficiently by the inhomogeneous electric fields to pass by the wire stop.

The possibility that the half frequency line could arise because of harmonics in the oscillator which provides the radiofrequency field is believed to be ruled out by the following studies. The intensity of the 6 Mc/sec. line was studied as a function of the r-f amplitude applied to the C field plates and it was found that the line was of zero intensity when the r-f amplitude was a factor of ten less than that used for the data of Fig. 4. Then a low pass filter, which attenuated the amplitude of signals near 6 Mc/sec. by a factor of about 30 relative to those near 3 Mc/sec., was inserted between the oscillator and the C field plates, but this had no effect on the intensity of the extra line group. Furthermore, the extra line was observed to be exactly the same when taken using two different oscillators—a GR signal generator 805C, and a home-made oscillator. A study of the harmonic output of these two oscillators was made; using a GR 724—A Precision Wavemeter it was found that the second harmonic output from the

---

[17] It is to be remarked that the reasons mentioned earlier in the analysis of the $Rb^{85}F$ spectrum for expecting the line intensities associated with $J=2$ state transitions to be weak are applicable to $Rb^{87}F$ as well. However, a $J=2$ state transition was observed for $Rb^{87}F$ whereas none was observed for $Rb^{85}F$. It is felt that the relative line intensity problem is not well enough understood so that the non-appearance of several presumably weak lines for the $J=2$ state of $Rb^{85}F$ should be regarded as surprising.

[18] The line for vibrational state $v=2$ (Fig. 1) is not shown in Fig. 4 because this very weak line was only observed when the radiofrequency field was larger than it was for the lines shown and no fine structure was resolved.

home-made oscillator was about 1 percent, and by a measurement technique involving a radio receiver it was found that the second harmonic output from the GR 805*C* oscillator was as high as 10 percent. Both the 6 Mc/sec. and the 3 Mc/sec. lines were taken with approximately the same amplitude of the radiofrequency field as is indicated in the captions to Figs. 4 and 9. Hence, the harmonic of the 3 Mc/sec. signal could not have caused the extra line, because the 6 Mc/sec. harmonic of the 3 Mc/sec. signal would have been too low to cause a 6 Mc/sec. line of appreciable intensity. Thus it seems clear that harmonics generated by the oscillator cannot account for the extra line. And there is no other place in our circuit at which harmonics of the required amplitude would be expected to be generated, since the circuit consists simply in the connection of the output from the oscillator to the two gold-plated plates of the *C*-field.

A chemical analysis[19] of the RbF was undertaken to determine whether there were appreciable amounts of other alkali halides in the sample used. By a spectrographic analysis it was found that no other alkali was present in a concentration of greater than 1 percent, and by a precipitation test the same conclusion was reached as to the presence of other halides. Such concentrations are too small to give rise to an observable line in our spectrum. Furthermore, of course, the refocussing procedure would in general eliminate the effects of any impurities.

It was suggested by Professor I. I. Rabi that this half-frequency line might arise as a double quantum transition in which two half frequency quanta supply the energy for the transition. Transitions occur normally only because there is a perturbation of the field free state and the matrix element for a transition is proportional to $EE_{rf}\sin\omega t$ in which E is the intensity of the static field and E_{rf} is the maximum intensity of the radiofrequency field. If we regard the radiofrequency field as causing the perturbation of the field-free state as well as the transition, then the matrix element will be proportional to $E_{rf}^2\sin^2\omega t$, and hence there will be a resonance at $\omega/2$ whereas in the normal case there was a resonance at ω. A more detailed theory of this process is found in reference 10. It is clear that this process will be important when E_{rf} and E are of comparable magnitude; this is indeed the case for our experiment in which computation indicated that both the static field and the radio-frequency field are of several volts/cm intensity.

The extra line was found to disappear at a somewhat higher value of E ($\sim$55 volts/cm which is still a weak field condition, since the field interaction amounts only to about 750 kc/sec.), whereas the upper frequency line is present and split up in the predicted manner at this same value of E. The theory of the double quantum

[19] We are indebted to Professor T. I. Taylor and Mr. Robert Anderson of the Columbia Chemistry Department for performing this analysis.

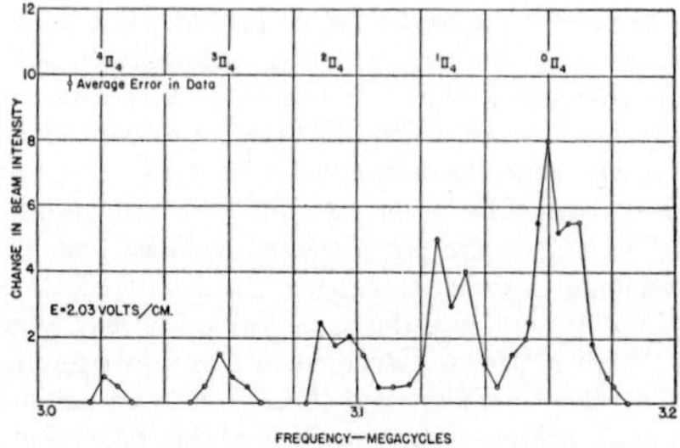

FIG. 9. Detailed spectrum of "extra" line group II_4. Radiofrequency field, $E_{rf}\sim 5$ volts/cm. Change in beam is expressed in cm of deflection on a galvanometer scale.

process does not predict the complete disappearance of the half-frequency line at such a low static field intensity. We were not able to determine whether the extra line was present at $E=0$ as would be expected from the theory of the double quantum transition. It should be remarked that it is often impossible to study resonance effects when the *C*-field voltage is near zero.[3]

An observation which is in apparent conflict with the double quantum theory of the line origin is that the intensities of both line groups were found to vary in about the same way with the amplitude of the radiofrequency field. This observation was made with amplitudes of the radiofrequency field from $\sim$1 to 8 volts/cm for which the corresponding line intensities (in arbitrary units) varied from $\frac{1}{2}$ (which is the minimum value observable) to 12. It would be expected that the lower frequency line, for which the transition matrix element is proportional to E_{rf}^2, would vary more rapidly in intensity with the amplitude of the radio-frequency field than would the upper frequency line, for which the transition matrix element is proportional to E_{rf}. Still it should be mentioned that the transition probability is high in these experiments, so that it is not true that the transition probability is simply proportional to the square of the connecting matrix element; indeed, a near unity transition probability may obtain for the higher amplitudes of radio-frequency fields used. A much more conclusive test should be made under conditions for which the transition probability is known to be very small, but such a test will necessitate a considerable improvement in sensitivity.

Nor is it clear from the theory of the double quantum transition why only one half-frequency line group is observed rather than one for each of the seven primary line groups.

VI. RESULTS

Table II shows the results for the internal interaction constants. The values of the quadrupole interaction constants and of $|c_2/h|$ for rotational state $J=1$ are taken from line groups II_1 and I_1 for which the field interaction is negligible. The quadrupole interaction constants for rotational state $J=2$ are obtained from line group I_3. The effective magnetic field produced at the F nucleus by the molecular rotation can be computed from the formula:

$$H = c_2/(\mu_N g_2)$$

in which H is the field per unit rotational quantum number and μ_N is 1 nuclear magneton. For $Rb^{87}F$, $H = 4.5 \pm 1.2$ gauss and for $Rb^{85}F$, $H = 3.8 \pm 0.9$ gauss. Within our experimental error there is no variation of H with vibrational state. The errors stated for the quadrupole interaction constant and $|c_2/h|$ arise primarily from the experimental inaccuracies in the determination of line positions.

The vibrational constant, ω_0, of the molecule can be determined from measured values of the relative intensities of the different vibrational lines of a line group.[2] The value obtained for $Rb^{85}F$ from a consideration of line groups II_1 and II_3 is:

$$\omega_0 = (340 \pm 68)\ \mathrm{cm}^{-1}.$$

The electric dipole moment, μ, of the molecule and the moment of inertia, A, can also be determined in principle by the electric resonance method, but the electrical quadrupole interactions for RbF are so large that the electric fields required for such a determination become very high. This results in a considerable decrease in line resolution, so that the accuracy of the determination would be quite poor.[3]

VII. DISCUSSION

A primary result of these studies is the determination of the electrical quadrupole interaction constant for Rb in RbF. This constant is determined for both molecular species for rotational state $J=1$ for several vibrational states, and for $Rb^{87}F$ also for rotational state $J=2$ for several vibrational states. These interaction constants are at least an order of magnitude larger than any other quadrupole interaction constant yet reported for an alkali halide molecule, and, incidentally, for this reason have not been determined by the magnetic resonance method. The absolute value of the quadrupole interaction constant is found to decrease by about 1.1 percent from one vibrational state to the next higher one. Within the experimental accuracy no difference was observed between the quadrupole interaction constants for rotational states $J=1$ and $J=2$ for $Rb^{87}F$.

It is expected that the ratio of the quadrupole interaction constants for the two molecular species $Rb^{85}F$ and $Rb^{87}F$ is largely independent of the molecular constants, q. Thus the perturbation treatment of the diatomic molecule yields for the molecular wave function $\psi = \psi_{\mathrm{electronic}}(x_i, R) \cdot \psi_{\mathrm{vibrational}}(R) \cdot \psi_{\mathrm{rotational}}(\theta, \phi)$ in which x_i are the electron coordinates, R is the internuclear distance, and θ, ϕ are the spherical coordinate angles specifying the direction of the internuclear axis with respect to a fixed coordinate system. $\psi_{\mathrm{electronic}}$ is independent of the reduced mass of the two nuclei. $\psi_{\mathrm{vibrational}}$ and $\psi_{\mathrm{rotational}}$ depend on the reduced mass of the system. The variation of q from one isotopic species to the other will be caused by the variation of $\psi_{\mathrm{vibrational}}$ and $\psi_{\mathrm{rotational}}$. The variation of q for a single molecular species has been found to be only about 1 percent from one vibrational state to another and no variation of q from rotational states $J=1$ to $J=2$ was found for $Rb^{87}F$. However, the variation of $\psi_{\mathrm{vibrational}}$ and $\psi_{\mathrm{rotational}}$ from one vibrational or rotational state to another is much larger than the variation of these functions from one molecular species to the other. Hence it seems to be a conservative estimate that q can vary by no more than 1 percent from one molecular species to the other. Thus we state $Q(Rb^{85})/Q(Rb^{87}) = 2.07 \pm 0.02$. In support of this argument there is evidence that the values of the field gradient, q, evaluated at the positions of the chlorine nuclei in the molecules $TlCl^{35}$ and $TlCl^{37}$ are the same.[20] For within the error of the experiment the ratio of the quadrupole interaction constants for chlorine in $TlCl^{35}$ and $TlCl^{37}$ is equal to the ratio of the electric quadrupole moments of the nuclei Cl^{35} and Cl^{37}, which have been measured by the method of atomic beams.[21] It is interesting to note that $\mu(Rb^{85})/\mu(Rb^{87}) = 0.492$ where μ is the nuclear magnetic dipole moment so that $(\mu Q)Rb^{85} \simeq (\mu Q)Rb^{87} \sim 1$. This is a further example of an empirical rule suggested by Gordy[22] and would indicate that the sign of the nuclear quadrupole moments of Rb is positive.

The constant $|c_2/h|$ characterizes the spin-orbit interaction between the spin of the fluorine nucleus and the rotational angular momentum of the molecule. Such an interaction in an alkali halide molecule was first reported by Nierenberg and Ramsey for LiF.[23] From their data we compute $|c_2/h| = 14$ kc/sec. Trischka found c_2/h to be $+16 \pm 2$ *kc*/sec for CsF. Our values for $|c_2/h|$ are 11 ± 3 kc/sec. for $Rb^{85}F$ and 14 ± 1 kc/sec for $Rb^{87}F$. The magnitudes of c_2/h for the RbF molecules are close to the values reported for LiF and CsF; unfortunately, the sign of c_2/h was not determined in our experiment. The theory of H. M. Foley[6] attributes this $c_2(\mathbf{I}_2 \cdot \mathbf{J})$ interaction to the effect of the bonding p electron of the halogen atom. This theory predicts an order of magnitude for the constants $|c_2/h|$ which is in agreement with the experimental results. The magnitudes of $|c_2/h|$ for the different alkali fluorides depend upon the moment of inertia of the molecules and upon the exact molecular wave functions, and have not been computed.

We are deeply indebted to our research director, Professor I. I. Rabi, for encouragement and for many enlightening discussions. We also wish to thank Professor H. M. Foley for several helpful discussions. One of the authors (VH) is indebted to the National Research Council for the grant of a predoctoral fellowship (1946–1949).

[20] H. Zeiger *et al.*, Bull. Am. Phys. Soc. **25**, No. 1, 35 (1950).
[21] L. Davis *et al.*, Phys. Rev. **76**, 1076 (1949).
[22] W. Gordy, Phys. Rev. **76**, 139 (1948).
[23] W. A. Nierenberg and N. F. Ramsey, Phys. Rev. **72**, 1075 (1947).

Reprinted from THE PHYSICAL REVIEW, Vol. 99, No. 6, 1842–1845, September 15, 1955
Printed in U. S. A.

Two-Quantum Transitions in the Microwave Zeeman Spectrum of Atomic Oxygen*

V. W. HUGHES AND J. S. GEIGER†‡
Yale University, New Haven, Connecticut

(Received May 20, 1955)

Two-quantum transitions have been observed in the microwave Zeeman spectrum of atomic oxygen in its ground $^3P_2$ state by the method of magnetic resonance absorption spectroscopy. The three lines observed originally by Rawson and Beringer are identified as arising from two-quantum transitions between Zeeman levels with the selection rule $\Delta M=\pm 2$. Each of the three lines is observed at the mean frequency of the two corresponding $\Delta M=\pm 1$ transitions to within the experimental accuracy of a few parts per million. The line width of the two-quantum transitions is approximately one-half that of the normal transitions, and the line intensity varies more rapidly with rf power, both in agreement with the theory. The principal features of these transitions are explained by second-order time-dependent perturbation theory. The two-quantum transitions reported here are essentially similar to the double-quantum transition reported by Hughes and Grabner in the electric quadrupole spectrum of $Rb^{85}F$ and to the multiple-quantum transitions seen by Kusch in the Zeeman spectra of K and O_2.

1. INTRODUCTION

A MULTIPLE quantum transition between two atomic or molecular energy levels is a transition in which the energy is supplied by two or more quanta. The transition is forbidden in first order perturbation theory and occurs through one or more intermediate states. In recent years, several examples of multiple-quantum transitions have been reported in atomic and molecular beam experiments. A double-quantum (or half-frequency) transition was first observed in the electric quadrupole spectrum of $Rb^{85}F$ by Hughes and Grabner using the molecular beam electric resonance method.[1] The transition was interpreted by second order perturbation theory, according to which two equienergetic quanta, each of one half the Bohr frequency for the transition, supply the energy for the transition.[2] Also it was pointed out that the line width for a double-quantum transition should be one-half that of a single-quantum transition in agreement with the experimental observation. Subsequently, it was shown experimentally and theoretically that the transition occurs also if two different frequencies are applied, provided that the sum of the two frequencies equals the Bohr frequency for the transition.[3] Recently Kusch observed double- and triple-quantum transitions in the Zeeman spectrum of K and O_2[4]; Braunstein and Trischka observed double-quantum transitions in the Stark spectrum of LiF molecules[5]; Hamilton *et al.* observed multiple-quantum transitions in the hyperfine spectrum of Au^{198}.[6] The theory of transitions involving the emission or absorption of two quanta (processes closely related to the Raman effect) was first discussed by M. Goeppert Mayer on the basis of the Dirac radiation theory.[7] Recently Salwen has discussed the line shapes of multiple-quantum transitions, including the effect of the velocity distribution in an atomic beam experiment.[8]

It is the purpose of this paper to point out an example of a double-quantum transition in the Zeeman spectrum of atomic oxygen observed in a microwave magnetic resonance absorption spectroscopy experiment. The double-quantum transitions being reported here were observed originally by Rawson and Beringer but remained unidentified.[9] Further, a slight variation of the second order perturbation theory originally proposed to explain double-quantum transitions in electric quadrupole spectra[2] is applied to double-quantum transitions in Zeeman spectra of the type observed by Kusch and reported here.

A preliminary report of this work has already appeared.[10]

2. EXPERIMENTAL DATA

The ground state of atomic oxygen is a $(2p)^4$ configuration with three fine-structure levels as shown in Fig. 1. In the presence of a magnetic field the $^3P_2$ level is split into five magnetic sublevels designated by the magnetic quantum number M with values from -2 to $+2$. Because of incipient Paschen-Back effect, i.e., the mixing in of $^3P_1$ state by the magnetic field, adjacent magnetic sublevels of the $^3P_2$ state are not equally spaced, and hence the magnetic resonance spectrum consists of four lines.

The method of observation was the microwave magnetic resonance absorption method developed by Beringer and his students,[11] and the apparatus was that

* This research has been supported in part by the Office of Naval Research.

† To be submitted by J. S. Geiger in partial fulfillment of the Ph.D. thesis requirement at Yale University.

‡ Loomis-Sheffield Fellow, 1954–55.

[1] V. W. Hughes and L. Grabner, Phys. Rev. 79, 314 (1950).

[2] V. W. Hughes and L. Grabner, Phys. Rev. 79, 828 (1950).

[3] L. Grabner and V. W. Hughes, Phys. Rev. 82, 561 (1951).

[4] P. Kusch, Phys. Rev. 93, 1022 (1954).

[5] R. Braunstein and J. W. Trischka, Phys. Rev. 98, 1092 (1955); private communication from J. W. Trischka on Li^6F^{19}

[6] Private communication from D. R. Hamilton.

[7] M. Goeppert Mayer, Ann. Physik 9, 273 (1931). See also G. Breit, Revs. Modern Phys. 4, 504 (1932).

[8] H. Salwen, Phys. Rev. 99, 1274 (1955).

[9] E. B. Rawson and R. Beringer, Phys. Rev. 88, 677 (1952).

[10] V. W. Hughes and J. S. Geiger, Bull. Am. Phys. Soc. 30, No. 3, 66 (1955).

[11] R. Beringer and J. G. Castle, Phys. Rev. 78, 581 (1950).

used by Beringer and Heald.[12] A spectrum of the type originally studied by Rawson and Beringer[9] is shown in Fig. 2. Essentially the derivative of the absorption is plotted as a function of the magnetic field for a fixed microwave frequency. The magnetic field was measured with the so-called regulator proton resonance probe,[12] which was placed adjacent to the microwave cavity. The inclined line indicates the zero drift of the galvanometer output and line centers are at the cross-over points on this line. The four spectral lines a, b, c, and d are the expected Zeeman transitions corresponding to the transitions indicated in Fig. 1. These lines have been shown to occur to within the experimental accuracy[9] of a few ppm (parts per million) at the fields predicted by elaborate Zeeman calculations of Abragam, Van Vleck, and Kambe[13] which take into account relativistic and quantum electrodynamic effects. The extra lines e, f, and g remained unidentified by Rawson and Beringer. The suggestion was made in their work that the extra lines might be due to argon impurity in the tank oxygen, since the metastable $^3P_2$ state of argon also has a g-value of about 1.5. However, lines of the type shown in Fig. 2 have been taken with oxygen which was 99.6% pure (Matheson Company extra dry oxygen), and any appreciable line intensity from an impurity could not be expected.

Inspection indicates first that to within the experimental accuracy of a few ppm the magnetic field at which line e occurs is $\frac{1}{2}$ of the sum of the fields at which lines a and b occur. Similarly, line f occurs at the mean field of lines b and c, and line g occurs at the mean field of lines c and d. Since the principal part of the Zeeman energy depends linearly on the magnetic field H, it can be concluded to sufficient accuracy that for a fixed magnetic field the frequency of line e would be $\frac{1}{2}$ of the sum of the frequencies of lines a and b, and similar statements apply for lines f and g. The second observation is that the widths of the unidentified lines e, f, and g are considerably less than—indeed approximately $\frac{1}{2}$ that of—the normal lines a, b, c, and d.

Figure 3 presents information on the relative line intensities of the extra and normal lines as a function of microwave power measured at the bolometer. It is noticed that the intensity of the extra lines decreases considerably more rapidly with decrease of microwave power than does the intensity of the normal lines for the range of powers used.

Finally, a search indicated no additional unexpected lines in the oxygen spectrum. The two lines associated with the $^3P_1$ Zeeman splittings were seen, of course, as has been reported in the work of Rawson and Beringer.[9]

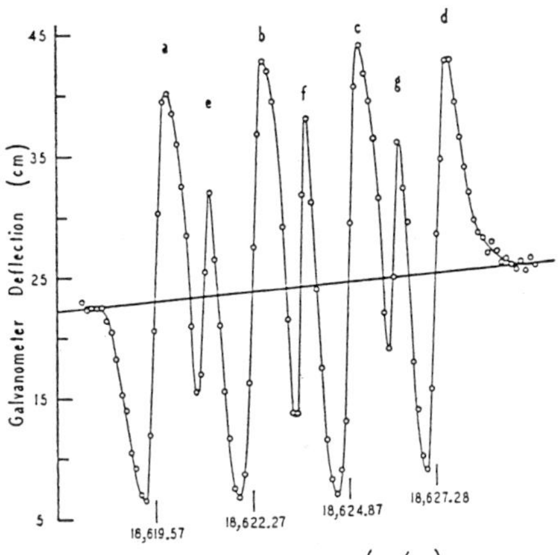

FIG. 2. Paramagnetic resonance absorption spectrum of ground state $^3P_2$ oxygen. Essentially the derivative of the absorption is plotted *vs* the magnetic field for a fixed microwave frequency of 9188.528 Mc/sec. The mean fields of single-quantum transitions and the fields of the extra lines are as follows:

$\frac{1}{2}(a+b)=18\,620.92$ kc/sec; $e=18\,620.93$ kc/sec,
$\frac{1}{2}(b+c)=18\,623.57$ kc/sec; $f=18\,623.58$ kc/sec,
$\frac{1}{2}(c+d)=18\,626.08$ kc/sec; $g=18\,626.08$ kc/sec.

The magnetic field was measured with the regulator proton resonance probe, which is placed adjacent to the microwave cavity. An additive correction would have to be applied to obtain the magnetic field in the cavity. This correction would not affect significantly the foregoing result that the extra lines occur at the mean field of corresponding single-quantum transitions. This curve was taken with ordinary tank oxygen.

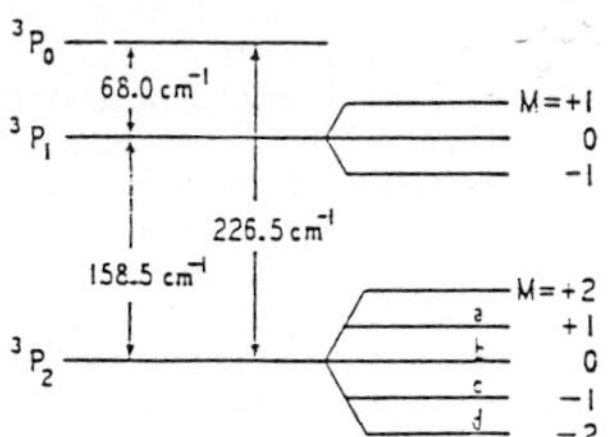

FIG. 1. Energy levels of atomic oxygen in a magnetic field.

3. THEORY OF TWO-QUANTUM TRANSITIONS

Consider an atomic state with three equally spaced Zeeman energy levels as, for example, the $^3S_1$ state of helium. The Zeeman energies are given by $W=\mu_0 g_J HM$, in which the symbols have their usual significance and $M=+1$, 0, or -1. It should be emphasized that the energy spacings between adjacent Zeeman energy levels are equal. It is well known that if such an atom is initially in the state $M=-1$ and a radio-frequency magnetic field with a component perpendicular to the static field and a frequency equal to $\mu_0 g_J H/h$ is applied, then after a time t the atom will have a certain probability of being in the states $M=0$ or $M=+1$.[14] The

[12] R. Beringer and M. A. Heald, Phys. Rev. 95, 1474 (1954).
[13] A. Abragam and J. H. Van Vleck, Phys. Rev. 92, 1448 (1953); K. Kambe and J. H. Van Vleck, Phys. Rev. 96, 66 (1954).
[14] V. W. Hughes *et al.*, Phys. Rev. 91, 828 (1953).

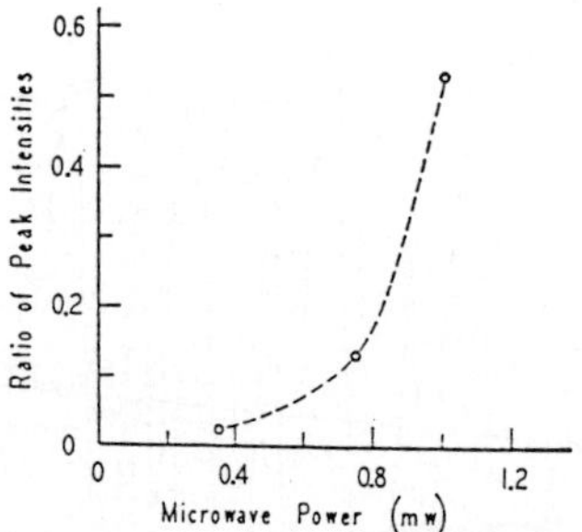

FIG. 3. Ratio of line intensities of extra to normal lines as a function of microwave power. The power is measured at the bolometer, and these data were taken with the Matheson Company extra dry oxygen.

transition to the $M=0$ state is the usual one predicted by first order time-dependent perturbation theory because the selection rule allows $\Delta M=\pm 1$. The transition to the $M=+1$ state occurs by virtue of the intermediate state $M=0$, and, since the energy levels are equally space, the usual interpretation would be that a resonance transition occurs from the $M=-1$ to $M=0$ state with the absorption of one quantum, and then a second resonance transition occurs from the $M=0$ to the $M=+1$ state with the absorption of a second quantum. The transition probability is given by the well known Majorana formula.[15]

The Zeeman spectrum of atomic oxygen presents a different situation because adjacent energy levels are not equally spaced, as was pointed out in the previous section. Thus, for example, if we consider an atom in the $M=-2$ magnetic substate of the $^3P_2$ level and apply a radio-frequency magnetic field whose frequency is the Bohr frequency for the transition from $M=-2$ to $M=-1$ the atom will, of course, have a substantial probability of making the transition to the $M=-1$ state, but the frequency will be far off resonance for a transition from the $M=-1$ to the $M=0$ state, so there will be negligible probability for the atom to reach the $M=0$ state.

It will now be pointed out that by a consideration of second-order time-dependent perturbation theory a substantial probability for the transition from the $M=-2$ to the $M=0$ state of $^3P_2$ oxygen is expected at a frequency one half the Bohr frequency for the $M=-2$ to $M=0$ transition. The Hamiltonian is:

$$\mathcal{H}=\mathcal{H}_0+\mathcal{H}',$$

$$\mathcal{H}'=\mu_0 g_J \mathbf{J}\cdot\mathbf{H}_{rf}=\mathcal{H}'(0)e^{-i\omega t};\quad \mathbf{H}_{rf}=\mathbf{H}_1 e^{-i\omega t},$$

in which $\mathcal{H}_0$ is the time-independent part of the Hamiltonian which leads to Zeeman levels as shown in Fig. 1. $\mathcal{H}'$ is the interaction with the rf magnetic field, and in this expression μ_0 is the Bohr magneton, g_J is the atomic g-value, $\mathbf{J}$ is the total atomic angular momentum operator, and $\mathbf{H}_{rf}$ is the applied radio-frequency magnetic field.[16]

First-order time-dependent perturbation theory gives the probability amplitude that an atom initially in the state n shall be in the state m after it has been subjected to the radio-frequency field for a time t [see Eq. (1)].

$$a_m^{(1)}(t)=(-i/\hbar)(m|\mathcal{H}'(0)|n)\times\frac{\{\exp[-i\omega t+i(E_m-E_n)t/\hbar]-1\}}{[-i\omega+i(E_m-E_n)/\hbar]}\simeq-(it/\hbar)(m|\mathcal{H}'(0)|n), \quad (1)$$

The usual resonance denominator appears for ω near the Bohr frequency. Provided t is sufficiently small so that the initial state amplitude can still be considered approximately 1, the expression is approximately that given on the right. Similarly, second-order time-dependent perturbation theory gives the probability amplitude:

$$a_m^{(2)}(t)=-\frac{1}{\hbar^2}\frac{(m|\mathcal{H}'(0)|l)(l|\mathcal{H}'(0)|n)}{[-i\omega+i(E_l-E_n)/\hbar]}\times\left\{\frac{\exp[-i2\omega t+i(E_m-E_n)t/\hbar]-1}{[-i2\omega+i(E_m-E_n)/\hbar]}-\frac{\exp[-i\omega t+i(E_m-E_l)t/\hbar]-1}{[-i\omega+i(E_m-E_l)/\hbar]}\right\}\simeq\frac{it}{\hbar^2}\frac{(m|\mathcal{H}'(0)|l)(l|\mathcal{H}'(0)|n)}{[-\omega+(E_l-E_n)/\hbar]}, \quad (2)$$

where l designates an intermediate state.

Consider the case realized for the $^3P_2$ magnetic substates of oxygen because of the unequal spacing of adjacent Zeeman levels, in which the applied frequency ω can be far off resonance from the Bohr frequencies for either the $M=-2$ to $M=-1$ or the $M=-1$ to $M=0$ transitions but approximately equal to one-half the Bohr frequency for the transition $M=-2$ to $M=0$. If the atom is initially in the $M=-2$ state, there is negligible transition probability to the $M=-1$ state which is the only allowed transition in first-order perturbation theory. In second-order perturbation theory the transition from the $M=-2$ to $M=0$ state is allowed, because of the occurrence of the intermediate state l corresponding to $M=-1$. Furthermore, it will be observed in the expression for the probability amplitude in second-order perturbation theory that one of the terms in the bracket involves a resonance denominator for $2\omega=(E_m-E_n)/\hbar$. This term will be dominant, and provided t is sufficiently small so that the amplitude of the initial state n can still be regarded as approximately 1, the expression is that given on the second line of Eq. (2). The occurrence of 2ω instead of ω in the

[15] E. Majorana, Nuovo cimento 9, 43 (1932); F. Bloch and I. I. Rabi, Revs. Modern Phys. 17, 237 (1945).

[16] As usual it is only necessary to consider a single rotating component of the radiofrequency field. F. Bloch and A. Siegert, Phys. Rev. 57, 522 (1940).

resonance denominator implies that two equienergetic quanta are involved in the transition. It also implies that the line width for these two-quantum transitions will be $\frac{1}{2}$ that of the normal transitions. It was pointed out above that this was one of the characteristics of the unidentified lines.

The principal surprise about these two-quantum transitions is their substantial intensity relative to the normal transitions. This fact can be understood by reference to Eqs. (1) and (2). It will be noticed that the probability amplitude for a two quantum transition at its resonance frequency differs from the probability amplitude for a single-quantum transition at its different resonance frequency by a factor which is the ratio of a matrix element of the interaction with the rf field to the energy difference between the half-frequency for the transition from $M=-2$ to $M=0$ and the Bohr frequency for the transition from $M=-1$ to $M=0$. In the experimental situation the rf magnetic field was of the order of 0.1 gauss and the energy difference between the half-frequency and the $M=-1$ to $M=0$ frequency corresponds to about 0.28 gauss. Thus the ratio is of the order of 1 and hence the intensity of a two-quantum transition is comparable to that of a single-quantum transition. It will be noticed, however, that the intensity of a two-quantum transition should decrease more rapidly with decrease of rf field than the intensity of a single-quantum transition. This prediction is in agreement with the experimental results shown in Fig. 3. It should be emphasized that these remarks on line intensities are only intended in a qualitative sense.[17] The actual experimental conditions are such that the transition probabilities are high so that perturbation theory is not a very good approximation. A better knowledge of the experimental factors influencing line intensity such as relaxation phenomena and saturation effects, and a careful comparison with Salwen's theory[8] would be required for a more quantitative understanding of the line intensities.

Lines e, f, and g are, of course, being identified with the double-quantum transitions from $M=0\to M=+2$, $M=-1\to M=+1$, and $M=-2\to M=0$, respectively. The question arises about the double-quantum transition predicted in connection with the $^3P_1$ Zeeman spectrum. This line would be expected to be much weaker than the double-quantum transitions associated with the $^3P_2$ levels, because of the greater difference between the half-frequency for the transition $M=-1$ to $M+1$ and the Bohr frequency from $M=0$ to $M=+1$. Furthermore, this half-frequency line is predicted to occur at the field value for which a normal line in the $^3P_2$ spectrum occurs. Hence it is understandable that this double-quantum line was not observed.

[17] E. B. Rawson reports in his thesis [Yale, 1952 (unpublished)] that the relative intensity of the unidentified lines to the normal lines depends on pressure. Indeed a decrease in pressure caused this relative intensity to change from a value of about $\frac{1}{4}$ to a value of about 3. Possibly relaxation phenomena affect the double-quantum lines and the single-quantum lines differently, but no careful study has been made of this question.

The extension of the perturbation theory given for double-quantum transitions to the case of triple-quantum transitions is apparent.[18] Since the separations between adjacent lines a, b, c, and d are equal, it will be appreciated that triple-quantum transitions will coincide in field with single-quantum or double-quantum transitions [e.g., $(a+b+c)/3=b$, etc.]. This situation differs from Kusch's case in which adjacent single-quantum transitions are not equally spaced.[4]

The essential similarity of the double-quantum transitions discussed in this paper and of the double-quantum transition in the electrical quadrupole spectrum of RbF is clear. In the RbF case also second order time-dependent perturbation theory explains the basic features of the transition—the necessity of an intermediate state, resonance at one-half the Bohr frequency, and the line width one-half that of a normal transition. In the RbF example, the normal electric quadrupole transition can occur only when a static electric field is present, and the two-quantum transition can occur in the absence of the static electric field. The factor by which the probability amplitude for a two-quantum transition differs from that of a single-quantum transition is the ratio of the matrix element of the interaction with the radio-frequency electric field to the matrix element of the interaction with the static electric field. The ratio is of the order of 1 when the radio-frequency electric field is nearly equal to the static electric field.

The microwave magnetic resonance absorption method has certain advantages compared with conventional atomic beam experiments for a further study of the line intensities of multiple-quantum transitions as given in Salwen's theory: first, a more definite knowledge of the microwave field in a cavity than in an "rf-hairpin" of the type used in atomic beams, and, secondly, perhaps a better knowledge of the velocity distribution of the atoms, since no loss of low velocity atoms is involved as in the atomic beam case.

Finally, it might be pointed out that in conventional microwave spectroscopy involving transitions between different rotational states of a molecule, the conditions for the occurrence of double-quantum lines are not met because the separations of adjacent rotational levels differ by large factors. It seems possible that in the field of nuclear paramagnetic resonance where the level spacings are equal, one might find appropriate perturbations in some cases so that the conditions for a double- or multiple-quantum transition of the type discussed in this paper might be realized.

It is a pleasure to thank Professor R. Beringer for several helpful discussions about the apparatus.

[18] The principal term characteristic of a triple-quantum transition is found by third-order time-dependent perturbation theory to be:

$$a_m^{(3)}(t)=+\left(\frac{i}{\hbar^3}\right)\frac{(m|\mathcal{H}'(0)|k)(k|\mathcal{H}'(0)|l)(l|\mathcal{H}'(0)|n)}{[-i\omega+i(E_l-E_n)/\hbar][-i2\omega+i(E_k-E_n)/\hbar]}\times\frac{\exp[-i3\omega t+i(E_m-E_n)t/\hbar]-1}{[-i3\omega+i(E_m-E_n)/\hbar]},$$

where l and k designate intermediate states.

Reprinted from THE PHYSICAL REVIEW, Vol. 105, No. 1, 170–172, January 1, 1957
Printed in U. S. A.

Experimental Limit for the Electron-Proton Charge Difference*

VERNON W. HUGHES
Yale University, New Haven, Connecticut
(Received August 22, 1956)

By a study of the deflection of a beam of CsI molecules in a homogeneous electric field an upper limit is established for the charge on a CsI molecule. This upper limit is 1.7×10^{-22} esu or 4×10^{-13} of the electronic charge, and implies that the magnitude of the electron-proton charge difference is less than 3×10^{-15} of the electronic charge or that the charge on a neutron is less than 2×10^{-15} of the electronic charge.

I. INTRODUCTION

THE equality of the magnitude of the charge on all "elementary" particles is a basic principle in physics. Yet there seems to be very little direct experimental evidence on a microscopic level to test this principle. In the oil drop experiment[1] Millikan verified that the unit positive ionic charge is equal to the unit negative electron charge with an accuracy of about 1 part in 1500. This result can be interpreted further from a macroscopic point of view to imply that there can be no difference in the magnitude of the electron and proton charges sufficient to yield a net charge of 1/1500 of a unit charge from all the electrons and protons comprising an oil drop. For a typical oil drop with a radius of 10^{-4} cm the charge difference can be no greater than about 3 parts in 10^{16}. Still higher accuracy is achieved in a macroscopic experiment done by Piccard and Kessler.[2] They established an upper limit to the charge on a mass of de-ionized carbon dioxide gas by observing with an electrometer that the potential of an iron sphere in which the gas is contained under about 8 atmospheres pressure does not change within the limit of experimental error when the gas is allowed to escape from the sphere. The charge per kilogram of gas is thus found to be less than 1.4×10^{-3} esu, which implies that the magnitude of the electron charge is equal to the magnitude of the proton charge to within 5 parts in 10^{21}.

In this paper a microscopic experiment will be described which tests the electrical neutrality of single molecules by a study of the deflection of a molecular beam in a homogeneous electric field. The upper limit of the charge on the molecule can be interpreted to yield an upper limit to the charge difference between an electron and a proton. A brief report on this experiment has appeared.[3]

II. DESCRIPTION OF EXPERIMENT

The apparatus was built by the author to do molecular beam electric resonance spectroscopy and has been described by Lee *et al.*[4] Since the apparatus was designed for an experiment quite different from the experiment being described here, many features of the apparatus are not optimum for the present application. A schematic diagram giving the features of the apparatus pertinent to the present experiment is shown in Fig. 1. The region l_2 of homogeneous electric field was formed by two parallel plates separated by about 0.65 cm. Because of its spectroscopic design history the spacing between the plates was uniform to 0.001 cm and the plates were gold-plated to minimize contact potentials, but no provision was present for reducing end effect field inhomogeneities. Standard techniques were employed for the production, collimation, and detection of a beam of alkali halide molecules.[5] In particular, the detector was a hot-wire, surface-ionization detector together with an FP54 tube electrometer circuit. The molecule used was CsI.

The beam intensity in the plane of the detector is observed with and without an electric field present in region l_2. In the absence of an electric field the theoretical beam shape in the plane of the detector is trapezoidal in shape[6] (Fig. 2). The integrated intensity on a detector of finite width w_d is easily computed as an integral over the detector width. A comparison of the

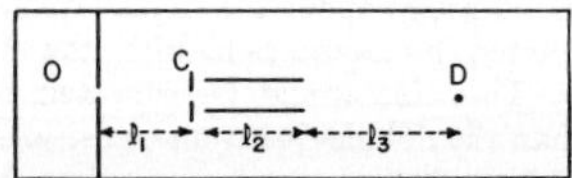

FIG. 1. Schematic diagram of apparatus, showing horizontal cross section through center of apparatus. Region O is the source chamber, which includes the oven with a source opening 1.1 cm high and 0.0025 cm wide. C represents the collimator slit which is 1.1 cm high and 0.005 cm wide. The region of length l_2 contains the homogeneous electric field which is produced by applying a voltage across two parallel plates. D is the hot-wire surface-ionization detector of width 0.0076 cm. The regions of length l_1 and l_3 are field free regions. $l_1=13$ cm, $l_2=12$ cm, and $l_3=18$ cm. The total distance from oven slit to detector wire is 44 cm.

* This experiment was performed in the molecular beams laboratory at Columbia University in 1948. This report, which has been delayed for the usual reasons, and its publication have been assisted by recent support from the Air Force Office of Scientific Research.

[1] R. A. Millikan, *Electrons (+ and −), Protons, Photons, Neutrons, and Cosmic Rays* (University of Chicago Press, Chicago, 1941), p. 85.

[2] A. Piccard and E. Kessler, Arch. sci. phys. et nat. **7**, 340 (1925).

[3] V. W. Hughes, Phys. Rev. **76**, 474 (A) (1949).

[4] Lee, Fabricand, Carlson, and Rabi, Phys. Rev. **91**, 1395 (1953).

[5] N. F. Ramsey, *Molecular Beams* (Clarendon Press, Oxford, 1956), Chap. XIV.

[6] See reference 5, Chap. II.

calculated and observed beam intensities is shown in Fig. 3; the agreement of about 10% in relative intensities is well within customary experience and is within the accuracy allowed by our knowledge of the geometry of the apparatus.

If a uniform electric field is applied in region l_2, and if the molecules are uncharged, the beam intensity at the detector should be unchanged from that of Fig. 3. On the other hand, if the molecules have a small charge q they experience a force **F** in the homogeneous electric field **E** given by

$$\mathbf{F}=q\mathbf{E},$$

and the integrated beam intensity on a detector of finite width is now different. The detector was placed on one side of the beam with its center at the half-intensity point of the trapezoidal pattern (Fig. 2), because at this position the signal is most sensitive to small deflections of the molecules. Furthermore, this unsymmetrical placement of the detector makes possible the distinction between an effect due to a net charge on the molecule and one due to a polarization force arising from the inhomogeneity of the applied electric field.

The beam intensity I at the detector in the absence of the field is simply

$$I=I_{00}(d-p)/2; \qquad (1)$$

the symbols are defined by Fig. 2, and the detector width w_d is taken equal to $d-p$. With the field present and a charge q on each molecule the deflection of a molecule with mass m and with velocity α, the most probable velocity for molecules in the source, is given by[7]

$$s_\alpha=\frac{qE}{2mv^2}(l_2^2+2l_2l_3), \qquad (2)$$

in which the lengths l_2 and l_3 are defined in Fig. 1. The deflected beam intensity pattern in the detector plane, $I(s)$, including the average over the molecular velocity

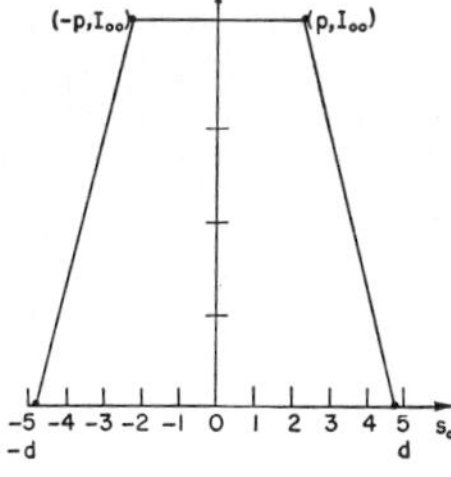

FIG. 2. Beam shape before deflection in plane of detector. Calculated curve. The maximum intensity is I_{00}. Each unit of s_0 is 0.0025 cm. The parameters p and d have the values: $p=0.0056$ cm, $d=0.012$ cm.

[7] See reference 5, Chap. IV for a discussion of formulas (2) and (3). Note that an error in sign is present in this reference. A plus sign appears before s_α in Eq. (IV.34), whereas a minus sign should appear as in the integrand of our Eq. (3).

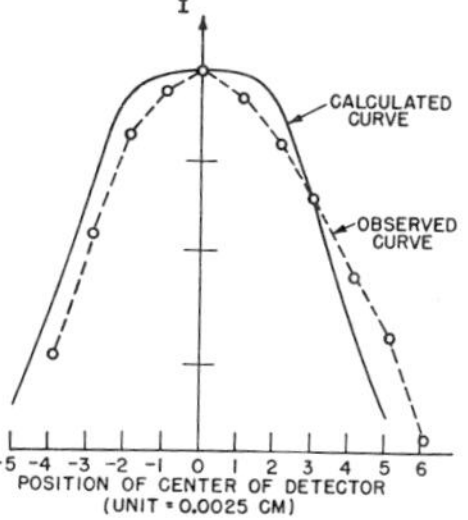

FIG. 3. Relative beam intensity I in arbitrary units seen by a detector of width $w_d=0.0076$ cm *vs* position of center of detector. Solid curve is calculated curve and dashed curve is curve through observed points.

distribution, is given by the formula

$$I(s)=I_{00}\int_{s_\alpha/(s+d)}^{\infty}\frac{(d+s-s_\alpha/x)xe^{-x}dx}{d-p}, \qquad (3)$$

in which $x=s_\alpha/(s-s_0)$, where s_0 is the position coordinate in the plane of the undeflected beam and s is the position coordinate in the plane of the deflected beam. Both s_0 and s are measured with respect to the zero position shown in Fig. 2. Equation (3) applies when s_α is positive and when $-d<s<-p$.

Hence, by integration,

$$I(s)=I_{00}\left(\frac{d+s}{d-p}\right)\left(\frac{s_\alpha}{s+d}+1\right)\exp[-s_\alpha/(s+d)]$$
$$-I_{00}\frac{s_\alpha}{d-p}\exp[-s_\alpha/(s+d)]. \qquad (3')$$

Integration of this expression over the region of the detector from $s=-d$ to $s=-p$ and utilization of the condition that s_α is small compared to the beam and detector dimensions yields the formula for the relative change in beam intensity due to the field

$$\Delta I/I=-2s_\alpha/(d-p). \qquad (4)$$

If s_α is negative, then, with the same assumptions used in deriving Eq. (4), the relative change in beam intensity will have the same magnitude but it will be an increase in intensity for this case.

The applied electric field of the parallel plates is, of course, not perfectly homogeneous principally because of end effects and nonparallelism. Hence a polarization force, $\mathbf{F}_p$, will act on the molecules.

$$\mathbf{F}_p=-(\partial W/\partial E)\nabla E, \qquad (5)$$

in which W is the energy of the molecule in the field of magnitude E. With the geometry fixed, this force will remain unchanged both in magnitude and in direction when the polarity of the potential applied to the plates is changed. Thus the effect of a polarization force can be distinguished from the effect of a force on a net charge q, because the change in beam intensity at the detector does not depend on polarity in the first case

TABLE I. Observed change in beam intensity when an electric field is applied. The unit of beam intensity is 1 cm of deflection on a galvanometer scale ($\sim 10^{-15}$ amp of beam current). With the detector centered to give maximum beam signal, the beam intensity was 15 000. For the data given in the table, the detector was moved about 7×10^{-3} cm off the centered position.

Beam intensity	Change in beam intensity Field on, negative direction	Field on, positive direction	Electric field intensity (volts/cm)
11 000	−40	−40	3100
11 000	−100	−100	3500

but does depend on polarity in the second case, provided the detector is unsymmetrically placed in the beam.

The experimental data are given in Table I. The over-all stability of the apparatus allowed the measurement of changes of about 0.1% in beam intensity. A small change in beam intensity is observed when the field is applied, but this change is independent of the direction of the field to within the experimental accuracy, and hence cannot be due to a force associated with a net molecular charge. This change in beam intensity agrees in order of magnitude with the estimated effect due to field inhomogeneity of the end regions of the parallel plates. It may be remarked that the ion content in the beam should be quite negligible. Thus it can be concluded that the relative change in beam intensity $\Delta I/I$ due to a force on charged molecules is less than 0.1% of the total beam intensity.

The upper limit of the allowed charge q on a molecule can be computed from Eqs. (4) and (2). The values of the experimental parameters were $p=5.6\times10^{-3}$ cm, $d=12\times10^{-3}$ cm, $w_d=7.6\times10^{-3}$ cm, T (oven temperature)$=650°$K. Hence by Eq. (4), $s_\alpha=3.2\times10^{-6}$ cm. The inaccuracy of Eq. (4) due to the fact that w_d ($=7.6\times10^{-3}$ cm) is not exactly equal to $d-p$ ($=6.4\times10^{-3}$ cm), as was assumed in its derivation, is negligible. Use of Eq. (2) yields the result that q must be less than 1.7×10^{-22} esu or 4×10^{-13} of the electronic charge.

III. DISCUSSION

Since the molecule CsI has an atomic number of 108, the above result for q implies that the difference between the magnitudes of the electron and proton charges is less than 3×10^{-15} of the electronic charge. The experimental result can also be interpreted as implying an upper limit to the charge of the neutron.[8] Thus if the electron and proton are assumed to have charges of equal magnitude, then the 152 neutrons in CsI must have a net charge of less than 1.7×10^{-22} esu or the charge of a single neutron is less than 2.0×10^{-15} electronic charge.[9]

The question of the equality in magnitude of the electron and proton charges is not only of importance to fundamental particle physics but also to the study of large scale matter. Indeed the experiment of Piccard and Kessler was performed to test Einstein's suggestion[2] that the magnetic field of the earth might be explained if the earth had a net charge due to an electron-proton charge difference of only 3 parts in 10^{19}. Apart from any other objections to this suggestion, the Piccard-Kessler experiment rules out this possibility. A further general remark can be made. It is well known that gravitational forces are much smaller than electrostatic forces, e.g., by a factor of about 10^{39} for the electron-proton system, and hence only a very slight deviation from charge equality would have important effects on the behavior of large scale matter where gravitational effects are important.

It was remarked at the beginning of this paper that the apparatus used was not well suited to do the optimum experiment on the electron-proton charge difference. It may be of interest to estimate the accuracy which might be attained with appropriate modern design. With an electric field 10 meters in length and of intensity 5×10^5 v/cm, with an ion-multiplier–mass-spectrometer detector to reduce noise background from the surface ionization detector and to increase the inherent signal-to-noise ratio of the current measuring circuit as compared with an FP54 electrometer circuit, and with careful oven design and vacuum technique to obtain a very stable beam, it is estimated that an electron-proton charge difference of about 1 part in 10^{21} could be detected.

ACKNOWLEDGMENT

The experiment reported in this paper was originally suggested to the author by Professor I. I. Rabi.

[8] B. T. Feld, in *Experimental Nuclear Physics*, edited by E. Segrè (John Wiley and Sons, Inc., New York, and Chapman and Hall, Ltd., London, 1953), Vol. 2, p. 214. The experiment described in the present paper is the one quoted in Feld's article. Feld states that the charge on a CsI molecule is proven to be less than 10^{-10} of the electronic charge; the correct limit should be 4×10^{-13} of the electronic charge, as given in the present paper. Hence the upper limit for the neutron charge is about a factor of 200 less than the value given in Feld's article.

[9] *Note added in proof.*—A recent experiment on the deflection of a neutron beam in an electric field puts an upper limit on the neutron charge of 6×10^{-12} of the electron charge. I. S. Shapiro and I. V. Estulin, Zhur. Eksptl'. i Teort. Fiz. **30**, 579 (1956); Soviet Phys. JETP (to be published).

Z. Phys. D – Atoms, Molecules and Clusters 10, 145–151 (1988)

Atoms, Molecules
Zeitschrift für Physik D and Clusters

The electrical neutrality of atoms*

V.W. Hughes, L.J. Fraser, and E.R. Carlson*****
Yale University, Physics Department, New Haven, CT 06520, USA

Received 4 August 1988

An atomic beam deflection experiment is described which establishes upper limits to the electrical charges of potassium and cesium atoms of $q(\mathrm{K}) < 2.4 \times 10^{-18}\, q_e$ and $q(\mathrm{Cs}) < 3.0 \times 10^{-18}\, q_e$ at the 90% confidence level, in which q_e is the electron charge magnitude. If we assume that the electron neutrino charge is zero and that charge conservation holds in neutron beta decay, then the difference δq in the electron and proton charge magnitudes has the upper limit $\delta q < 1.3 \times 10^{-20}\, q_e$ at the 90% confidence level. The possibility of a more sensitive atomic beam deflection experiment using a laser-cooled atomic beam is suggested. A brief review of the topic of the electrical neutrality of atoms is given.

PACS: 35.10

I. Introduction

All experimental evidence to date indicates that atoms are electrically neutral, i.e. have zero charge, or equivalently that the electron and proton charge magnitudes are equal and that the neutron charge is zero. Although present experimental sensitivity is very high, still only a very tiny charge on an atom would have very important consequences for particle physics, astrophysics and cosmology.

In this paper we give a brief review of this topic and report some details on the most recent atomic beam experiment to measure the electrical charge of an atom [1, 2]. The atomic beam type experiment is conceptually very simple involving only observation of the deflection of an atomic beam by an applied static electric field; still its sensitivity is comparable to that of bulk matter experiments and its interpretation is quite clean and unambiguous. The beautiful atomic beam techniques developed by Stern in his classic papers on Untersuchungen zur Molekularstrahlmethode provide the full background for the atomic beam type experiment on the electrical neutrality of atoms [3, 4].

The modern standard theory of particle physics [5], which includes atomic physics, has leptons and quarks as the elementary constituents and it does require that particle and antiparticle have charges of equal magnitude but opposite sign. The charges for the electron and proton are assigned so that the electron charge magnitude q_e equals the proton charge magnitude q_p and the neutron charge $q_n = 0$. However, this assignment is arbitrary, and since baryon conservation and lepton conservation are assumed in the standard theory, there are no reactions allowed by the theory which would be forbidden if $q_e \neq q_p$ or $q_n \neq 0$, and the changes in atomic energy levels would be unobservably tiny. In the speculative grand unified theories (GUT) [5, 6], however, baryon and lepton number are not separately conserved so that proton decay in a reaction such as $p \to e^+ + \pi^0$ is predicted. A difference in q_p and q_e (provided π^0 has no charge) together with the basic assumption of charge conservation would forbid proton decay and hence nullify the GUT theories [7]. Recently it has been speculated that atomic charge neutrality might be violated by a tiny amount determined by an energy scale identified as the mass of the magnetic monopole [8].

For large scale matter as treated in astrophysics and cosmology, a slight difference δq between the

* Research supported in part by the Department of Energy and the National Science Foundation

** Xerox Corporation, J.C. Wilson Center for Technology, 800 Phillips Rd., 311-211K, Webster, NY 14580, USA

*** AT & T Bell Laboratories, Communications Sciences, Room 4E-605 Crawfords Corner Rd., Holmdel, NJ 07733, USA

electron and proton charge magnitudes, $\delta q = q_e - q_p$, could have important consequences on gravitation, on the magnetic fields of astronomical bodies, and on cosmology [9]. It is well known that the ratio of the magnitudes of the electrical force between two protons to their gravitational force is the large number

$$\frac{F_{\mathrm{el}}}{F_{\mathrm{grav}}} = \frac{e^2/r^2}{GM_p^2/r^2} = 1.2 \times 10^{3b}.$$

Hence if $\delta q = y q_e$ the electrical force between two hydrogen atoms would equal in magnitude the gravitational force provided $y = 0.9 \times 10^{-18}$ [10].

The origin of the magnetic fields of astronomical bodies is another problem for which a slight charge difference may be relevant. Einstein remarked that a slight difference δq could lead to a net volume charge for matter [11]; hence a rotating body such as the earth would have an associated magnetic field similar to that of a magnetic dipole and a value of $y = 2 \times 10^{-19}$ would account for the earth's magnetic field. Einstein's suggestion did lead to an early sensitive measurement of δq [12], and there was other interest as well in this possible origin of the magnetic field of astronomical bodies [13, 14]. However modern evidence indicates that the earth's magnetic field is due to an internal dynamo mechanism [15] and, furthermore, experiments have found that δq is too small to account for the magnetic fields of astronomical bodies.

In cosmology, Lyttleton and Bondi suggested that an electron-proton charge difference might account for the observed expansion of the universe, which could be due to an electrical repulsion between atoms with a small net charge [16]. Several discussions of their suggestion followed [17, 18, 19]. In their model, the observed expansion of the universe as obtained from the measured Hubble constant would require $y = 2 \times 10^{-18}$. Their imaginative suggestion stimulated several experiments and the results were sensitive enough to rule out this speculation.

Various measurements of the electrical neutrality of atoms have been made since the earliest days of atomic physics. The classical Millikan oil drop experiment [20] established directly that electronic and ionic charges are equal to about 1 part in 1000. Furthermore, from the electrical neutrality of an oil drop composed of some 10^{13} electron-proton pairs it could be concluded that $|\delta q| < 10^{-16}\, q_e$. From other bulk matter experiments and in particular from current searches for free quarks using magnetically levitated balls [21, 22] it is established that $|\delta q| < 10^{-21}\, q_e$ (see Table 1).

II. The atomic beam experiment

II.1. Principle of experiment

The simplest experiment conceptually on the electrical neutrality of atoms is an atomic beam experiment

Table 1. Some measurements of electrical neutrality of atoms

	Molecule	Upper limit[a] (for q)	Upper limit[b] (for q/M)	Date	Ref.
Millikan	oil drop		$\pm 10^{-16}$	1935	[31]
Stover, Moran and Trischka	iron sphere		$\pm 0.8 \times 10^{-19}$	1967	[32]
Piccard and Kessler	CO_2	$\pm 2 \times 10^{-19}$	$\pm 5 \times 10^{-21}$	1925	[12]
Hillas and Cranshaw	Ar	$(4 \pm 4) \times 10^{-20}$	$(1 \pm 1) \times 10^{-21}$	1959	[33]
	N_2	$(6 \pm 6) \times 10^{-20}$	$(2 \pm 2) \times 10^{-21}$		
King	H_2	$(1.8 \pm 5.4) \times 10^{-21}$	$+(0.9 \pm 2.7) \times 10^{-21}$	1960	[34]
	He	$(-0.7 \pm 4.7) \times 10^{-21}$	$(-0.2 \pm 1.2) \times 10^{-21}$		[35]
	SF_6	$(0 \pm 4.3) \times 10^{-21}$	$(0 \pm 3.0) \times 10^{-23}$		[35]
Hughes	CsI	4×10^{-13}	2×10^{-15}	1957	[24]
Zorn, Chamberlain, and Hughes	Cs	$(1.3 \pm 5.6) \times 10^{-17}$	$(1.0 \pm 4.2) \times 10^{-19}$	1963	[25]
	K	$(-3.8 \pm 11.8) \times 10^{-17}$	$(-1.0 \pm 3.0) \times 10^{-18}$		
	H_2	$\pm 2 \times 10^{-15}$	$\pm 1 \times 10^{-15}$		
	D_2	$\pm 2.8 \times 10^{-15}$	$\pm 7 \times 10^{-16}$		
Fraser, Carlson and Hughes	Cs	$\pm 1.7 \times 10^{-18}$	$\pm 1.3 \times 10^{-20}$	1968	[1]
	K	$\pm 1.3 \times 10^{-18}$	$\pm 3.3 \times 10^{-20}$		
Shapiro and Estulin	n	6×10^{-12}	6×10^{-12}	1956	[36]
Shull, Billman and Wedgwood	n	$(-1.9 \pm 3.7) \times 10^{-18}$	$(-1.9 \pm 3.7) \times 10^{-18}$	1967	[37]
Gähler, Kalus and Mampe	n	$(-1.5 \pm 2.2) \times 10^{-20}$	$(-1.5 \pm 2.2) \times 10^{-20}$	1982	[30]
Marinelli and Morpurgo	steel ball		$(0.8 \pm 0.8) \times 10^{-21}$	1984	[22]

[a] Measured charge per molecule in units of the electronic charge
[b] Measured charge per molecule divided by the total number of nucleons

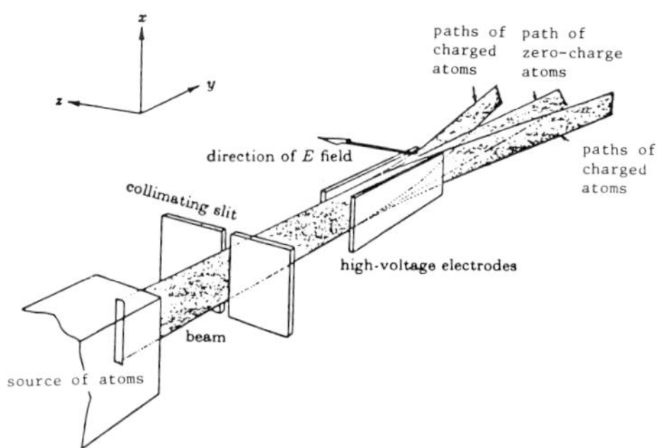

Fig. 1. Molecular beam measurement of atomic or molecular charge

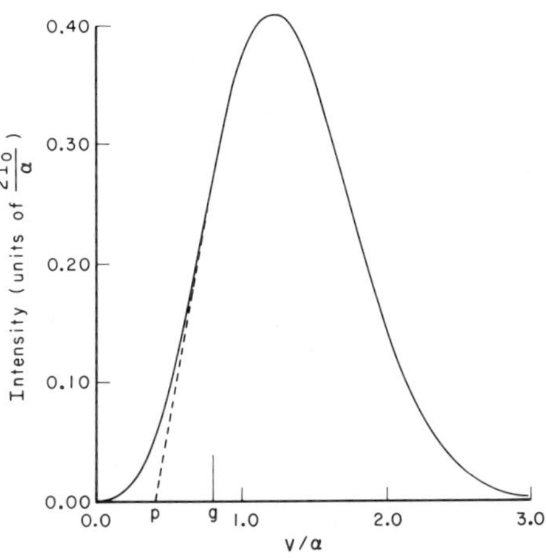

Fig. 3. Velocity distributions for the atomic beam. The solid line is an ideal Maxwellian distribution and the dashed line is a modified distribution with a deficiency of slow atoms due to scattering and gravitational deflection of the beam. The velocity v is measured in units of α, the most probable velocity for atoms in the oven; p is the velocity below which no atoms are found in the modified distribution, g is the velocity above which the modified distribution is identical to the Maxwellian distribution, and I_0 is the total integrated beam intensity

in which the deflection of an atomic beam by a static electric field is studied. It will turn out that very high sensitivity is possible with this simple method. This experiment was suggested to the author (VWH) in 1947 by I.I. Rabi following his discussion with A. Einstein.

The method of the experiment is shown in Fig. 1. Three versions of this experiment have been done [23, 24, 25, 1, 2]. The geometry of the latest version is shown in Fig. 2. An atom with the most probable velocity α of atoms in the oven source ($\alpha=\sqrt{2kT/m}$, in which m = mass of atom, k = Boltzmann's constant, T = oven source temperature) will undergo a deflection at the detector plane of magnitude s_α due to the force of the applied static electric field E on its charge q

$$s_\alpha=\frac{qE}{4kT}\,1_1(1_1+21_2). \tag{1}$$

With the geometry of Fig. 2, with $E=180$ kV/cm, $T=480$ °K and $s_\alpha=10^{-6}$ cm, $q\simeq 5\times10^{-18}\,q_e$.

The velocity distribution of the atoms in the beam is a Maxwellian distribution applicable to effusion from a source at temperature T, modified somewhat at low velocities by atom-atom collisions in the beam. The gravitational force also removes some slow atoms from the useful height of the beam. Figure 3 shows the calculated velocity distributions where the effect of collisions is based on measured scattering cross sections. The solid curve is the ideal Maxwellian distribution, and the dashed curve is the modified Maxwellian distribution.

In practice the applied electric field is only approximately homogeneous and hence a force arises on an

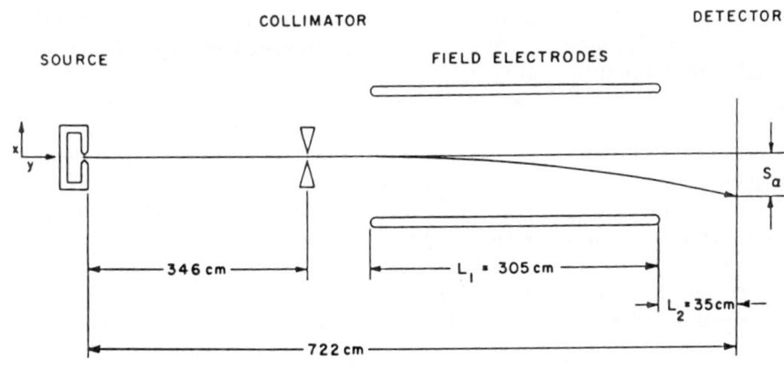

Fig. 2. Geometry of the apparatus, showing trajectory of an atom which has been deflected by the electric field. The dimensions are $l_1=3.05$ m, $l_2=0.35$ m, $l_{sc}=3.46$ m, $l_{sd}=7.22$ m

atom with zero charge due to its polarizability and the field inhomogeneity. The leading correction term is associated with the atomic electric dipole polarizability [26] and the electric field gradient. The force is given by

$$F = q\mathbf{E} + +\nabla(\boldsymbol{\mu}\cdot\mathbf{E}) \tag{2}$$

in which $\boldsymbol{\mu} = p\mathbf{E}$ where p = atomic dipole polarizability. The force components associated with the atomic quadrupole polarizability and higher order terms are negligibly small. The magnitude of the force associated with the field gradient is quite large so that the beam is deflected when the static electric field is applied with $s_\alpha \simeq 10^{-4}$ cm. However the direction of the force associated with the field gradient does not depend on the direction of E. Hence if the voltage polarity across the parallel plates is reversed, the deflecting force should be unchanged, in distinction to a deflection associated with a charge q on an atom.

II.2. Apparatus and procedure [1]

The vacuum chamber was a stainless steel can 8.1 m in length and 30 cm in diameter. The chamber was divided into three separately pumped chambers – an oven chamber, a separating chamber to provide differential pumping, and a main chamber containing the electrodes and the detector. The pumping system consisted of mercury diffusion pumps with Freon-cooled baffles and liquid nitrogen traps, backed by an oil-filled mechanical forepump. Gold and aluminum metal 0-rings were used. The main chamber was wrapped with heating tape and fiberglass insulation to permit bakeout to a maximum temperature of 150 °C. Typical operating pressures were 5×10^{-8} Torr in the source chamber, 5×10^{-9} Torr in the separating chamber and 2×10^{-9} Torr in the main chamber.

In order to isolate the apparatus from building vibrations, the main chamber was firmly mounted on three concrete-filled steel pillars which extended through the laboratory floor to separate concrete pads on the ground directly below. The upstream end of the source chamber was connected to the laboratory floor.

The atomic beam source was a nickel oven (Fig. 4) containing two glass ampules for the alkali. The oven slit was 0.04 mm wide and 5 mm high, formed from 0.05 mm thick stainless steel shims. Potassium or cesium of 99.95% purity were used. Together with the oven slit, a collimator slit 0.04 mm wide by 2.5 cm high, placed at about the midpoint of the main chamber, and vertical beam stops defined the ribbon-shaped atomic beam.

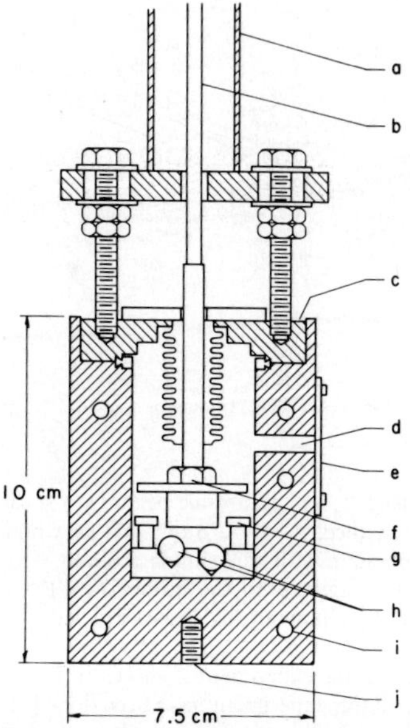

Fig. 4. Source oven: *a* is a support tube; *b*, plunger rod; *c*, removable flange; *d*, beam channel; *e*, slit jaws; *f*, plunger; *g*, baffle; *h*, ampules; and *j*, thermocouple well

The surface ionization detector consisted of a 0.025 mm irridium wire heated to about 1300 °K, a collector electrode for the ion current, and a vibrating reed electrometer. It had a detection efficiency close to 100%.

The electric field was provided by two parallel plates 305 cm in length shaped as shown in Fig. 5a to produce a uniform electric field over a height of 3 cm and constructed of cold-drawn annealed Invar 36 bars ground to a 16 to 32 µin finish. The straightness of the plates was measured against a 4.5 m granite surface plate. The plates were mounted with a separation of (1.000 ± 0.025) mm on insulators made of steatite blocks with precision ground internal faces (Fig. 5b). A 30 kV dc power supply capable of rapid polarity reversal and with a short term stability of 0.01% and an absolute calibration of 2% was used.

An observed atomic beam shape at the detector plane compared to the calculated shape based on apparatus geometry is shown in Fig. 6, where the agreement is regarded as satisfactory, particularly in view of small angle scattering of the long atomic beam. Figure 7 shows observed atomic beam shapes with no applied electric field and with an applied electric field of 180 kV/cm, where the broadening of the beam

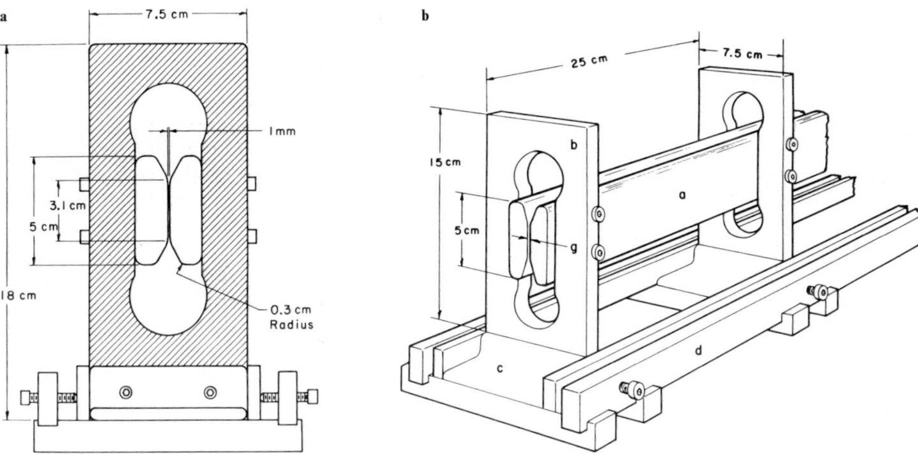

Fig. 5a. Front view of the electrode system; **b** Electrode system: *a*, Invar electrode; *b*, steatite insulator; *c*, insulator mounting block; *d*, Invar support structure with positioning screws; *g*, interelectrode gap of (1.000 ± 0.025) mm

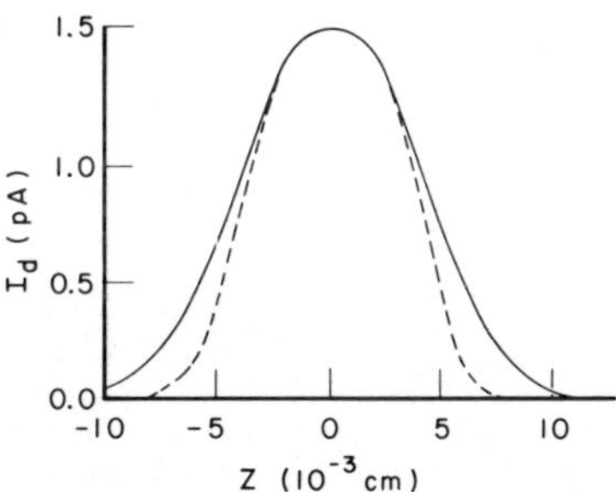

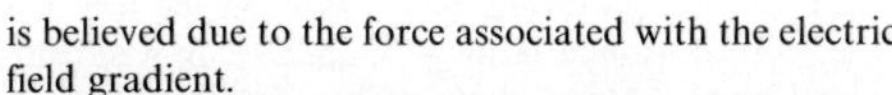

Fig. 6. Beam shape: detector current, I_d, versus detector position, z. The dotted line is calculated on the basis of the apparatus geometry and the solid line is a typical measured curve

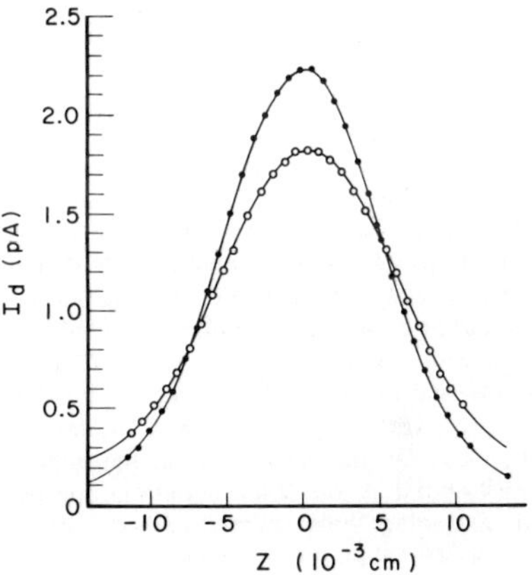

Fig. 7. Detector current, I_d, as a function of detector position, z. The closed dots show the beam shape with no applied field and the open circles give the beam broadened by a 180 kV/cm field

is believed due to the force associated with the electric field gradient.

The procedure of the experiment is to look for small deflections of the atomic beam due to the applied electric field and in particular to identify a deflection due to a net charge on the atom. Since the dependence of the signal on detector position is particularly strong at the points of half maximum intensity at the sides of the beam, very small deflections of the beam may be most sensitively measured with the detector at these points, and most of the data were taken in this condition. Of course the difference between the beam deflections for the two voltage polarities across the plates is the quantity related to atomic charge q. It can be shown [25] that the change in intensity ΔI for a beam with a Maxwellian velocity distribution measured by a detector placed near the half intensity point is given by

$$\Delta I = -\left(\frac{\partial I}{\partial z}\right) s_\alpha \qquad (3)$$

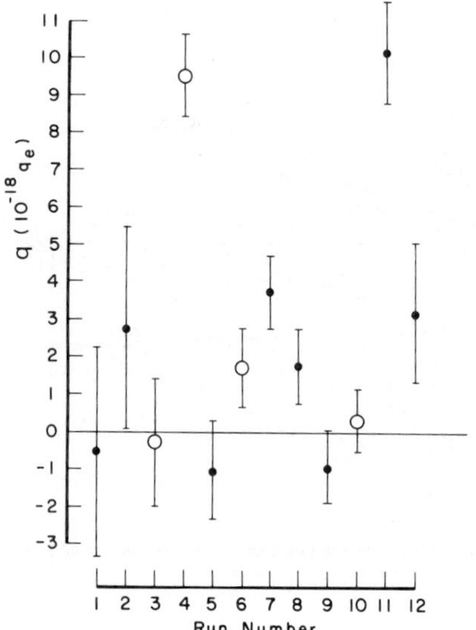

Fig. 8. Apparent atomic charge, q, versus run number. The closed circles are results for cesium and the open circles for potassium

in which $\frac{\partial I}{\partial z}$ is the slope of the measured beam intensity. A correction to (3) can be made to account for deviation from a Maxwellian velocity distribution.

The basic data obtained consisted of beam intensity measurements with the detector set at a half intensity point alternately with one voltage polarity across the plates and then the opposite polarity. The polarity was modulated manually with a period of about 1 min and typically some 70 pairs of data points were obtained in a one hour period. Then a similar data set is obtained with the detector set at the opposite half intensity point. About 300 h of data-taking were involved, including both potassium and cesium atoms. From these data, using (1) and (3), the charge on an atom can be determined.

II.3. Results and discussion [1]

The results on atomic charge are shown in Fig. 8. Except for the two runs nos. 4 and 11, the data points are self-consistent statistically and the point-to-point variations are believed due to noise in the detector and in the atomic beam. The disagreement of runs 4 and 11 are correlated with insufficient bakeout of the apparatus and conditioning of the electric field plates. These runs are discarded from the data sample used to deduce our final result. This type of systematic error associated with the condition of the apparatus and in particular of the electric field plates is believed to be due to insulating hydrocarbon films formed on the electric field plates (patch effect). These films can become electrically charged and produce polarity dependent asymmetries in the electric field. This effect provides the limiting systematic error in the present experiment. Omitting runs 4 and 11, the results from the data are:

$$q(\mathrm{K}) = (0.84 \pm 0.78) \times 10^{-18}\, q_e$$
$$q(\mathrm{Cs}) = (1.62 \pm 0.70) \times 10^{-18}\, q_e, \tag{4}$$

in which the error is the one standard deviation statistical error added in quadrature with a smaller systematic error associated with drifts in beam shape and instrumental errors.

If we assume that the charge q on an atom is given by $q = Z\delta q + N q_n$ in which Z equals the number of electron-proton pairs and N equals the number of neutrons, then from the values for $q(\mathrm{K})$ and $q(\mathrm{Cs})$ given in (4) we obtain

$$\delta q = (0.9 \pm 2.0) \times 10^{-19}\, q_e,$$
$$q_n = (0.4 \pm 1.5) \times 10^{-19}\, q_e. \tag{5}$$

With the assumption of charge conservation in neutron beta decay this implies that the charge on the neutrino is $q_{\bar{\nu}_e} = (0.5 \pm 3.5) \times 10^{-19}\, q_e$. If we assume that the neutrino charge is zero so that $\delta q = q_n$, then we obtain $\delta q = (1.28 \pm 0.52) \times 10^{-20}\, q_e$.

Improvement in the sensitivity of this type of atomic beam experiment could certainly be achieved through the use of a cleaner ultrahigh vacuum system free of all hydrocarbon contaminants and through greater attention to the patch effect. In addition, more rapid polarity reversal, automation of data-taking, use of a mass spectrometer with the surface ionization detector, and improved electronics should provide an improvement in sensitivity by a factor of 10 to 100.

Another type of atomic beam deflection experiment appears promising now based on the use of a laser-cooled atomic beam [27, 28, 29]. A well-collimated alkali atomic beam with a mean velocity of 10^3 cm/s and a velocity spread of about 25% could be produced. Such a slow beam would allow use of a short electric field region and have relatively high beam intensity.

III. Discussion

Table 1 gives the results presently available on the electrical neutrality of atoms. In addition to the atomic beam type experiment there has been a similar type of experiment on a neutron beam [30]. The droplet method, first used in the Millikan oil drop experiment, has been employed in current searches for free quarks with magnetically levitated balls. In the gas efflux method the total charge on a large number of gas molecules is measured by observing the change in potential on a metal container relative to the surroundings as gas effuses from the container. All of the experiments find that atoms are electrically neutral and they can be interpreted to indicate that $\delta q \lesssim (10^{-21}$ to $10^{-20})\, q_e$.

Although all experimental evidence indicates that atoms are electrically neutral, it is still of great importance to extend the sensitivity of the experiments because of the great significance of this matter to modern physics. There seem to be excellent possibilities for much improved experiments.

References

1. Fraser, L.J., Carlson, E.R., Hughes, V.W.: Bull. Am. Phys. Soc. **13**, 636 (1968);
   Fraser, L.J.: Ph.D. Thesis, Yale University (1968)
2. Hughes, V.W.: In: Atomic physics 10. Narumi, H., Shimamura, I. (eds.), p. 1. Amsterdam: North Holland 1987
3. Kusch, P., Hughes, V.W.: Atomic and molecular beam spectroscopy in encyclopedia of physics, 37/1. Berlin, Göttingen, Heidelberg: Springer 1959
4. Ramsey, N.F.: Molecular beams, Oxford: Clarendon Press 1956
5. Quigg, C.: Gauge theories of the strong weak and electromagnetic interactions. Menlo Park, CA: Benjamin/Cummings 1983
6. Ross, G.G.: Grand unified theories. Menlo Park, CA: Benjamin/Cummings 1984
7. Feinberg, G., Goldhaber, M.: Proc. Natl. Acad. Sci. **45**, 1301 (1959)
8. Chu, S.-Y.: Phys. Rev. Lett. **59**, 1390 (1987)
9. Hughes, V.W.: In: Gravitation and relativity. Chiu, H.-Y., Hoffmann, W.F. (eds.), p. 259. New York: Benjamin, W.A. 1964
10. Swann, W.F.G.: Astrophys. J. **133**, 733 (1961)
11. Einstein, A.: Schw. Naturf. Ges. Verh. **105**, Pt. 2, 85(1924)
12. Piccard, A., Kessler, E.: Arch. Sci. Phys. Nat. (Geneva) **7**, 340 (1925)
13. Blackett, P.M.S.: Nature **159**, 658 (1947)
14. Angel, J.R.P.: Ann. Rev. Astron. Astrophys. **16**, 487 (1978)
15. Cowling, T.G.: Ann. Rev. Astron. Astrophys. **19**, 115 (1981)
16. Lyttleton, R.A., Bondi, H.: Proc. Roy. Soc. London, A **252**, 313 (1959)
17. Chamber, Ll.G.: Nature **191**, 1082 (1961)
18. Lyttleton, R.A., Bondi, H.: Proc. R. Soc. London, A **257**, 442 (1960)
19. Hoyle, F.: Proc. R. Soc. London, A **257**, 431 (1960)
20. Millikan, R.A.: The electron. Chicago: University of Chicago Press 1917
21. Marinelli, M., Morpurgo, G.: Phys. Rep. **85**, 161 (1982)
22. Marinelli, M., Morpurgo, G.: Phys. Lett. **137** B, 439 (1984)
23. Hughes, V.W.: Phys. Rev. **76**, 474 (1949)
24. Hughes, V.W.: Phys. Rev. **105**, 170 (1957)
25. Zorn, J.C., Chamberlain, G.E., Hughes, V.W.: Phys. Rev. **129**, 2566 (1963)
26. Van Vleck, J.H.: The theory of electric and magnetic susceptibilities. London: University Press 1932
27. Phillips, W.D.: Laser-cooled and trapped atoms. NBS 653, U.S. Department of Commerce, Washington, D.C. (1983)
28. Phillips, W.D.: Ann. Phys. Fr. **10**, 717 (1985)
29. Chu, S. et al.: In: Atomic physics 10. Narumi, H., Shimamura, I. (eds.), p. 377. Amsterdam: North Holland 1987
30. Gähler, R., Kalus, J., Mampe, W.: Phys. Rev. D **25**, 2887 (1982)
31. Millikan, R.A.: Electrons (+and−), protons, photons, neutrons, and cosmic rays. Chicago: University of Chicago 1935
32. Stover, R.W., Moran, T.I., Trischka, J.W.: Phys. Rev. **164**, 1599 (1967)
33. Hillas, A.M., Cranshaw, T.E.: Nature **184**, 892 (1959)
34. King, J.G.: Phys. Rev. Lett. **5**, 562 (1960)
35. Dylla, H.F., King, J.G.: Phys. Rev. A **7**, 1224 (1973)
36. Shapiro, I.S., Estulin, I.V.: Soviet Phys.-JETP **3**, 626 (1957)
37. Shull, C.G., Billman, K.W., Wedgwood, F.A.: Phys. Rev. **153**, 1415 (1967)

Reprinted from
QUANTUM ELECTRONICS
A Symposium
edited by C. H. Townes

Columbia University Press, New York

NARROW LINEWIDTHS FOR DECAYING STATES BY THE METHOD OF SEPARATED OSCILLATING FIELDS*

V. W. HUGHES
Gibbs Laboratory, Yale University

THE REMARKS I should like to make were stimulated by the recent experiment of Madansky and Owen(1) in which an intense beam of fast hydrogen atoms in the metastable $2^2S_{\frac{1}{2}}$ state was produced. The beam was produced by passing protons from an rf ion source having an energy of some 10 kev through low-pressure hydrogen gas ($\sim 10^{-4}$ mm of Hg). A proton has a sufficiently high probability for capturing an electron from a H_2 gas molecule to form the metastable state of atomic H so that about 1 percent of the initial proton beam from the ion source forms the metastable state of H. The resultant beam intensity of metastable H atoms is many orders of magnitude greater than the metastable beams used by Lamb and his colleagues in their experiments(2) on the fine structure of hydrogen. Furthermore, the present beam can be highly directional and hence is not subject to the inverse square law loss. The beam energy of 10 kev can be controlled to about 0.1 percent without difficulty, so that a highly monochromatic rather than a Maxwellian beam is available.

I should like to discuss one possible application of such a beam to the measurement of the fine structure of hydrogen by use of the method of separated oscillating fields. Since the velocity of an atom in the beam is about 10^8 cm/sec, the atom will travel an appreciable distance in the 10^{-8} sec to 10^{-7} sec spent in the transition region, so separated oscillating fields

*This work has been supported by the National Science Foundation.

can be realized physically. In the experiments done thus far on the fine structure of hydrogen, the linewidth is the natural one determined by the optical decay rate of the $2p$ state and amounts to 100 Mc/sec. In principle, linewidths narrower than this natural width can be obtained by use of the method of separated oscillating fields.

In Fig. 1, a schematic diagram illustrating the method of separated oscillating fields[(3)] is shown. The radiofrequency field is applied over two short regions of length l separated by a relatively long region of length L in which no rf field is present.

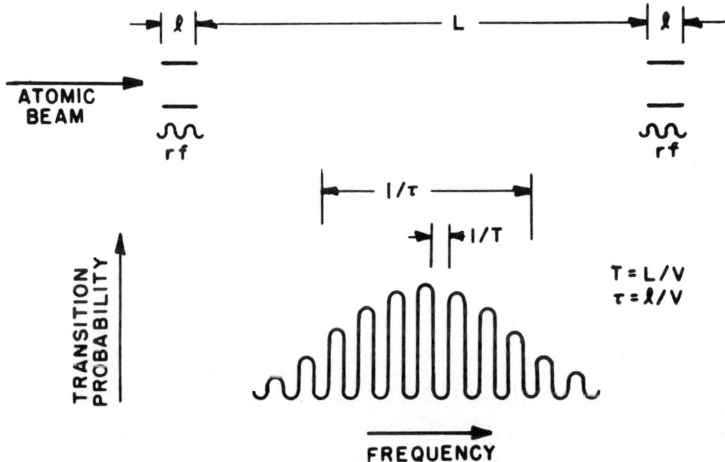

Fig. 1. Method of separated oscillating fields

The principal features of the line shape are indicated for a transition between two stable energy levels for a monoenergetic atomic beam. The overall width corresponds to the transit time τ through the short regions l. The narrower interference peaks correspond to the transit time T through the long region L.

The problem under consideration is the extension of the theory of transitions by the method of separated oscillating fields to include a decay rate for one of the two states. Figure 2 indicates two energy levels p and q; the level q has a decay rate γ_q. This diagram would apply to the $2^2S_{\frac{1}{2}}$ and $2^2P_{\frac{1}{2}}$ levels in hydrogen, in which only the decay rate of the $2p$ level need be considered.

$$p \text{ ———————— } \gamma_p = 0$$

$$q \text{ ———————— } \gamma_q$$

$$H = H_o + v(t)$$

$$H_o \phi_n = w_n \phi_n$$

$$V_{pq} = h\, b e^{+i\omega t}; \quad V_{qp} = \hbar\, b^* e^{-i\omega t}; \quad V_{pp} = V_{qq} = 0$$

$$\psi = \sum_n c_n(t)\,\phi_n$$

$$c_n(t) = a_n(t)\, e^{-\frac{i w_n}{\hbar} t}$$

$$\dot{a}_p = a_q(-i\,b)\, e^{+i(\omega_{pq}+\omega)t}$$

$$\dot{a}_q = a_p(-i\,b^*)\, e^{+i(\omega_{qp}-\omega)t} - \underline{\gamma}_q\, a_q$$

Fig. 2. Equations of motion

The usual formulation[4] of the transition process is given in which $V(t)$ is the time-dependent interaction with the rf field. The time-dependent Schrödinger equation for the state amplitudes a_p and a_q are given. The a's are so defined that a_p is constant if $V(t)$ is zero. The optical decay rate of state q is treated in the usual phenomenological manner[5] through the term $-(\gamma_q/2)\,a_q$.

$$a_p = A_1 e^{-\delta_1 t} + A_2 e^{-\delta_2 t}$$

$$a_q = -\frac{i}{b}\left[\delta_1 \Delta_1 e^{-\delta_1 t + i\lambda t} - \delta_2 A_2 e^{-\delta_2 t + i\lambda t}\right]$$

$$\delta_{1,2} = \left(+\frac{i\lambda}{2} + \frac{\gamma_q}{4}\right) \pm \frac{i}{2}\left[\lambda + \frac{\gamma_q}{2}^{\,2} + 4\,bb^*\right]^{\frac{1}{2}}$$

$$\lambda = \omega_{qp} = \omega$$

Fig. 3. General solution

Figure 3 gives the general solution of these coupled differential equations. The quantities A_1 and A_2 are determined by the initial conditions. The quantities δ_1 and δ_2 depend on the energy separation ω_{qp}, on the decay rate γ_q, on the transition matrix element b, and on the applied frequency ω.

Figure 4 gives the solution for a_p for the case of separated oscillating fields. The atom is assumed to be in the state p at

$$(a_p = 1, \quad a_q = 0 \text{ at } t = 0)$$

$$a_p(2\tau + T) = \left(\frac{\delta_2}{\delta_2 - \delta_1} e^{-\delta_1 \tau} - \frac{\delta_1}{\delta_2 - \delta_1} e^{-\delta_2 \tau}\right)^2$$

$$+ \frac{\delta_1 \delta_2}{\delta_2 - \delta_1}(e^{-\delta_1 \tau} - e^{-\delta_2 \tau}) e^{-i\lambda T}$$

$$- \frac{\gamma_q}{2} T \left[\frac{1}{\delta_2 - \delta_1}(e^{-\delta_2 \tau} - e^{-\delta_1 \tau})\right]$$

$$a_p(2\tau + T) = A_{pp}^2(\tau) + A_{pq}(\tau) A_{qq}(T) A_{qp}(\tau)$$

$$P_{pp}(2\tau + T) = a_p(2\tau + T) a_p^*(2\tau + T)$$

$$P_{pp}(2\tau + T) = |A_{pp}^2(\tau)|^2 + e^{-\frac{\gamma_q}{2}} \{A_{pp}^2(\tau) A_{pq}^*(\tau) A_{qp}^*(\tau) e^{+i\lambda T}$$

$$+ A_{pp}^2(\tau)^* A_{pq}(\tau) A_{qp}(\tau) e^{-i\lambda T}\}$$

$$+ e^{-\gamma_q T} |A_{pq}(\tau) A_{qp}(\tau)|^2$$

Fig. 4. Solution for separated oscillating fields case

time $t = 0$ as it enters the first oscillating field region and the probability amplitude $a_p(2\tau + T)$ that it is in the state p at the end of the second oscillating field region is given. This probability amplitude is expressible as the sum of two terms. The first term, $A_{pp}^2(\tau)$ is the product of the probability amplitude, $A_{pp}(\tau)$, that the atom be in the state p at the end of its traversal of the first oscillating field region (at time τ) times the probability amplitude that the atom remain in state p during the time T

(which is 1) times the probability amplitude that the atom remain in state p during its traversal of the second oscillating field, which is again $A_{pp}(\tau)$. The second term is the product of the probability amplitude $A_{pq}(\tau)$ that the atom is in the state of q at the time τ times the probability amplitude $A_{qq}(\tau)$ that the atom remain in the state q during the time T, which is the exponential optical decay factor $e^{-\gamma_q T/2}$, times the probability amplitude for returning to the state p in the traversal of the second oscillating field region.

The observable probability, $P_{pp}(2\tau + T)$, that the atom be in the state p at the time $2\tau + T$ is $|a_p(2\tau + T)|^2$, which is expressed as the sum of three terms in the last equation. The first term is simply the absolute square of the first term in the expression for $a_p(2\tau + T)$, and the third term is the absolute square of the second term for $a_p(2\tau + T)$. The middle term is an interference term which arises from a cross product of the first and second terms for $a_p(2\tau + T)$. In the limiting case of $\gamma_q = 0$, which applies for two stable energy levels, the expression for $P_{pp}(2\tau + T)$ reduces to the usual expression for the transition probability in the separated oscillating field method, and the middle term is responsible for the sharp interference peaks of frequency width approximately $1/T$. In the case under consideration with $\gamma_q \neq 0$, the middle term also gives rise to interference peaks of width $1/T$, but their amplitude is reduced by the exponential factor $e^{-\gamma_q T/2}$. Hence the achievement of a narrow interference line is severely limited by the attendant reduction in intensity of the interference peak.

I estimate that in an experiment which seeks to study the $2s \longrightarrow 2p$ transition in hydrogen it may be possible to obtain an interference peak $\frac{1}{3}$ of the usual natural width of 100 Mc/sec with reasonable signal to noise. In order to achieve this it will be necessary to utilize the facts that the transition probability for all atoms is closely the same because of the monoenergetic character of the beam and that the interference term of interest depends on the relative phases of the radiofrequency in the two oscillating field regions whereas the other terms for the transition probability do not.

I should like to suggest that apart from this possibility of achieving lines narrower than the natural width with a fast beam of hydrogen metastable atoms, the high intensity, unidirectionality, and monoenergetic character of the beam may by themselves prove useful in fine structure measurements. Also with a fast atomic beam time dilation experiments may be possible. Finally it may be possible to form high-beam intensities of He^+ and of Li^{++} in the $2^2S_{\frac{1}{2}}$ state and hence to do fine structure measurements on these ions. At Yale we are undertaking some experiments to investigate these possibilities.

In conclusion, I might say that the principal motivations at present for higher precision fine-structure experiments seem to be the determination of the fine-structure constant α to a higher precision (a more precise value of α would be most valuable in correction with the interpretation of the hyperfine structure of hydrogen for information on the structure of the proton[(6)]) and the further study of the substantial discrepancy between the experimental and theoretical values of the Lamb shift in singly ionized helium.[(7)]

REFERENCES

1. L. Madansky and G. E. Owen, Phys. Rev. Lett. *2*, 209 (1959).
2. W. E. Lamb, Jr., and R. C. Rutherford, Phys. Rev. *79*, 549 (1950).
3. N. F. Ramsey, Phys. Rev. 78, 698 (1950).
4. N. F. Ramsey, *Molecular Beams*, Oxford University Press (1956).
5. W. E. Lamb, Jr., Phys. Rev. *85*, 259 (1952).
6. A. C. Zemach, Phys. Rev. *104*, 1771 (1956).
7. E. Lipworth and R. Novick, Phys. Rev. *108*, 1434 (1957).

DISCUSSION

C. O. ALLEY: I should like to point out the similarity between the method using spatially separated oscillatory fields described by Professor Hughes and the method using a single coherently pulsed oscillatory field discussed in my paper presented at this Symposium.

VOLUME 4, NUMBER 7 PHYSICAL REVIEW LETTERS APRIL 1, 1960

UPPER LIMIT FOR THE ANISOTROPY OF INERTIAL MASS FROM NUCLEAR RESONANCE EXPERIMENTS*

V. W. Hughes, H. G. Robinson, and V. Beltran-Lopez
Gibbs Laboratory, Yale University, New Haven, Connecticut
(Received March 2, 1960)

Mach's principle states that the inertial mass of a body is determined by the total distribution of matter in the universe; if the matter distribution is not isotropic, it is conceivable that the mass of a body depends on its direction of acceleration and is a tensor rather than a scalar quantity. Thus the matter in our galaxy is not distributed isotropically with respect to the earth, and hence the mass of a body on the earth may depend on the direction of its acceleration with respect to the direction towards the center of our galaxy. Cocconi and Salpeter[1] have proposed that the total inertial mass of a body on the earth be considered the sum of an isotropic part m and an anisotropic part Δm, and that the contribution to the mass of a body on the earth due to a mass $\mathfrak{M}$ a distance r away from the body is proportional to $\mathfrak{M}/r^\nu$ $(0 \leq \nu \leq 1)$. The ratio of Δm, due to a mass $\mathfrak{M}$ a distance r away, to m, due to the total mass in the universe, is

$$\frac{\Delta m}{m} = \frac{\mathfrak{M}}{r^\nu} \frac{3-\nu}{4\pi\rho R^{(3-\nu)}}, \tag{1}$$

in which ρ = average density of matter in the universe (10^{-29} g/cm^3) and R = radius of the universe (3×10^{27} cm).[2] If Δm is ascribed to our own galaxy, then $r = 2.5\times10^{22}$ cm and $\mathfrak{M} = 3\times10^{44}$ g, where the total mass of the galaxy is considered concentrated at its center. Hence for $\nu = 1$, $\Delta m/m = 2\times10^{-5}$ and for $\nu = 0$, $\Delta m/m = 3\times10^{-10}$.

Cocconi and Salpeter have suggested several experiments to test for this anisotropy of mass based on the observation that the contribution to the binding energy of a particle in a Coulomb potential due to the anisotropic mass term Δm is

$$\Delta E = (\Delta m/m)\overline{T}\,\overline{P}_2(\cos\theta). \tag{2}$$

Here $\overline{T}$ is the average kinetic energy of the particle, P_2 is the Legendre polynomial of order 2, and θ is the angle between the direction of acceleration of the particle (determined by the direction of an external magnetic field $\vec{H}$ and by the magnetic quantum state) and the direction to the galactic center. This equation is based on the assumption that Δm varies as $P_2(\cos\theta)$. The first experiment suggested was to observe the Zeeman splitting in an atom[1] and the second was to observe the Zeeman splitting in the excited nuclear state of Fe^{57} by use of the Mössbauer effect.[3] [The change in binding energy due to Δm will not be given exactly by Eq. (2) in the nuclear case, but if the nucleus is idealized as a single particle in a spherically symmetric square well potential a similar type of equation applies.] For both experiments effects are to be measured as a function of the angle θ. The models used for the atom and the nucleus are adequate for the order of magnitude estimate we require for ΔE.

In this Letter we report an experiment using nuclear magnetic resonance in the ground state of nuclei to test for the anisotropy of mass. This method gives a sensitivity some factor of 10^6 greater than could be achieved in the experiment suggested by Cocconi and Salpeter using the Mössbauer effect. In addition, we report experiments on the Zeeman effect in atoms of the first type suggested by Cocconi and Salpeter.

We discuss first the atomic Zeeman experiments. There are at least two methods of searching for the effect of mass anisotropy in an atomic state with orbital angular momentum quantum number $L \geq 1$ and with total angular momentum quantum number $J \geq 3/2$. One method is to observe, for example, the frequency of the Zeeman transition $M_J = +3/2 \leftrightarrow M_J = +1/2$ in the ${}^2P_{3/2}$ state as a function of the relative orientation of the direction of the magnetic field and the direction to the galactic center. In our experiment with our electromagnet fixed to the earth and with the magnetic field pointing approximately in the north-south direction, the change in relative orientation is achieved as a function of time due to the rotation of the earth. In New Haven at 41° latitude at a certain time in the sidereal day the south direction in the horizontal plane points within 22 degrees towards the center of our galaxy; 12 hours later this same direction along the earth's horizontal plane points 104 degrees away from the galactic center. It is important, of course, that the frequency standard with respect to which the Zeeman transition frequency is compared shall itself exhibit no mass anisotropy effect. Most Zeeman transition frequency measurements such as those quoted by Cocconi and Salpeter[1] have been referred to crystal oscillator secondary standards calibrated occasionally against the signal from WWV, and hence because of possible mass anisotropy effects in the crystal oscillator are not suitable for the present purpose. In our experiment a frequency derived from the cesium atomic frequency standard (National Company Atomichron) was used. The transition $(F, M_F) = (4,0) \leftrightarrow (3,0)$ used in this frequency standard[4] will not exhibit any mass anisotropy effect. The magnetic field is maintained constant with a proton resonance probe whose resonance frequency is compared with the Atomichron frequency. For a constant magnetic field the proton resonance frequency will exhibit no mass anisotropy effect. The transition $M_J = +3/2 \leftrightarrow M_J = +1/2$ with $M_I = +3/2$ in the ${}^2P_{3/2}$ state of Cl^{35} was observed over a twelve-hour period. No variation with time of the Zeeman transition frequency occurring at about 9190 Mc/sec in a magnetic field of 4730 gauss was observed within the experimental error of 30 kc/sec. If the electronic structure of chlorine, which is an atom lacking one electron to complete the outer $3p$ shell, is treated as a hole moving in a Coulomb potential due to the nucleus and electrons, the upper limit to $\Delta m/m$ of 10^{-10} is obtained.

A second method is to observe the frequencies of two Zeeman transitions where the intervals would be affected differently by any mass anisotropy. It is only necessary to observe the two transitions at the time at which the direction to the galactic center is such as to maximize the mass anisotropy effect. Simple experimental data of this type are available from the Zeeman transitions $\Delta M_J = \pm 1, \pm 2$ in the ${}^3P_2$ state of atomic oxygen.[5] From these data it can be deduced that $\Delta m/m \leq 10^{-10}$.

By far the most sensitive test is obtained from a nuclear magnetic resonance experiment on an appropriate nucleus in its ground state. Consider the Li^7 nucleus in its ground state which has nuclear spin $I = 3/2$. In a magnetic field there will be four energy levels corresponding to the allowed values of the magnetic quantum number M_I. In the absence of any mass anisotropy, adjacent levels are equally spaced and a single nuclear resonance line will be observed. If the mass anisotropy effect is present, there will be three different intervals which will lead to a triplet nuclear resonance line, if the structure is resolved, or to a single broadened line if the structure is unresolved. Over a twelve-hour period, the resonance line for Li^7 was observed in a $1N$ water solution of $FeCl_3$ saturated with LiCl. The magnetic field of about 4700 gauss was stabilized against the proton resonance frequency with the Atomichron as a frequency standard. Only a single line was observed. The line width of 4.3 parts per million is due primarily to the inhomogeneity of the magnetic field and from the width of the proton resonance line should be (5.0 ± 1) parts per million. Hence a broadening of no greater than 8 cps could be due to the effect of mass anisotropy. If the nuclear structure of Li^7 is treated as a single $P_{3/2}$ proton in a central nuclear potential,[6] the limit $\Delta m/m \leq 10^{-20}$ is obtained. The increase in sensitivity over that which one could obtain from the Mössbauer effect is due to the far narrower line width obtainable for a transition with a nucleus in its ground state as compared with a nucleus in an excited state.

The limit here obtained of $\Delta m/m \leq 10^{-20}$ is far less than the value of 3×10^{-10} obtained by setting $\nu = 0$ in Eq. (1). Hence it seems that within the framework of the Mach theory as discussed by Cocconi and Salpeter one should conclude that there is no anisotropy of mass of the type which varies as $P_2(\cos\theta)$ associated with effects of mass in our galaxy.

In the interest of completeness we intend to improve the sensitivity of the nuclear resonance experiment at least one order of magnitude by obtaining a narrower line width. We shall study the nuclear resonance signal of nuclei consisting of closed shells plus or minus one nucleon. Also we shall search for an anisotropic mass which varies other than as $P_2(\cos\theta)$ by studying nuclear resonance signals of nuclei with spin greater than 3/2. Finally, since in the spirit of this investigation it is not necessarily excluded that mass anisotropy could be associated with a point in the universe other than the center of our galaxy, we shall study the nuclear resonance signal with respect to any arbitrary direction.

It is a pleasure to thank Professor H. Smith and Professor R. Wildt of the Yale Astronomy Department and Professor E. E. Salpeter for helpful and enlightening discussions.

---

*This research has been supported in part by the National Aeronautics and Space Administration.

[1]G. Cocconi and E. E. Salpeter, Nuovo cimento 10, 646 (1958).

[2]C. W. Allen, Astrophysical Quantities (The Athione Press, London, 1955).

[3]G. Cocconi and E. E. Salpeter, Phys. Rev. Letters 4, 176 (1960).

[4]P. Kusch and V. W. Hughes, Encyclopedia of Physics (Springer-Verlag, Berlin, 1959), Vol. 37, Part 1.

[5]H. E. Radford and V. W. Hughes, Phys. Rev. 114, 1274 (1959).

[6]M. G. Mayer and J. H. D. Jensen, Elementary Theory of Nuclear Shell Structure (John Wiley & Sons, Inc., New York, 1955).

Reprinted from
Proc. Nat. Acad. Sci. USA
Vol. 71, No. 8, pp. 3287–3289, August 1974

Atomic Regime in Which the Magnetic Interaction Dominates the Coulomb Interaction for Highly Excited States of Hydrogen

(energy levels/transition probabilities/autoionization/proposed atomic beam absorption spectroscopy experiment)

RONALD O. MUELLER AND VERNON W. HUGHES

Yale University, Gibbs Laboratory, Physics Department, New Haven, Connecticut, 06520

Contributed by Vernon W. Hughes, May 29, 1974

ABSTRACT **The atomic regime in which the interaction of the electron with an external magnetic field dominates the Coulomb interaction with the nucleus, relevant to pulsars, can be realized at laboratory magnetic fields for discrete autoionized states of hydrogen, at energies above the ionization limit. Approximate wave functions, energy levels, and electric dipole transition probabilities are presented for hydrogen, and an atomic beam absorption spectroscopy experiment at 50 kG is proposed to study this new regime.**

Recent astrophysical evidence indicates the existence of intense magnetic fields B of the order of 10^{13} G (1 G = 10^{-4} T, tesla) on pulsars (1), and in this connection* the behavior of atoms has been discussed theoretically (2–11). From the atomic physics viewpoint these intense magnetic fields provide a new atomic regime in which the magnetic interaction dominates the Coulomb interaction. For example, the energy spectrum of hydrogen in an intense magnetic field is that of an electron in the magnetic field perturbed by the relatively small Coulomb interaction. This regime is achieved for the ground state of hydrogen for B of order 5×10^9 G, at which value the cyclotron radius is less than the Bohr radius. It would be interesting to study this new atomic regime in the laboratory. At any value of the magnetic field this regime is realized, e.g., for hydrogen, in a state of sufficiently high excitation (12–17).

The nonrelativistic Schrödinger equation for the relative motion of a hydrogen-like atom or ion in a uniform magnetic field taken along the z axis is approximately (18)†

$$H\Psi = \left[\frac{p^2}{2\mu} + \frac{1}{2}\omega_c l_z + \frac{1}{8}\mu\omega_c^2\rho^2 - \frac{Ze^2}{r}\right]\Psi = E\Psi, \quad [1]$$

where ρ and ϕ are cylindrical coordinates, e is the magnitude of the electron charge, Ze is the nuclear charge, μ is the reduced mass, ω_c is the angular Larmor frequency ($\omega_c \equiv eB/\mu c$) and l_z is the z component of the angular momentum. l_z commutes with H and hence the magnetic quantum number m is a good quantum number. Parity is also conserved. Apart from the ϕ coordinate this equation is nonseparable.

For the intense magnetic field regime, the well known adiabatic approach, described initially by Schiff and Snyder (13), is applicable in which the fast motion in the plane perpendicular to the z-direction is considered separable from the relatively slow z-motion and the Coulomb interaction is neglected with regard to the perpendicular motion. The normalized eigenfunctions for the perpendicular motion of the electron in the magnetic field are (19)

$$\Phi_{n_\rho m}(\rho,\phi) = \left[\frac{n_\rho!}{2\pi r_c^2(|m| + n_\rho)!}\right]^{1/2} e^{-\rho^2/4r_c^2} \times \left(\frac{\rho^2}{2r_c^2}\right)^{|m|/2} L_{n_\rho}^{|m|}(\rho^2/2r_c^2)e^{im\phi}, \quad [2]$$

in which $r_c \equiv (\hbar c/eB)^{1/2}$ is the cyclotron radius, $L_{n_\rho}^{|m|}$ is the generalized Laguerre polynomial, $n_\rho = 0,1,2\ldots$, and $m = 0, \pm1, \pm2, \ldots$. The corresponding infinitely degenerate eigenenergies are

$$E^{(0)}_{n_\rho m} = (\hbar\omega_c/2)(2n_\rho + |m| + m + 1). \quad [3]$$

Solutions of Eq. **1** are obtained in the form

$$\Psi = \Phi_{n_\rho m}(\rho,\phi)f(z). \quad [4]$$

Substitution of **4** into Eq. **1**, multiplication by $\Phi^*_{n_\rho m}$, and integration over the ρ, ϕ variables leads to a one-dimensional Schrödinger equation for f

$$\left[\frac{p_z^2}{2\mu} + V_{n_\rho|m|}(z)\right]f(z) = \epsilon f(z), \quad [5]$$

in which

$$V_{n_\rho|m|}(z) = -Ze^2\iint|\Phi_{n_\rho m}(\rho,\phi)|^2(\rho^2 + z^2)^{-1/2}\rho d\rho d\phi. \quad [6]$$

The discrete eigenfunctions and eigenenergies of Eq. **5** are written as $f_{n_\rho m n_z}(z)$ and $\epsilon_{n_\rho m n_z}$ where $n_z = 0,1,2,\ldots$ orders the solutions of Eq. **5** in increasing energy and z nodes. They are infinite in number since asymptotically the leading term in $V_{n_\rho|m|}$ approaches $-Ze^2/z$. There is also a continuum of solutions of Eq. **5** for $\epsilon > 0$, for unbounded motion along z. The discrete energy eigenvalues of Eq. **1** in the intense magnetic field regime are then written approximately

$$E_{n_\rho m n_z} = E^{(0)}_{n_\rho m} + \epsilon_{n_\rho m n_z}. \quad [7]$$

If we start with the Pauli equation rather than the Schrödinger equation, then for the magnetic field strengths of interest here (10^4–10^5 G) there is a complete Paschen-Back effect with the spin orbit interaction small relative to the interaction of the spin with the magnetic field and all the eigenfunctions of the Pauli equation can be represented in a first approximation by a product of a spatial and spin wave-

* This problem has also received considerable attention in a solid state context. See, in particular, refs. 10 and 11, and references therein.

† Although in the presence of the magnetic field there is no longer translational invariance, the center of mass and relative motions still separate under certain conditions, and if we neglect a term of order of the electron to proton mass we obtain Eq. 1. See ref. 18.

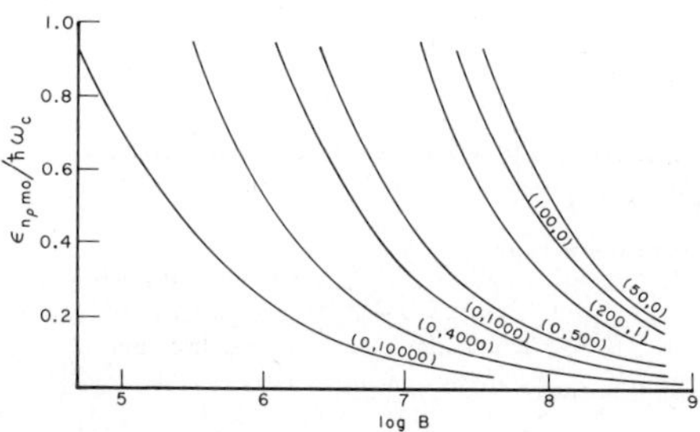

FIG. 1. Variational estimates of the eigenvalues $\epsilon_{n_\rho m0}$ in units of $\hbar\omega_c$ for particular values of (n_ρ, $|m|$) as a function of magnetic field strength B in Gauss.

function. The electron spin magnetic quantum number $m_s = \pm 1/2$ is then a good quantum number. Other relativistic corrections to the energies are negligible for this intense magnetic field regime. We then obtain approximately an additional energy term $\hbar\omega_c m_s$ in the total energy eigenvalues.

Variational estimates of the eigenvalues $\epsilon_{n_\rho m n_z}$ for the lowest even ($n_z = 0$) and odd ($n_z = 1$) states of Eq. (**5**) for very high values of n_ρ and $|m|$ have been obtained with the trial function‡

$$f_t = N_z^P(z)^P e^{-\alpha z^2}(1 + \beta z^2), \qquad [8]$$

where α and β are variational parameters, $P = 0,1$ for even and odd solutions and N_z^P are the normalization factors. [The parity of solutions **4** is $(-1)^{m+P}$] Results for $\epsilon_{n_\rho m0}$ with $Z = 1$ are shown in Fig. 1. The adiabatic approximation is valid, that is $\epsilon_{n_\rho m0} < \hbar\omega_c$, provided n_ρ and $|m|$ are large enough. A rough estimate of this limit is obtained by requiring the extension of wave function **4**, approximated by the expectation $\langle\rho^2\rangle^{1/2}$, to be larger than the radius of the Coulomb state with a binding energy equal in magnitude to $\hbar\omega_c$, leading to the condition

$$(2n_\rho + |m| + 1) > (Z^2 r_c^2/8a_o^2). \qquad [9]$$

Fig. 2 is a schematic energy spectrum of hydrogen, Eq. **1**, with B taken to be 50 kG. The lowest levels represent conventional Coulomb states, with n, l, m taken as good quantum numbers and with small linear Zeeman corrections to the energies. As n increases the diamagnetic term $^1/_2\mu\omega_c\rho^2$ eventually causes significant mixing of states with differing l and n. Perturbative treatments break down when $\hbar\omega_c$ is comparable to the energy separation between Coulomb states with adjacent values of n. At 50 kG, $\hbar\omega_c$ is about 6×10^{-4} eV and the breakdown occurs at n of about 40.

At higher energies, near the B = 0 ionization limit, an intermediate regime of states exists where magnetic and Coulomb interactions are comparable. The electron's motion along the z axis and along the ρ direction are presumably strongly coupled. The energies of these highly excited states are now being estimated using a two-dimensional WKB method for nonseparable potentials. A recent experiment (14) measured the principal series of barium with B = 25 kG, and reached the intermediate regime for this complex atom. The results have also been considered theoretically (15).

‡ Discrete solutions $f_{n_\rho m n_z}$ fall off exponentially for $z \to \infty$, but the Gaussian form permits analytic evaluation of most integrals.

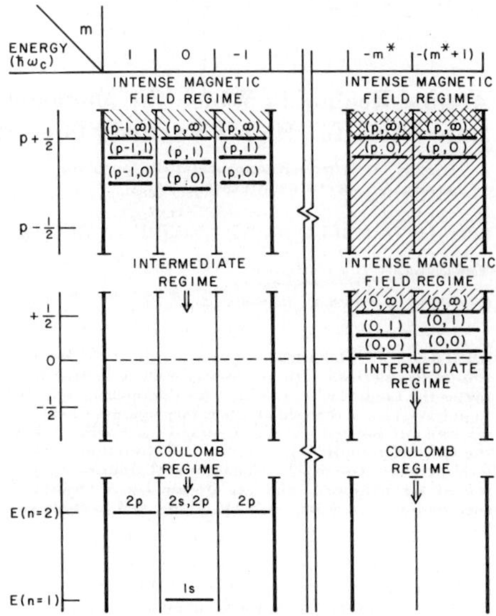

FIG. 2. Schematic diagram of the energy spectrum of hydrogen in a magnetic field. Values of m greater than 1 are not presented since the results are analogous to those for m negative, except that the magnetic eigenenergies, $E^{(0)}_{n_\rho m}$, increase with positive m. m^* is a value of $|m|$ that satisfies the inequality **9** with $n_\rho = 0$. Levels in the Coulomb regime are designated by (n,l), while in the intense magnetic field regime by (n_ρ,n_z). p is a positive integer large enough so that all values of n_ρ and m that give this magnetic eigenenergy also satisfy inequality **9**. The *shaded regions* refer to the continua of unbounded motion along the z axis.

The ionization continuum begins at $\hbar\omega_c/2$ (the zeropoint interaction energy of an electron with B) with the electron first becoming unbound along z. The electron's motion is confined normal to B, which drastically changes the character of continuum solutions from those with B = 0. At energies above $\hbar\omega_c/2$ there are discrete and continuum states which for $z \to \infty$ lead to excited solutions $\Phi_{n_\rho m}$ for motion normal to B; for each m, an infinite number of such discrete states converge on the levels $E^{(0)}_{n_\rho m}$ and each level marks the onset of a corresponding continuum. Such discrete states are hence embedded in continua with less excited asymptotic boundary conditions and will autoionize.

The intense magnetic field regime sets in at energies $E^{(0)}_{n_\rho m}$ where n_ρ and m satisfy the adiabatic inequality **9**. If m is large enough and negative this occurs at $\hbar\omega_c/2$. If $\hbar\omega_c > e^2/2a_o$ (pulsar case) the intense regime encompasses the entire spectrum and all states are beyond the B = 0 ionization limit. In higher order the Coulomb interaction couples adiabatic states of the same parity and m. Rough estimates of the autoionization width (20), $\Gamma = 2\pi|(\Psi', - Ze^2/r\Psi)|^2$, for the discrete state $\Psi_{n_\rho m0}$ due to configuration interaction with each underlying continuum $\Psi'_{n_\rho^* m p_z}$, where $n_\rho^* < n_\rho$ and the continuum solution for $f(z)$ has the appropriate energy

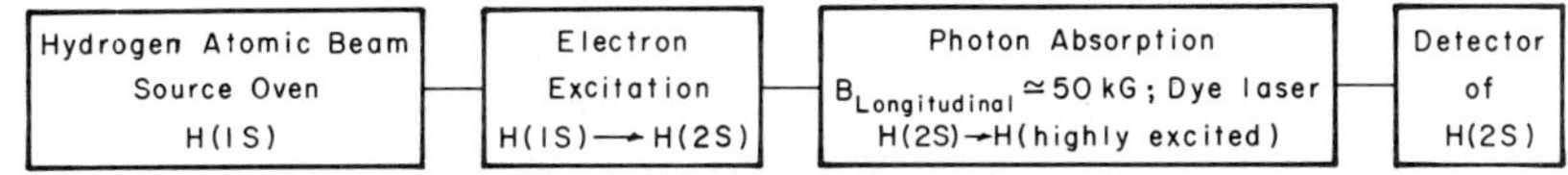

FIG. 3. Block diagram of experimental method of atomic beam absorption spectroscopy to study hydrogen in the intense magnetic field regime.

$p_z^2/2u$, indicate that even for the largest case ($n_\rho^* = n_\rho - 1$) the widths are extremely small relative to $\hbar\omega_c$.

The probability of excitation, by whatever process, to states in the intense magnetic field regime depends, as a function of energy, on interference between the discrete states and underlying continua. We have evaluated the electric dipole transition matrix element§, $\gamma = (\Psi_{n_\rho m n_z}, \boldsymbol{\varepsilon}\cdot\mathbf{r}\ \Psi_{nlm'})$, where, since ϕ symmetry is preserved, the basic polarizations are taken to be $\sigma(\Delta m = \pm 1)$ and $\pi(\Delta m = 0)$. The excitation probability is then¶ $w = 4\pi^2\alpha\ \omega\ N(\omega)\ |\gamma|^2\delta(m_{s'},m_s)$, where α is the fine structure constant, $\omega = (E_{n_\rho m n_z} - E_{nlm'})/\hbar$, the delta function specifies that m_s is unchanged and $N(\omega)$ is the number of incident photons per unit angular frequency, area, and time, which for simplicity here is assumed constant over the line width.

We present approximate results for the probability of excitation from the 2S state ($l = m' = 0$) to states $\Psi_{n_\rho m n_z}$ with $n_z = 0,1$. The corresponding functions $f_{n_\rho m n_z}$ are approximated by f_t, and the value of α that minimizes the energy is taken to be approximately $1/(4n_\rho r_c^2)$. The extension of $\Psi_{n_\rho m n_z}$ is much larger than the 2S state, and if we expand $\Psi_{n_\rho m n_z}$ about the origin the leading terms for the transition probabilities are

$$w^\sigma = (10^3)(a_0/r_c)^5(n_\rho)^{1/2}N(\omega)\ \delta(m, \pm 1)\delta(P,0)\ \text{sec}^{-1} \quad \textbf{[10a]}$$

$$w^\pi = (6 \times 10^5)(a_0/r_c)^7(n_\rho)^{-1/2}N(\omega)\delta(m,0) \times \delta(P,1)\ \text{sec}^{-1}, \quad \textbf{[10b]}$$

where $N(\omega)$ now has inverse cm² units, and the delta functions specify selection rules. w^π is reduced by the factor $(a_0^2/n_\rho r_c^2)$ relative to w^σ, which comes about because the odd solution $f_{n_\rho 01}$ approaches zero near the origin. Similarly, transition probabilities to higher states ($n_z \geq 2$) are greatly reduced.

A schematic diagram of a proposed atomic beam absorption spectroscopy experiment to measure precisely the spectrum of energy levels of hydrogen in the intermediate and intense magnetic field regimes is shown in Fig. 3. The initial hydrogenic state is chosen to be the metastable 2S state. A longitudinal magnetic field of 50 kG is provided by a superconducting solenoid about 10 cm in length. To avoid Stark quenching of the 2S state due to the motional electric field $(\mathbf{v} \times \mathbf{B})/c$, the atom's velocity must be parallel to $\mathbf{B}$ within about 3 mrad. The energy needed to excite from the 2S state to a state in the intermediate or intense magnetic field regimes is about 3.4 eV, and the upper level separations are about 6×10^{-4} eV in the magnetic field of 50 kG. This wavelength range with the required resolution of 10^{-6} to 10^{-5} is available from tunable dye lasers together with frequency doubling. For a transition to a state $(n_\rho,|m|) = (4000,1)$ in the intense magnetic field regime a transition rate of about 10^6 sec^{-1} is obtained with an achievable continuous wave (cw) energy flux of 10 μW/cm² incident on a 2S state H atom. For an atom with a velocity of 8×10^5 cm/sec an irradiation time of about 10^{-6} sec seems reasonable, corresponding to a transition probability of unity. To avoid Doppler line broadening the radiation is incident perpendicular to the beam direction. From the upper state the atom will either radiatively decay or autoionize into a continuum state yielding a free electron. The observed signal would be a decrease in the intensity of the detected 2S metastable H beam. In view of the enormous dimensions of the intense magnetic field states ($\sim 10^{-4}$ cm), it will be necessary to use ultrahigh vacuum of 10^{-10} torr or less to avoid excessive level shifts and collision broadening.

This proposed experiment could also be done on helium, using either the metastable 2^1S_0 or 2^3S_1 state as the initial state, or indeed with many other atoms or molecules that have suitable metastable states.

This research was supported in part by the Air Force Office of Scientific Research, AFSC, under Contract no. F44620-70-C-0091.

§ The basic dipole velocity matrix element is $(\Psi_{n_\rho m n_z}, \boldsymbol{\varepsilon}\cdot(\mathbf{p} + (e/c)\mathbf{A})\Psi_{nlm})$, where $\mathbf{A}$ is the vector potential. With the commutation relation $[r,H] = -i\hbar/\mu\ (\mathbf{p} + e\mathbf{A}/c)$ this matrix element can be rewritten

$$(\Psi_{n_\rho m n_z}, \boldsymbol{\varepsilon}\cdot\mathbf{r}[(E_{nlm} + (1/2)\mu\omega_c^2\rho^2) - (E_{n_\rho m n_z} + Ze^2/r - V_{n_\rho|m|}(z))]\Psi_{nlm}),$$

and to lowest order this reduces to the conventional dipole length matrix element.

¶ This matrix element actually represents the excess transition probability of the discrete level over that of the underlying continua (see ref. 20). The discrete transition probabilities that we have calculated are particularly large for intense field states with $n_z = 0$; rough estimates of the transition probabilities to the underlying continua, averaged over the level width, are much less in comparison.

1. Ruderman, M. (1972) *Annu. Rev. Astron. Astrophys.* **10,** 427–476.
2. Cohen, R., Lodenquai, J. & Ruderman, M. (1970) *Phys. Rev. Lett.* **25,** 467–469.
3. Kodomtsev, B. B. (1970) *Zh. Eksp. Theor. Fiz.* **58,** 1765–1769. [*Sov. Phys.-JETP* **31,** 945–947.]
4. Mueller, R. O., Rau, A. R. P. & Spruch, L. (1971) *Phys. Rev. Lett.* **26,** 1136–1139.
5. Edmonds, A. R. (1973) *J. Phys. Sect. B* **6,** 1603–1615.
6. Barbieri, R. (1971) *Nucl. Phys.* **A161,** 1–11.
7. Praddaude, H. C. (1972) *Phys. Rev.* **A6,** 1321–1324.
8. Callaway, J. (1972) *Phys. Lett.* **40A,** 331–332.
9. Smith, E. R., Henry, R. J. W., Surmelian, G. L., O'Connell, R. F. & Rajagopal, A. K. (1972) *Phys. Rev.* **D6,** 3700–3701.
10. Cabib, D., Fabri, E. & Fiorio, G. (1972) *Nuovo Cimento* **10B,** 185–199.
11. Hasegawa, H. & Howard, R. E. (1961) *J. Phys. Chem. Solids* **21,** 179–198.
12. Jenkins, F. A. & Segrè, E. (1939) *Phys. Rev.* **55,** 52–58.
13. Schiff, L. I. & Snyder, H. (1939) *Phys. Rev.* **55,** 59–63.
14. Garton, W. R. S. & Tomkins, F. S. (1969) *Astrophys. J.* **158,** 839–845.
15. Starace, A. F. (1973) *J. Phys. Sect. B* **6,** 585–590.
16. O'Connell, R. F. (1974) *Astrophys. J.* **187,** 275–277.
17. Hughes, V. W. & Mueller, R. O. (1973) *Bull. Amer. Phys. Soc.* **18,** 1509.
18. Lamb, W. (1952) *Phys. Rev.* **85,** 259–276.
19. Landau, L. D. & Lifshitz, E. M. (1965) *Quantum Mechanics-Non-Relativistic Theory* (Pergamon Press, London), 2nd ed., p. 426.
20. Fano, U. (1961) *Phys. Rev.* **124,** 1866–1878.

Volume 67B, number 4 PHYSICS LETTERS 25 April 1977

PARITY NONCONSERVATION IN HYDROGEN INVOLVING MAGNETIC/ELECTRIC RESONANCE★

E.A. HINDS and V.W. HUGHES

Gibbs Laboratory, Physics Department, Yale University, New Haven, Conn., USA

Received 3 January 1977

Parity nonconservation effects in the $n = 2$ state of hydrogen involving a simultaneous magnetic and electric microwave resonance transition are discussed as a possible test of the Weinberg–Salam theory.

The modern intense search for parity nonconserving (PNC) interactions in atomic physics [1] and in electromagnetic interactions has been stimulated by the Weinberg–Salam unified theory of weak and electromagnetic interactions, which predicts PNC effects of first order in the neutral current weak interaction.

An important type of experiment on hydrogen, which utilizes the level crossings of the 2S and 2P states at two magnetic fields and seeks to observe a PNC effect in laser absorption to the 3S state, has been discussed by Lewis and Williams [2]. We discuss here a different type of experiment on hydrogen which utilizes only microwave induced transitions within the $n = 2$ state and is of a general type suggested by Sandars [3].

The relevant portion of the energy level diagram for hydrogen in the $n = 2$ state [4] is shown in fig. 1. The general idea is to induce a transition from state α to state β as a magnetic dipole (M1) transitions and also as an electric dipole (E1) transition associated with an admixture of 2P state due to a PNC interaction. Interference between these two transitions is the quantity to be observed. The PNC interaction Hamiltonian is:

$$\mathcal{H}_{\mathrm{PNC}} = \frac{G}{\sqrt{2}}(C_1^{\mathrm{p}} \bar{\psi}_{\mathrm{e}}\gamma_\alpha\gamma_5\psi_{\mathrm{e}}\bar{\psi}_{\mathrm{p}}\gamma_\alpha\psi_{\mathrm{p}} + C_2^{\mathrm{p}} \bar{\psi}_{\mathrm{e}}\gamma_\alpha\psi_{\mathrm{e}}\bar{\psi}_{\mathrm{p}}\gamma_\alpha\gamma_5\psi_{\mathrm{p}}), \qquad (1)$$

in the notation of Feinberg and Chen [5] where C_1^{p} and C_2^{p} are dimensionless constants characterizing the strengths of the interactions. In the Weinberg/Salam model $C_1^{\mathrm{p}} = (1/2 - 2\sin^2\theta_{\mathrm{W}})$ and $C_2^{\mathrm{p}} = 1.2(1/2 - 2\sin^2\theta_{\mathrm{W}})$.

In order to discuss the method we consider specifically states $|1\rangle$ and $|4\rangle$. The perturbed states due to $\mathcal{H}_{\mathrm{PNC}}$ are:

$$|1\rangle' = |1\rangle + \delta_{1,5}|5\rangle; \; |4\rangle' = |4\rangle + \delta_{4,6}|6\rangle + \delta_{4,8}|8\rangle, \qquad (2)$$

in which, for example

$$\delta_{4,8} = \langle 8|\mathcal{H}_{\mathrm{PNC}}|4\rangle \left[E_4 - E_8 - \frac{i\hbar}{2}(\Gamma_4 - \Gamma_8)\right],$$

where E_4, E_8 are the energies and Γ_4, Γ_8 are the inverse lifetimes of states $|4\rangle$ and $|8\rangle$. At a field of 1100 gauss where levels 4 and 8 cross, $\delta_{1,5}$ and $\delta_{4,6}$ can be neglected and $\delta_{4,8} = -2.6 \times 10^{-10}(C_1^{\mathrm{p}} + 1.1\, C_2^{\mathrm{p}})$.

In the presence of applied oscillating magnetic and electric fields $B_x \cos\omega t$ and $E_y \cos(\omega t + \phi)$ the matrix elements for M1 and E1 transitions from state 1 to state 4 are approximately

$$b_{\mathrm{M}} = \mu_{\mathrm{B}} g_J B_x / 4\hbar = 4.4 \times 10^6 B_x \text{ (gauss) sec}^{-1};$$

$$b_{\mathrm{E}} = \delta_{4,8} \mathrm{i} \mathrm{e}^{-\mathrm{i}\phi}\sqrt{3} e a_0 E_y / 2\hbar$$

$$= -0.54\, \mathrm{i}\mathrm{e}^{-\mathrm{i}\phi}(C_1^{\mathrm{p}} + 1.1\, C_2^{\mathrm{p}}) E_y \text{ (esu) sec}^{-1}.$$

If an atom is in state 1 at time $t = 0$, the probability that it be in state 4 at time T is $P \simeq |b_{\mathrm{M}} + b_{\mathrm{E}}|^2 T^2$ at resonance and for small transition probability [6]. A method of measuring the M1–E1 interference term is to observe the change δP in P when ϕ is changed from $-\pi/2$ to $+\pi/2$.

$$\left(\frac{\delta P}{P}\right) = \frac{4|b_{\mathrm{E}}|}{|b_{\mathrm{M}}|} = 4.9 \times 10^{-7}(C_1^{\mathrm{p}} + 1.1\, C_2^{\mathrm{p}})\frac{E_y}{B_x}, \qquad (3)$$

in which only the dominant term $|b_{\mathrm{M}}|^2$ in P is retained.

★ Research supported in part by the National Science Foundation under Grant No. PHY75-02376 A01.

Volume 67B, number 4 PHYSICS LETTERS 25 April 1977

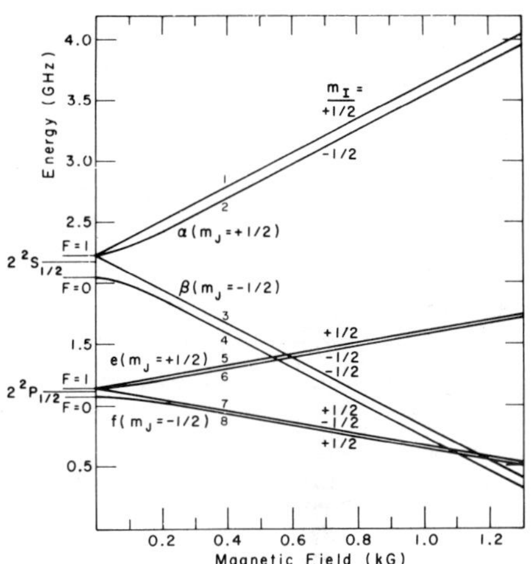

Fig. 1. The relevant energy levels of hydrogen in a magnetic field, drawn approximately to scale. The zero field hyperfine splittings are $a(2\,^2S_{1/2}) = 177.56$ MHz and $a(2\,^2P_{1/2}) = 59.2$ MHz. The Lamb shift is 1057.9 MHz.

Another method is to reverse the direction of the static magnetic field. The linewidth $\delta\nu$ of this transition will be $\delta\nu \sim 1/T$ and hence the transition can be resolved from the nearby $2 \rightarrow 3$ transition.

Other transitions in H at the level crossings can also be used, in particular $2 \rightarrow 3$ at the 3, 7 crossing for which $\delta P/P \sim 4.9 \times 10^{-7}(C_1^p - C_2^p)E_y/B_x$ and $2 \rightarrow 4$ at the 4, 6 crossing for which $\delta P/P = 9.1 \times 10^{-6}$ $C_2^p E_z/B_z$. In the case of transition $2 \rightarrow 4$, $\delta P/P$ is enhanced because the M1 transition is small.

A possible experimental method uses an intense beam of metastable hydrogen atoms travelling parallel to a static magnetic field. The β states can be removed by quenching, for example, in a transverse electric field at the β, e crossing. The beam then enters a microwave cavity where B_x and E_y induce the desired $\alpha \rightarrow \beta$ transition. A large ratio E_y/B_x will enhance $\delta P/P$ in (3); with a suitable design of microwave cavity one can obtain $E_y/B_x \sim 15$. We also wish to make B_x and hence E_y as large as possible in order to maximize P. We find that an effective upper limit on E_y is set by the rate at which atoms in the state α are quenched; hence P will be small.

In principle this type of experiment can determine both C_1^p and C_2^p. For hydrogen, of course, there are no ambiguities associated with the atomic calculations. The experiment can also be considered for deuterium to determine the constants C_1^n and C_2^n for the neutron. This experimental approach applies in principle to any hydrogen-like atom, and, in particular, the experimental parameters would be of the same general order of magnitude for $^4He^+$ and $^3He^+$. For a muonic atom such as μ^-p or $(\mu^-\,^4He)^+$ the magnetic fields required for level crossings are too large to be attainable in the laboratory.

We gratefully acknowledge clarifying discussions with W.L. Williams.

References

[1] M.A. Bouchiat and C. Bouchiat, J. Phys. (Paris) 35 (1974) 899.
[2] R.R. Lewis and W.L. Williams, Phys. Lett. 59B (1975) 70.
[3] P.G.H. Sandars, in: Atomic Physics, vol. 4, ed. G. zu Putlitz, E.W. Weber and A. Winnacker (Plenum Press, New York, 1975) p. 71.
[4] W.E. Lamb, Jr., Phys. Rev. 85 (1952) 259; W.E. Lamb, Jr., and R.C. Retherford, Phys. Rev. 79 (1950) 549.
[5] G. Feinberg and M.Y. Chen, Phys. Rev. D10 (1974) 190; Phys. Rev. D10 (1974) 3145; Phys. Rev. D10 (1974) 3789.
[6] P. Kusch and V. Hughes, in: Handbuch der Physik, vol. 37/1, ed. S. Flugge/Marburg (Springer-Verlag, Berlin, 1959) p. 1.

THE FINE STRUCTURE CONSTANT α

Vernon W. Hughes

Gibbs Laboratory
Physics Department
Yale University
New Haven, Connecticut 06520

I very much appreciate the opportunity to contribute an article for a volume in honor of Professor Rabi. Since so much of Professor Rabi's research involved the electromagnetic interaction both as a study of its own foundations and as a tool to learn about nuclei and elementary particles, I hope it is appropriate to review the history and status of our knowledge of the famous fine structure constant α, which characterizes the strength of the electromagnetic interaction.

Early History

The fine structure constant α can be defined by the relation:

$$\alpha = \frac{e^2}{\hbar c} \tag{1}$$

in which e is the magnitude of the charge on an electron, $\hbar$ is Planck's constant h divided by 2π, and c is the velocity of light. The quantity α is dimensionless and is approximately equal to $\frac{1}{137}$. Clearly α involves constants that characterize the discreteness of electric charge (e), quantum theory (h), and relativity theory (c).

Apparently the first suggestions[1,2] that this constant might be interesting were implied by Planck and Einstein in the period from 1905 to 1910. They noted that the elementary quantum of energy h has the same dimensions as e^2/c and approximately the same order of magnitude. They suggested that in a comprehensive theory there might be a relationship between the quantum structure of radiation and the elementary charge.

The fine structure constant was first introduced explicitly and related usefully to atomic theory by Sommerfeld.[3-6] He extended the old Bohr quantum theory of a hydrogen-like atom to include relativistic effects by solving the problem of relativistic Kepler motion together with the old quantum conditions. Equation 2 shows the resulting Sommerfeld formula for the energy levels, in which k is the azimuthal quantum number ($k = 1, 2, \ldots$), n_r is the radial quantum number ($n_r = 0, 1, 2, \ldots$), and n is the total quantum number. The quantity $\alpha(\alpha = e^2/\hbar c)$ occurs in the formula and was called the fine structure constant because it appears in the term of order $\alpha^2 Ry$ in the expansion of W which accounts for atomic fine structure. The Dirac theory gives the same formula (Equation 3) for the energy levels, but the interpretation of the quantum numbers is different. The quantum number k is given by: $k = j + \frac{1}{2}, (j = \frac{1}{2}, \frac{3}{2}, \ldots)$,

Reprinted from
TRANSACTIONS OF THE NEW YORK ACADEMY OF SCIENCES
Series II Volume 38 Pages 62–76
November 4, 1977 25448

$$E = W + mc^2$$

$$W = mc^2\left\{\left[1 + \frac{\alpha^2 Z^2}{(n - k + \sqrt{k^2 - \alpha^2 Z^2})^2}\right]^{-1/2} - 1\right\}$$

(Sommerfeld Formula) (**2**)

$$\oint_0^{2\pi} P_\phi \mathrm{d}\phi = kh; \oint P_r \mathrm{d}r = n_r h$$

$$n = n_r + k$$

$$W \simeq -Ry\frac{Z^2}{n^2} - \alpha^2 Ry \frac{Z^4}{n^3}\left(\frac{1}{k} - \frac{3}{4n}\right)$$

$$W = mc^2\left\{\left[1 + \frac{\alpha^2 Z^2}{(n - k + \sqrt{k^2 - a^2 Z^2})^2}\right]^{-1/2} - 1\right\} \quad \text{(Dirac Formula)} \quad (3)$$

$$k = j + \frac{1}{2},$$

in which j is the total electronic angular momentum quantum number, including electron spin, which was not present in the old quantum theory.

We consider the comparison of Sommerfeld's fine structure theory with experiment. At the time of Sommerfeld's theory (1915), the constant α was known[7,8] to about 3 parts in 10^3 from independent measurements of c, e, and h/e, and the Rydberg constant Ry was known with much greater accuracy, so that the fine structure intervals could be calculated with an accuracy of better than 1%. Optical measurements[9] of fine structure intervals, principally for H and He^+, were of comparable accuracy and were in excellent agreement with the theoretical predictions from Sommerfeld's formula both as to the number and the spacing of the fine structure components. Later, after the development of quantum mechanics and the Dirac theory of the electron, the theory of fine structure was assumed to be correct, and α was determined from optical measurements of the fine structure of hydrogenic atoms.

Since World War II, the value of α has been based primarily on high precision radiofrequency and microwave spectroscopy measurements of atoms or solids. The original method of radiofrequency spectroscopy was, of course, Professor Rabi's molecular beam magnetic resonance method,[10] and much of the subsequent work in this field of spectroscopy is closely related in viewpoint and in spirit to the molecular beam magnetic resonance method.

Principal Measurements of α

The six principal modern measurements that determine the present-day value of α are the ac Josephson effect, the fine structure of hydrogen, the fine structure of helium, the hyperfine structure of muonium, the hyperfine structure of hydrogen, and the g-value of the electron.

The most precise value of α is obtained at present from the measurements of e/h through the ac Josephson effect and of the gyromagnetic ratio of the proton, γ_p, using the identity

$$\alpha^{-1} = \left(\frac{c}{4R_\infty\gamma_p} \frac{\mu_p}{\mu_B^e} \frac{2e}{h}\right)^{1/2} \tag{4}$$

in which c = velocity of light, R_∞ = Rydberg constant, μ_p = proton magnetic moment, μ_B^e = electron Bohr magneton, e = electronic charge, and h = Planck's constant. The ac Josephson relation applies to a junction consisting of two superconductors weakly coupled by an insulator and reads[11-15]

$$h\nu = 2\,\text{eV}, \tag{5}$$

where V is the dc potential difference across the junction and ν is the frequency of the alternating supercurrent. This relation is based on the viewpoint that the superconducting state is a highly correlated phase-coherent quantum state of macroscopic scale and that the supercurrent is due to the tunnelling of bound electron pairs. Theoretically, Equation 5 is believed to be general, exact, and independent of the detailed properties of the materials, geometry, or environment of the system. The experimental measurement of e/h involves application of a microwave frequency to the junction in order to zero beat with the ac supercurrent and observation of the resulting structure in the I-V dc curve. Measurements with various junctions of different materials and geometries have shown that the relative values e/h agree within the experimental error of about 1 part in 10^8. The value obtained for e/h is

$$2e/h = 4.835\,934\,20 \times 10^{14}\,\text{Hz}\,\text{V}_{\text{NBS}}^{-1}\,(0.030\,\text{ppm}) \tag{6}$$

in which the error is mainly associated with voltage measurements.

Recently a high precision measurement[16] of the gyromagnetic ratio of the proton in H_2O, γ_p', has been made by a nuclear induction experiment using a weak magnetic field of $12G$ produced by a solenoid. Electrical methods were employed to measure the effective dimensions of the solenoid to ppm level accuracy. The result quoted is

$$\gamma'_{p\text{NBS}} = 2.675\,131\,4(11) \times 10^8\,\text{rad}\,\text{s}^{-1}\,\text{T}_{\text{NBS}}^{-1}(0.42\,\text{ppm}). \tag{7}$$

The magnetic shielding correction is well known for H_2O so γ_p for the free proton can be given with essentially the same accuracy. We find by using in Equation 4 the values of e/h and γ_p from Equations 6 and 7 together with values of the other fundamental constants[15]

$$\begin{aligned} c &= 2.997\,924\,58(1.2) \times 10^{10}\,\text{cm/sec}\,(0.004\,\text{ppm}) \\ R_\infty &= 1.097\,373\,143(10) \times 10^5\,\text{cm}^{-1}(0.01\,\text{ppm}) \\ \mu_p/\mu_B^e &= 1.521\,032\,209(16) \times 10^{-3}(0.01\,\text{ppm}) \end{aligned} \tag{8}$$

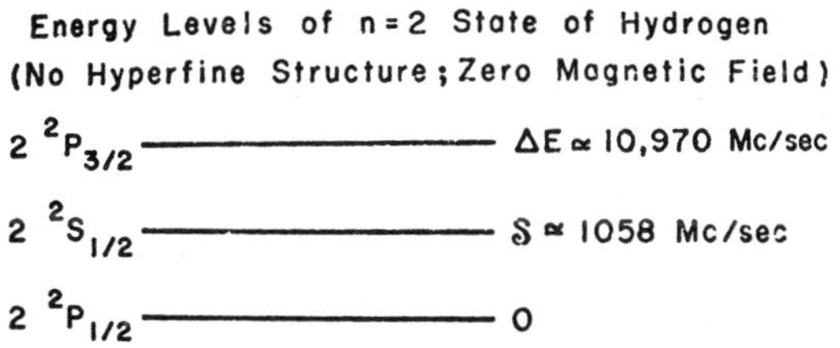

Energy Levels of n=2 State of Hydrogen
(No Hyperfine Structure; Zero Magnetic Field)

$2\,^2P_{3/2}$ ———————— $\Delta E \approx 10{,}970$ Mc/sec

$2\,^2S_{1/2}$ ———————— $S \approx 1058$ Mc/sec

$2\,^2P_{1/2}$ ———————— 0

FIGURE 1. Fine structure of hydrogen.

gives

$$\alpha^{-1} = 137.035\ 987(29)\ (0.21\ \text{ppm}). \tag{9}$$

One of the earliest precision determinations of α came from the classic experiments of Lamb and his colleagues on the fine structure in the $n = 2$ state of hydrogen (FIGURE 1). The theoretical formula for the $2^2P_{3/2}$–$2^2P_{1/2}$ fine structure interval for a hydrogen-like atom is given by

$$\Delta E(2^2P_{3/2} - 2^2P_{1/2}) = \frac{R_\infty(Z\alpha)^2}{16}\left\{\left[1 + \frac{5}{8}(Z\alpha)^2\right]\left(1 + \frac{m}{M}\right)^{-1} - \left(\frac{m}{M}\right)^2 \cdot \left(1 + \frac{m}{M}\right)^{-3} + 2a_e\left(1 + \frac{m}{M}\right)^{-2} + \frac{\alpha}{\pi}(Z\alpha)^2 \ln Z\alpha\right\}, \tag{10}$$

in which m is the electron mass, M is the nuclear mass, and a_e is the anomalous g-value of the electron. The most accurate direct measurement of this interval in hydrogen has been made by an optical-pumping level-crossing method[15,17] giving

$$\Delta E(\text{H}, 2^2P_{3/2} - 2^2P_{1/2}) = 10\ 969.127(87)\ \text{MHz}\ (7.9\ \text{ppm}) \tag{11}$$

Using Equations 10 and 11 we find[15]

$$\alpha^{-1} = 137.035\ 44(52)\ (3.9\ \text{ppm}) \tag{12}$$

Other indirect determinations of this interval have been made using measured values of both the ($2^2P_{3/2}$–$2^2S_{1/2}$) and the ($2^2S_{1/2}$–$2^2P_{1/2}$) intervals.[15]

Study of the fine structure of helium in the 2^3P state has led to a precision determination of α.[18] The relevant energy levels for helium are shown in FIGURE 2. The 2^3S_1 state is 19.8 eV above the ground $^1S_0$ state and is metastable. The 2^3P state lies about 1 eV or 9231 cm^{-1} above the 2^3S_1 state. The fine structure is inverted with the $J = 0$ level being highest; the interval between the $J = 0$ and $J = 1$ levels is about 30 GHz and that between the $J = 1$ and $J = 2$ levels is about 2.3 GHz. The principal reason that a study of helium fine structure may have an advantage relative to hydrogen fine structure for determining α is that the lifetime

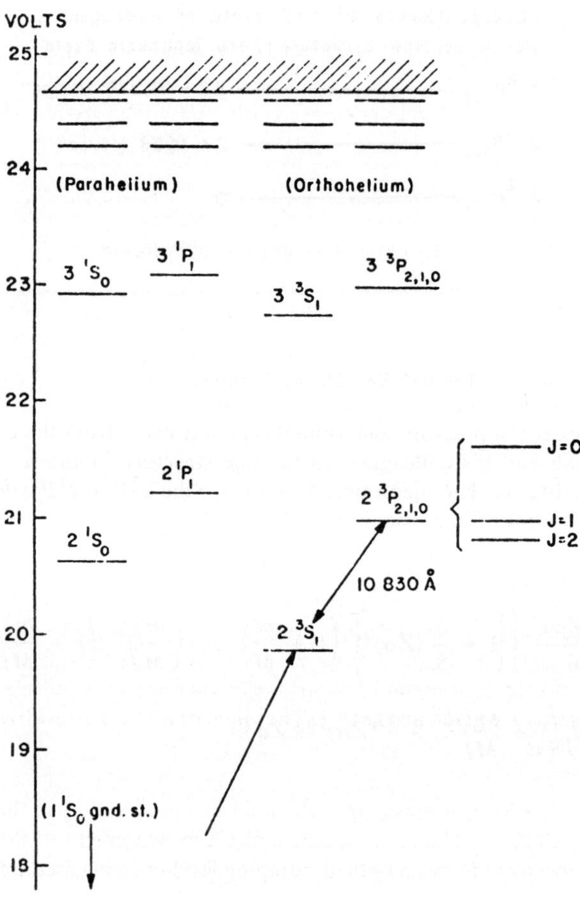

FIGURE 2. Helium energy levels.

of the 2^3P state of helium is about 100 times longer (10^{-7} sec) than that of the $2P$ state of hydrogen, and hence the ratio of the fine structure interval to the natural line width is about 100 times larger for He in the 2^3P state than for H in the $2P$ state. The helium fine structure interval $J = 0$ to $J = 1$ in the 2^3P state has been measured to a precision of 1.2 ppm, which is about an order of magnitude better than that achieved for the hydrogen fine structure interval $2^2P_{3/2}$–$2^2P_{1/2}$.

The experimental method, indicated in FIGURE 3, is the atomic beam magnetic resonance method adapted to optically excited states, which was first proposed by Professor Rabi and was first applied by Perl *et al.*[19] to measure the hyperfine structure of Na in the $3P$ state. In the He measurements done at Yale[20–22] He atoms in the ground 1^1S_0 state are excited by electron bombardment to the metastable 2^3S_1 state, and then pass through a conventional atomic beam magnetic resonance apparatus with inhomogeneous A- and B-fields and a homogeneous

C-field. In the C-region resonance optical radiation induces the transition from the 2^3S to the 2^3P state and, in addition, a microwave magnetic field induces a transition between different 2^3P fine structure levels. The microwave-induced resonance line has a microwave power-broadened width of about 5 MHz. The measured values for the $J = 0$ to $J = 1$ and $J = 1$ to $J = 2$ intervals in the 2^3P state are given in TABLE 1.

The theory for He fine structure is of course much more difficult than that for H fine structure. The energy levels in helium can be expressed by a power series in α

$$E_J = E_0 + \alpha^2 \langle \mathcal{H}_2 \rangle_J + 0(\alpha^3) + \alpha^4 \langle \mathcal{H}_2 \frac{1}{E_0 - \mathcal{H}_0} \mathcal{H}_2 \rangle_J$$

$$+ \alpha^4 \langle \mathcal{H}_4 \rangle_J + 0\left(\alpha^2 \frac{m}{M_{\mathrm{He}}}\right), \tag{13}$$

in which E_0 is the Schroedinger term, $\mathcal{H}_2$ is the Breit interaction which contributes in first and second order perturbation theory, $\mathcal{H}_4$ is a higher-order spin dependent operator derived from the covariant two-particle Bethe-Salpeter equation, the term $O(\alpha^3)$ arises from the electron anomalous magnetic moment, and the term $O(\alpha^2 m/M_{\mathrm{He}})$ arises from the nuclear recoil. Elaborate theoretical calculations have been made using extensive Hylleraas-type wave functions. A survey of these calculations is given by Lewis[23] and the latest theoretical results are given by Lewis *et al.*[24] and by Lewis and Serafino.[25] The theoretical values of the He fine structure intervals are given in TABLE 1. The value of α^{-1} obtained from the $J = 0$ to $J = 1$ interval is

$$\alpha^{-1} = 137.036\,08(13)\ (0.94\ \mathrm{ppm}). \tag{14}$$

The error is determined about equally from the experimental and theoretical inaccuracies. Efforts are in progress to improve both the experimental and theoreti-

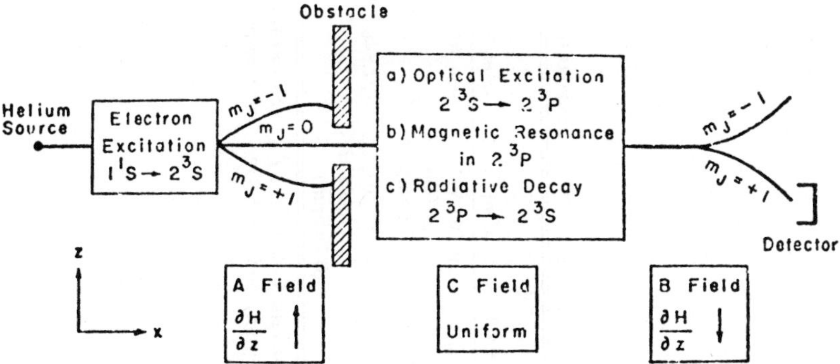

FIGURE 3. Schematic diagram of helium fine structure experiment.

TABLE 1

THEORETICAL CONTRIBUTIONS TO THE FINE STRUCTURE OF 2^3P HELIUM (IN MHz)*

Interval	$\alpha^4 mc^2$	$\alpha^5 mc^2$	$\left(\frac{m}{M}\right)\alpha^4 mc^2$	Second Order	$\alpha^6 mc^2$	ν_{theory}	ν_{exp}	$\nu_{theory} - \nu_{expt}$
ν_{01}	29 564.577 ±0.006 (0.21 ppm)	54.708	−10.707 ±0.000 44 (0.015 ppm)	11.657 ±0.042 (1.42 ppm)	−3.331 ±0.003 9 (0.13 ppm)	29 616.904 ±0.043 (1.44 ppm)	29 616.864 ±0.036 (1.2 ppm)	0.040 = 1.35 ppm
ν_{12}	2 317.203 ±0.001 8 (0.76 ppm)	−22.548	1.952 ±0.000 88 (0.39 ppm)	−6.866 ±0.081 (35 ppm)	1.542 ±0.006 8 (3.0 ppm)	2 291.283 ±0.081 (35 ppm)	2 291.196 ±0.005 (2.2 ppm)	0.087 = 37 ppm

*The values of α^{-1}, c, R_∞, and (m/M) are 137.035 987(29) (0.21 ppm), 2.997 924 58(12) × 10^{10} cm/sec (0.004 ppm), 109 737.314 3 cm^{-1} (0.009 ppm), and 1.370 934 × 10^{-4}, respectively. Thus $(1/2)\,\alpha^2 c R_\infty$ = 87.594 28 GHz (0.42 ppm).

cal values of He fine structure intervals, so that a still more precise value of α will be obtained.

Muonium (μ^+e^-) is the simple hydrogen like atom consisting of a positive muon and an electron. Since the muon and the electron are structureless Dirac particles, the theory of the energy levels of muonium is a well-defined quantum electrodynamic problem. Hence study of its hyperfine structure interval provides an opportunity for a precise determination of α. FIGURE 4 shows the energy level diagram for the ground $n = 1$ state hfs levels in a magnetic field as given by the famous Breit-Rabi formula.[26] The microwave magnetic resonance method has been used to measure various transitions between these levels at both weak and strong magnetic field.[27] The most recent and precise measurement has been done

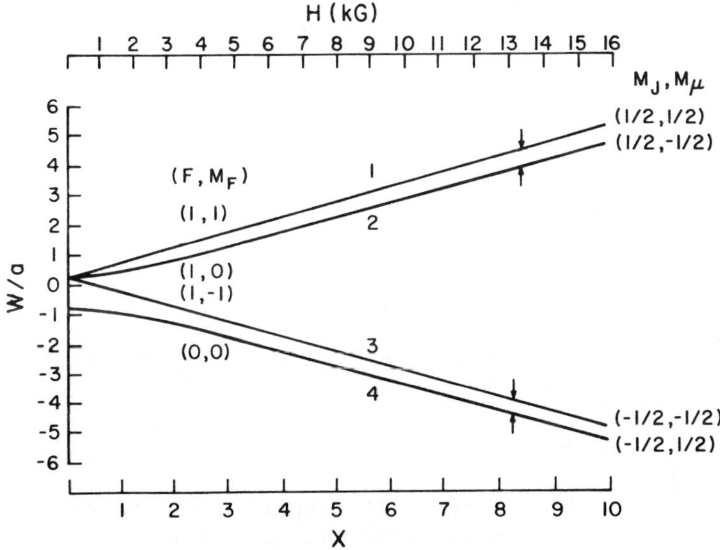

FIGURE 4. Energy levels for muonium in its $1^2S_{1/2}$ ground state in a magnetic field. The energy difference between $F = 1$ and $F = 0$ levels at $H = 0$ is the hfs interval $\Delta\nu$.

at a strong magnetic field of 13.6 kG, using the intense muon beam at the Los Alamos Meson Factory (LAMPF).[28] A resonance line is shown in FIGURE 5. The values of the hfs interval $\Delta\nu$ and of the ratio of muon to proton magnetic moments μ_μ/μ_p obtained in this experiment are:

$$\Delta\nu = 4\,463\,302.35(52)\ \text{kHz}\ (0.12\ \text{ppm}),$$
$$\mu_\mu/\mu_p = 3.183\,340\,3(44)\ (1.4\ \text{ppm}). \qquad \textbf{(15)}$$

The theoretical formula for $\Delta\nu$ is given in TABLE 2.[29-31] This formula is based on conventional quantum electrodynamics and the assumption that the muon is a heavy structureless Dirac particle. The two-body bound state equation is solved in perturbation theory up to relative order α^3 in the radiative corrections and $\alpha^2 \ln\alpha$ m_e/m_μ in the relativistic recoil terms. The estimated theoretical error in the evaluation of the α^3 term is 0.6 ppm. The value obtained for α is

$$\alpha^{-1} = 137.036\ 02(9)\ (0.7\ \text{ppm}) \tag{16}$$

where the error is due principally to the 1.2 ppm uncertainty in μ_μ/μ_p and to the estimated theoretical error of about 0.6 ppm. It seems likely that improvements in both experiment and theory will soon reduce the error in α^{-1} from muonium by a factor of about 3.

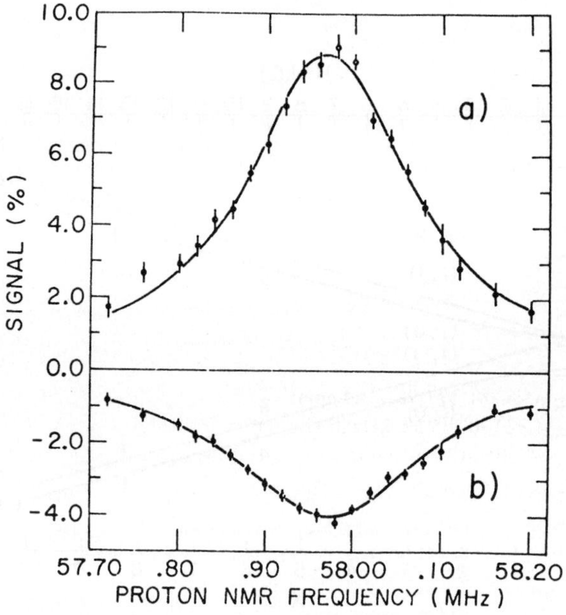

FIGURE 5. A resonance line for transition ν_{12} in a 1.7-atm Kr target, for (a) backward and (b) forward positron decay. The solid line is a least-squares fit of a Lorentzian curve to the data. The linewidth is 55 G and arises from the muon decay and power broadening. The data shown were obtained in 11 h.

The hyperfine structure interval of hydrogen in its gound state, $\Delta\nu$, can also be used to determine a value for α. The experimental value of $\Delta\nu$ is one of the most precisely known quantities in physics, and indeed serves as a frequency standard. The value of $\Delta\nu$ is given[15] in TABLE 3 to an accuracy of 1 part in 10.[12] The theoretical formula is also given in TABLE 3 and is very similar to that of muonium. The principal difference occurs in the correction term δ_p, which includes not only the relativistic recoil terms present in δ'_μ for muonium but also contributions associated with proton structure, given by form factors of the proton for elastic electron-proton scattering as well as the spin dependent form factors char-

TABLE 2

THEORETICAL FORMULA FOR MUONIUM $\Delta\nu$ AND COMPARISON WITH EXPERIMENT

$$\Delta\nu_{\text{theor}} = [\tfrac{16}{3}\,\alpha^2 cR_\infty(\mu_\mu/\mu_B{}^e)][1 + (m_e/m_\mu)]^{-3}[1 + \tfrac{3}{2}\alpha^2 + a_e + \epsilon_1 + \epsilon_2 + \epsilon_3 - \delta'_\mu]$$

$$a_e = \frac{\alpha}{2\pi} - 0.32848\frac{\alpha^2}{\pi^2} + (1.181 \pm 0.015)\frac{\alpha^3}{\pi^3}$$

$$\epsilon_1 = \alpha^2(\ln 2 - 5/2)$$

$$\epsilon_2 = -\frac{8\alpha^3}{3\pi}\ln\alpha\left(\ln\alpha - \ln 4 + \frac{281}{480}\right)$$

$$\epsilon_3 = \frac{\alpha^3}{\pi}(18.4 \pm 5)$$

$$\delta'_\mu = \frac{m_e}{m_\mu}\left\{\frac{3\alpha}{\pi}\left[1 - \left(\frac{m_e}{m_\mu}\right)^2\right]^{-1}\ln\frac{m_\mu}{m_e} + 2\alpha^2\ln\alpha\left[1 + \left(\frac{m_e}{m_\mu}\right)\right]^{-2}\right\}$$

$\alpha^{-1} = 137.035\ 987(20)$ (0.21 ppm)

$R_\infty = 1.097\ 373\ 143(10) \times 10^5\ \text{cm}^{-1}$ (0.01 ppm)

$c = 2.997\ 924\ 58(1.2) \times 10^{10}$ cm/sec (0.004 ppm)

$\mu_\mu/\mu_B{}^e = (\mu_\mu/\mu_p)(\mu_p/\mu_B{}^e)$

$\mu_p/\mu_B{}^e = 1.521\ 032\ 209(16) \times 10^{-3}$ (0.01 ppm)

$a_e = 0.001\ 159\ 656\ 7(35)$ (3 ppm)

$m_\mu/m_e = 206.768\ 51(25)$ (1.2 ppm)

$\mu_\mu/\mu_p = 3.183\ 341\ 7(39)$ (1.2 ppm) $\begin{cases} 3.183\ 340\ 3(44)\ [1.4\ \text{ppm}]\ \mu^+e^- \\ 3.183\ 346\ 7(82)\ [2.6\ \text{ppm}]\ \mu^+ \text{ in } H_2O \end{cases}$

$\Delta\nu_{\text{theor}} = \alpha^2[8.381\ 577\ 78 \pm 1.4\ \text{ppm}]$

$\Delta\nu_{\text{exp}} = 4\ 463\ 302.35(52)$ kHz (0.12 ppm)

$\alpha^{-1} = 137.036\ 02(9)$ (0.7 ppm)

TABLE 3

HYPERFINE STRUCTURE INTERVAL FOR HYDROGEN*

$$\Delta\nu_{\text{theor}} = \left(\frac{16}{3}\alpha^2 cR_\infty\frac{\mu_p}{\mu_B{}^e}\right)\left(1 + \frac{m_e}{m_p}\right)^{-3}\left(1 + \frac{3}{2}\alpha^2 + a_e + \epsilon_1 + \epsilon_2 + \epsilon_3 + \delta_p\right)$$

$$a_e = \frac{\alpha}{2\pi} - 0.32848\frac{\alpha^2}{\pi^2} + (1.181 \pm 0.015)\frac{\alpha^3}{\pi^3};\ \epsilon_1 = a^2\left(\ln 2 - \frac{5}{2}\right)$$

$$\epsilon_2 = -\frac{8\alpha^3}{3\pi}\ln\alpha\left(\ln\alpha - \ln 4 + \frac{281}{480}\right);\ \epsilon_3 = \frac{\alpha^3}{\pi}(18.4 \pm 5)$$

δ_p = Proton recoil and proton structure term

$\delta_p = \delta_p(\text{rigid}) + \delta_p(\text{polarizability})$

$\delta_p = -34.6(9) \times 10^{-6} + \delta_p(\text{pol})\ [\ |\delta_p(\text{pol})| < 4\ \text{ppm}]$

$\Delta\nu_{\text{expt}} = 1420\ 405\ 751.7667(10)$ Hz

*From Cohen and Taylor[15] and Brodsky and Drell.[52]

acterizing deep inelastic electron-proton scattering. The latter are associated with the proton polarizability correction in the term δ_p. The uncertainty in δ_p, principally that associated with the polarizability correction, is estimated to be about 4 ppm. Hence the uncertainty in α from H hfs as given in TABLE 3 is about 1.6 ppm.

The anomalous g-value of the electron, $a_e = (g_e - 2)/2$ provides another source for a precise value of α. The current theoretical value of a_e is[27,32-34]

$$a_e = \frac{\alpha}{2\pi} - 0.328\ 479\left(\frac{\alpha^2}{\pi^2}\right) + (1.181 \pm 0.015)\left(\frac{\alpha^3}{\pi^3}\right), \tag{17}$$

where the numerical uncertainty in the sixth-order contribution is given. Efforts are beginning to calculate the eighth-order $(\alpha/\pi)^4$ radiative correction. Hadronic corrections will be at most of order $(\alpha/\pi)^4$ in magnitude.

Two modern measurements provide the experimental value of a_e. The first, which has until recently for many years given the most precise value, is the well-known g-2 experiment in which the difference frequency ω_a between the spin precession frequency and the cyclotron frequency in a magnetic field is measured.

$$\omega_a = \left(\frac{e}{mc}\right) a_e B. \tag{18}$$

The value obtained for a_e was[35,36]

$$a_e = (1\ 159\ 656.7 \pm 3.5) \times 10^{-6}. \tag{19}$$

Recently a second method based on observation of the spin-cyclotron beat frequency and the cyclotron frequency of isolated 1 meV electrons in an electric-magnetic field trap has given the still higher precision value[37]

$$a_e = (1\ 159\ 652.41 \pm 0.20) \times 10^{-6}. \tag{20}$$

Use of Equations 17 and 20 gives

$$\alpha^{-1} = 137.035\ 979(60)(0.43\ \text{ppm}) \tag{21}$$

in which the error is about equally due to theoretical and experimental uncertainties.

TABLE 4 and FIGURE 6 summarize the various determinations of α. They all

TABLE 4

VALUES* OF THE FINE STRUCTURE CONSTANT

Method	α^{-1} Value
Josephson e/h and γ_p	137.035 987(29) (0.21 ppm)
Hydrogen Fine Structure, $2^2P_{3/2}$–$2^2P_{1/2}$	137.035 44(52) (3.9 ppm)
Helium Fine Structure, 2^3P_0–2^3P_1	137.036 08(13) (0.94 ppm)
Muonium Hyperfine Structure	137.036 02(9) (0.7 ppm)
Hydrogen Hyperfine Structure	137.035 97(22) (1.6 ppm)
Electron g-Value	137.035 979(60) (0.43 ppm)

*From Mohr.[53]

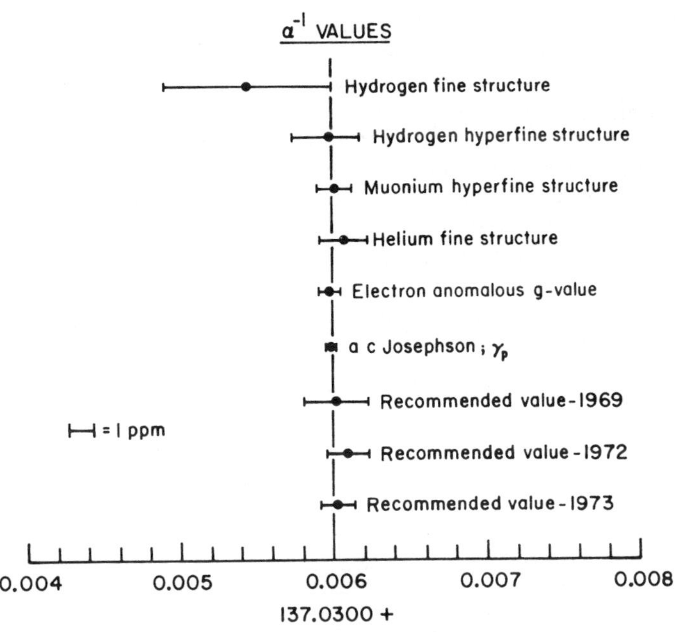

FIGURE 6. Values of α from different methods.

agree reasonably well within their errors. The most precise value is that based on the measurements of e/h by the ac Josephson effect and of γ_p.

The agreement of the values of α from these different sources can be regarded as confirming our quantitative understanding of the physics involved in these different systems.

A precise value for α is required in the overall determination of the fundamental atomic constants[15] and is essential for a comparison of theory and experiment for many basic quantities such as the Lamb shift, positronium fine structure, the muon g-value, and muonic atom energy levels. Our faith in QED is based in substantial part on these successful comparisons.

Theory of the Value of α

The fine structure constant α is a dimensionless pure number, which of course characterizes the strength of the electromagnetic interaction. It is reasonable to expect that the numerical value of α can be calculated in a comprehensive theory of the elementary particles, or perhaps of the universe.

Attempts to provide a theory for α date back at least to Eddington's famous proposal[38] of 1936, which relates α^{-1} to the number of dimensions of a certain wavefunction. In Heisenberg's unified field theory of elementary particles, which is characterized as a nonlinear spinor theory, quantum electrodynamics is incorporated and a value for α is calculated.[39] If magnetic monopoles exist[40] of strength g, one might expect the relationship $\alpha = e/(2g)$.

A general modern viewpoint is that unified physical theory will only provide meaningful, finite answers if α has its observed value. In particular, extensive studies have been made within the framework of quantum electrodynamics only for spin 0, spin $\frac{1}{2}$, and spin 1 particles in order to find finite, convergent solutions of the field equations which may occur only for a single value of α or of the unrenormalized coupling constant α_0.[41-44] More recent unified theories including not only the electromagnetic but also the weak and strong and even the gravitational interactions have been applied to the calculation of α.[45-48] As yet none of these theoretical approaches to calculating α can be regarded as successful or as a generally accepted viewpoint.

A proposal based on a group theoretical argument without obvious physical content suggested[49] that

$$\alpha^{-1} = 2^{19/4}3^{-7/4}5^{1/4}\pi^{11/4} = 137.036\ 082.$$

At the time this suggestion was made agreement with the experimental value of α was excellent. Several strictly arithmetic or numerological formulae were de-

TABLE 5

NUMERICAL EXPRESSIONS FOR α^{-1}

$\alpha^{-1} = 137.035\ 987(29)$ (0.21 ppm) experimental		
$\alpha^{-1} = 2^{19/4}3^{-7/4}5^{1/4}\pi^{11/4}$	$= 137.036\ 082$	Wyler
$\alpha^{-1} = 2^{-19/4}3^{10/3}5^{17/4}\pi^{-2}$	$= 137.035\ 938$	Roskies
$\alpha^{-1} = 2^{-13/4}3^{17/4}5^{2/3}\pi^{5/4}$	$= 137.036\ 163$	
$\alpha^{-1} = 2^{2/3}3^{7/3}5^{11/3}\pi^{-7/2}$	$= 137.036\ 120$	
$\alpha^{-1} = 2^{5/3}3^{-8/3}5^{5/2}\pi^{7/3}$	$= 137.036\ 007$	
$\alpha^{-1} = 2^{8/3}3^{3/4}5^{-1/2}\pi^{8/3}$	$= 137.036\ 289$	
$\alpha^{-1} = (5/2)^{1/2}(2)^{2/3}e^{4}$	$= 137.035\ 97$	Giaever
$\alpha^{-1}_{\text{Wyler}} - \alpha^{-1}_{\text{exp}} = 0.000\ 095 \pm 0.000\ 029$		

veloped by computer searches[50] that also gave values of α in agreement with the experimental value and hence lessened the marvel of the agreement of the group theoretical value of α with the experimental value. The current best experimental value for α does not agree well with the group theoretical value. TABLE 5 summarizes this situation.

All in all we do not yet have a convincing theory for α, or even a generally accepted viewpoint for such a theory. We must apparently await further understanding of physics before a good theory of the value of α can be given.[51]

References

1. EINSTEIN, A. 1909. Phys. Z. **10:** 185.
2. KLEIN, M. J. 1966. Phys. Today **19**(November): 23.
3. SOMMERFELD, A. 1915. Munchener Ber. 425, 459; 1916. Munchener Ber. 131.

4. SOMMERFELD, A. 1916. Ann. Physik **51:** 125.
5. SOMMERFELD, A. 1934. Atomic Structure and Spectral Lines. Vol. 1. 3rd edit. Methuen and Co. Ltd. London.
6. RUBINOWICZ, A. 1933. Handbuch der Physik. Vol. 24/1:1 J. Springer. Berlin.
7. MILLIKAN, R. A. 1917. The Electron. University of Chicago Press. Chicago, Ill.
8. COHEN, E. R., K. M. CROWE & J. W. M. DUMOND. 1957. Fundamental Constants of Physics. Interscience. New York, N.Y.
9. PASCHEN, F. 1916. Ann. Physik **50:** 901.
10. RABI, I. I., *et al.* 1939. Phys. Rev. **55:** 526.
11. JOSEPHSON, B. D. 1962. Phys. Lett. **1:** 251.
12. JOSEPHSON, B. D. 1965. Adv. Phys. **14:** 419.
13. PARKER, W. H., *et al.* 1969. Phys. Rev. **177:** 639.
14. FINNEGAN, T. F., *et al.* 1971. Phys. Rev. B **4:** 1487.
15. COHEN, E. & B. N. TAYLOR. 1973. J. Phys. Chem. Ref. Data **2:** 663.
16. OLSEN, P. T. & E. E. WILLIAMS. 1976. Atomic Masses and Fundamental Constants 5. J. H. Sanders & A. H. Wapstra, Eds.: 538. Plenum Press. New York, N.Y.
17. BAIRD, J. C. *et al.* 1972. Phys. Rev. A **5:** 564.
18. HUGHES, V. W. 1970. Facets of Physics. D. A. Bromley & V. W. Hughes, Eds. Academic Press. New York, N.Y.
19. PERL, M. L., I. I. RABI & B. SENITSKY. 1955. Phys. Rev. **98:** 611.
20. PICHANICK, F. M. J., *et al.* 1968. Phys. Rev. **169:** 55.
21. LEWIS, S. A., F. M. J. PICHANICK, & V. W. HUGHES. 1970. Phys. Rev. A **2:** 86.
22. KPONOU, A., *et al.* 1971. Phys. Rev. Lett. **26:** 1613.
23. LEWIS, M. L. 1975. Atomic Physics 4. G. zu Putlitz, E. W. Weber, and A. Winnacker, Eds.: 105. Plenum Press, New York, N.Y.
24. LEWIS, M. L., P. H. SERAFINO & V. W. HUGHES. 1976. Phys. Lett. **58A:** 125.
25. LEWIS, M. L. & P. H. SERAFINO. 1977. Second-order contribution to the fine structure of helium from all intermediate states. Phys. Rev. A (December).
26. BREIT, G. & I. RABI. 1931. Phys. Rev. **38:** 2082.
27. HUGHES, V. W. & T. KINOSHITA. 1977. Muon Physics. Vol. I. Chapt. 2. Academic Press. New York, N.Y.
28. CASPERSON, D. E., *et al.* 1977. Phys. Rev. Lett. **38:** 956, 1504.
29. BRODSKY, S. J. & G. W. ERICKSON. 1966. Phys. Rev. **148:** 26.
30. LAUTRUP, B. E., *et al.* 1972. Phys. Rep. Phys. Lett. C (Netherlands) **3C:** 193.
31. LEPAGE, G. P. 1977. SLAC PUB-1900. Phys. Rev. A (September).
32. CVITANOVIC, P. & T. KINOSHITA. 1974. Phys. Rev. D **10:** 3991, 4007.
33. LEVINE, M. J. & R. ROSKIES. 1976. Phys. Rev. D **14:** 2191.
34. CALMET, S. *et al.* 1977. Rev. Mod. Phys. **49:** 21.
35. WESLEY, J. C. & A. RICH. 1971. Phys. Rev. A **4:** 1341.
36. RICH, A. & J. C. WESLEY. 1972. Rev. Mod. Phys. **44:** 250.
37. VAN DYCK, R. S., JR., P. B. SCHWINBERG & H. G. DEHMELT, 1977. Phys. Rev. Lett. **38:** 310.
38. EDDINGTON, A. Relativity Theory of Protons and Electrons. Cambridge University Press, Cambridge, England.
39. HEISENBERG, W. 1966. Introduction to the Unified Field Theory of Elementary Particles. Intersciences. New York, N.Y.; DÜRR, H. P., *et al.* 1965. Nuovo Cimento **38:** 1220.
40. DIRAC, P. A. M. 1948. Phys. Rev. **74:** 817.
41. GELL-MANN, M. & F. E. LOW. 1954. Phys. Rev. **95:** 1300.
42. SALAM, A. 1963. Phys. Rev. **130:** 1287; SALAM, A. & R. DELBOURGO. 1964. Phys. Rev. **135:** B1398.
43. JOHNSON, K., *et al.*, 1964, 1967. Phys. Rev. **136:** B1111. **163:** 1699.
44. ADLER, S. L. 1972. Phys. Rev. D **5:** 3021.
45. ROY, S. M. & A. S. VENGURLEKAR. 1976. Nucl. Phys. **B114:** 449.
46. ARNOLD, R. C. 1977. Phys. Lett. **67B:** 91.
47. MOTZ. L. 1977. Nuovo Cimento **37A:** 13.

48. Terazawa, H., *et al.* 1977. Phys. Rev. D **15:** 1181.
49. Wyler, M. A. 1969. C. R. Acad. Sci. Paris A **269:** 743; 1971. C. R. Acad. Sci. Paris, **272:** 186.
50. Roskies, R. 1971. Phys. Today **24**(November), 9.
51. Hughes, V. W. 1968. A Tribute to I. I. Rabi. Columbia University Symposium, 1967.
52. Brodsky, S. J. & S. D. Drell. 1970. Ann. Rev. Nucl. Sci. **20:** 147.
53. Mohr, P. J. 1977. Atomic Physics 5. R. Marrus, M. Prior & H. Shugart, Eds.: 37. Plenum Press, New York, N.Y.

PHYSICAL REVIEW LETTERS

VOLUME 15 | 5 JULY 1965 | NUMBER 1

POLARIZED ELECTRONS FROM A POLARIZED ATOMIC BEAM*

R. L. Long, Jr., W. Raith, and V. W. Hughes

Gibbs Laboratory, Yale University, New Haven, Connecticut

(Received 25 May 1965)

A source of polarized electrons would have many uses in atomic, nuclear, and elementary-particle physics for the study of the spin dependence of interactions. Although polarized electrons are available from beta decay[1] and can also be produced by Mott scattering,[2] these sources of polarized electrons are not particularly convenient or useful as regards intensity, degree of polarization, energy variability, or suitability for injection into an accelerator. Many conceivable schemes for producing polarized electrons have been discussed.[2,3] All experiments which sought to utilize the electronic polarization in ferromagnetic materials have had negative results.[4] Utilization of optical pumping[5] and of elastic scattering of electrons with energies in the keV range from heavy atoms[6] are presently under investigation. In this Letter we report on the successful production of polarized electrons by photoionization of a polarized atomic beam of alkali atoms. We believe that this method can provide useful sources of polarized electrons.

The principles of the atomic beam method have been discussed in detail.[7] The first experimental attempt to use this method was made by Friedmann.[8] Ground-state alkali atoms with the electronic magnetic quantum number $m_J=+\frac{1}{2}$ are selected by deflection in a strong inhomogeneous magnetic field. The selected atoms having an electronic polarization P_a close to unity enter the photoionization region where there is a weaker magnetic field H along the axis of propagation. They change adiabatically into the states characteristic of H for which the electronic polarization P is smaller than P_a due to the hfs coupling of the electronic spin and the nuclear spin $\mathbf{I}$:

$$P=P_a f(H),$$

where

$$f(H)=(2I+1)^{-1}\sum_{m=-I+\frac{1}{2}}^{I+\frac{1}{2}}[x+m(I+\tfrac{1}{2})^{-1}]\times[x^2+2xm(I+\tfrac{1}{2})^{-1}+1]^{-1/2},$$

and

$$x=(g_J-g_I)\mu_0 H/\Delta W$$

(ΔW = hfs energy interval; μ_0 = Bohr magneton; g_J and g_I are the electronic and nuclear g values).[9] Photoionization is predominantly an electric dipole transition and the spin-orbit interaction in the final state is not large enough to cause a spin flip for the outgoing electron during the time it spends in the field of the positive ion[10]; hence, the polarization of the photoelectrons should be equal to the electronic polarization P of the atoms. Potassium was chosen as the alkali atom because of its relatively small hfs interaction and relatively low photoionization threshold energy corresponding to

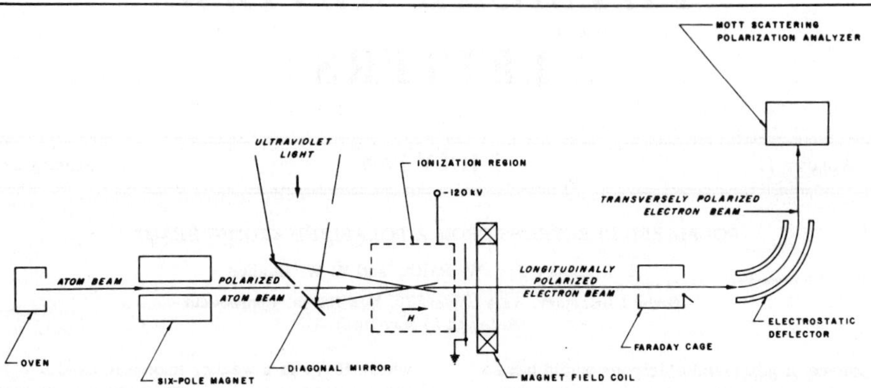

FIG. 1. Schematic diagram of the polarized electron source and the arrangements for measuring current and polarization (not to scale).

$\lambda = 2856$ Å.

A schematic diagram of the polarized electron source and of the Mott scattering apparatus used to measure the electron polarization is shown in Fig. 1. The inhomogeneous magnetic field for the state selection is provided by a permanent six-pole magnet[11] for which the value P_a was calculated to be 0.96. The light source is a mercury high-pressure arc (Osram HBO 200). The arc image is formed with a spherical mirror and a plane diagonal mirror in the ionization region, which lies between two cylindrical electrodes with a small bias voltage between them suitable for electron extraction, and with the electrodes at about -120 kV, in order that the analysis of the electron polarization can be made by the Mott scattering method.[12] The axial magnetic field H of 90 G is provided by a single coil, and the focusing properties of the electric and magnetic fields serve to discriminate against all photoelectrons emitted from the electrodes. The photoelectron current is measured in the Faraday cage. For measurement of the electron polarization, the electron beam is passed through a 112° cylindrical electrostatic deflector, which converts the longitudinal polarization into a transverse polarization, and then scatters in a gold foil. The electrons scattered through 118° were detected by solid-state junction counters having an energy resolution of 15 keV, thus providing a good discrimination against inelastically scattered electrons. A comparison of scattering from aluminum and gold foils was made to eliminate instrumental asymmetry. After extrapolation of the measured asymmetry to zero scattering-foil thickness, theoretical values for the scattering asymmetry from screened gold nuclei[13] were used to compute the polarization. For $H = 90$ G the predicted value for the electron polarization P is given by

$$P = P_a f(H) = 0.96 \times 0.58 = 0.56,$$

with an estimated uncertainty of ±0.02 chiefly due to the measurement and inhomogeneity of the magnetic field.

The first measurements showed a rapid decrease of the electron polarization with increase of oven temperature.[14] This effect was due to photoelectrons from K_2 molecules in the atomic beam. In saturated potassium vapor the molecular fraction increases with temperature,[15] and atomic beam measurements by Mehran at Harvard[16] gave fractions of 1.3×10^{-3} to 2.0×10^{-3} in the relevant temperature range of from 600 to 650°K. The observed decrease of polarization indicates that the photoionization probability of K_2 under our conditions is about 4000 times bigger than that of K, which can be understood considering that the photoabsorption cross section of K_2 is very large[17] and that the ionization potential of K_2 is at least 0.24 V lower than that of K.[18] The K_2 content in the beam was eliminated by thermal dissociation of the molecules in a tube attached to the oven exit and heated to about 1000°K.

Even with the elimination of the K_2 content in the beam, the observed electron polarization was about 11% smaller than the expected value. This difference decreased when the light intensity was reduced, which indicated that the depolarization was caused by a two-photon process. The two-step photoionization process was identified as the transition $4\,^2S \to 5\,^2P \to$ continuum. The resonance line doublet occurs at wavelengths of 4045 and 4048 Å, and is overlapped by a strong broad line in the mercury arc spectrum from $7\,^3S_1 \to 6\,^3P_0$ which occurs at about 4050 Å. With the full intensity and full spectrum of the light source, photoelectrons from this two-step photoionization process accounted for about 30% of the total current. Due to the fine structure interaction in the $5\,^2P$ state, these photoelectrons are partially depolarized. When the resonance excitation was suppressed by use of a nickel-sulfate-solution filter,[19] the measured polarization was $P = 0.58 \pm 0.03$, which agrees with the expected value.

The maximum current of electrons with the expected polarization was 10^{-12} A (6×10^6 electrons/sec). This current corresponded to a flux of polarized atoms in the ionization region of 1.2×10^{14} sec^{-1} and a photoionization probability of 5×10^{-8}. The latter figure agrees with a rough calculation based on lamp data, photoionization cross section, atomic velocity, and geometry.

Both the electron polarization and the current achieved do not represent the ultimate capacities of this method. Since the polarization obtained with the rather low ionizer field of 90 G agrees with the theoretical value, polarizations of at least 0.9 can be expected with magnetic fields greater than 400 G. The polarized electron-beam intensity depends most critically on the photoionization efficiency, and increase in beam intensity can be expected with the use of different light sources. For a pulsed polarized electron source flash lamps may have sufficiently high radiance at short wavelengths so that lithium, with an ionization threshold of 2300 Å but a large photoionization cross section, can be used instead of potassium for the atomic beam. Utilization of the two-step photoionization process involving polarized light (and perhaps no magnet at all) also appears promising.

The authors want to thank Mr. D. Lazarus for preparing the scattering foils and Dr. M. Posner for his assistance with the electronics and for valuable discussions.

---

*The research has been supported in part by th U. S. Office of Naval Research and by the National Aeronautics and Space Administration.

[1]H. Frauenfelder and R. M. Steffen, in Alpha-, Beta-, and Gamma-Ray Spectroscopy, edited by K. Siegbahn (North-Holland Publishing Company, Amsterdam, 1965), Vol 2., p. 1431.

[2]H. A. Tolhoek, Rev. Mod. Phys. 28, 277 (1956).

[3]E. Fues and H. Hellmann, Physik. Z. 31, 465 (1930).

[4]H. A. Fowler and L. Marton, Bull. Am. Phys. Soc. 4, 235 (1959); R. L. Long, V. W. Hughes, J. S. Greenberg, I. Ames, and R. L. Christensen, Bull. Am. Phys. Soc. 6, 265 (1961); and Phys. Rev. 138, A1630 (1965); W. T. Pimbley and E. W. Müller, J. Appl. Phys. 33, 238 (1962); H. v. Issendorff and R. Fleischmann, Z. Physik 167, 11 (1962); W. Raith and R. Schliepe, Z. Physik 170, 185 (1962).

[5]W. Raith, in Electron Microscopy, Fifth International Congress, Philadelphia, 1962, edited by S. S. Breese, Jr. (Academic Press, Inc., New York, 1962), paper AA-6; H. Boersch, J. Lemmerich, and R. Schliepe, Z. Physik 182, 166 (1964).

[6]G. Holzwarth and H. J. Meister, Nucl. Phys. 59, 56 (1964); J. Kessler and H. Lindner, Z. Physik 183, 1 (1965).

[7]V. W. Hughes, R. L. Long, Jr., and W. Raith, in Proceedings of the International Conference on High-Energy Accelerators, Dubna, 1963, edited by A. A. Kolomensky (Atomizdat., Moscow, 1964), p. 988.

[8]H. Friedmann, Sitzber. Math.-Naturw. Kl. Bayer. Akad. Wiss. München 1961, 13. However, his reported results could not be confirmed; see R. L. Long, Jr., W. Raith, and V. W. Hughes, Bull. Am. Phys. Soc. 10, 28 (1965); and F. Bopp, D. Maison, G. Regenfus, and H. Chr. Siegmann, "Experimente zur Herstellung Polarisierter Electronen aus Atomstrahlen" (to be published).

[9]P. Kusch and V. W. Hughes, in Handbuch der Physik, edited by S. Flügge (Springer-Verlag, Berlin, 1959), Vol. 37/1, p. 83.

[10]N. F. Mott and H. S. W. Massey, The Theory of Atomic Collisions, (Clarendon Press, Oxford, 1949), 2nd ed., p. 65.

[11]R. L. Christensen and D. R. Hamilton, Rev. Sci. Instr. 30, 356 (1959).

[12]J. S. Greenberg, D. P. Malone, R. L. Gluckstern, and V. W. Hughes, Phys. Rev. 120, 1393 (1960).

[13]S. Lin, Phys. Rev. 133, A965 (1964).

[14]Long, Raith, and Hughes, reference 8.

[15]R. W. Ditchburn, Proc. Roy. Soc. (London) 117, 486 (1927).

[16]F. Mehran, private communication.

[17]R. W. Ditchburn, J. Tunstead, and J. G. Yates,

Proc. Roy. Soc. (London) 181, 386 (1943).

[18]This value follows from the work of Y. Lee and B. H. Mahan, J. Chem. Phys. 42, 2893 (1965).

[19]M. Kasha, J. Opt. Soc. Am. 38, 929 (1948).

Reprinted from THE PHYSICAL REVIEW, Vol. 185, No. 4, 1251–1255, 20 September 1969
Printed in U. S. A.

Magnetic Moment and hfs Anomaly for He^3†

W. L. WILLIAMS* AND V. W. HUGHES
Gibbs Laboratory, Yale University, New Haven, Connecticut

(Received 2 January 1969)

The ratio of the nuclear magnetic moments of He^3 and H^1 has been measured by NMR techniques using a gaseous sample. The result is $-\mu_{He^3}(He)/\mu_p(H_2)=0.76178685$ (±0.1 ppm), uncorrected for magnetic shielding in He and H_2, and with these shielding corrections, $-\mu_{He^3}/\mu_p=0.76181237\pm0.6$ ppm. With this new value for the magnetic moment ratio the hfs anomaly is found to be $\delta_{He^3}=+(217\pm3)\times10^{-6}$. By comparison with the theory for δ_{He^3} the contribution of exchange currents to δ_{He^3} is calculated to be $+(15\pm5)\times10^{-6}$.

I. INTRODUCTION

A COMPLETE theory of the three-nucleon system must comprehend all the experimental information about the He^3 nucleus. Conversely, data about He^3 should aid in the development of the theory of the three-nucleon system and in the examination of the magnitude of a three-nucleon force. Recently,[1–3] studies have been made of He^3 and H^3 through the elastic scattering of high-energy electrons, and values for the rms charge and magnetic moment radii have been determined. Further relevant quantities are the nuclear masses, spins, magnetic dipole moments, and the hfs anomalies.

In this paper we report a new high-precision measurement of the magnetic moment ratio μ_{He^3}/μ_p by an NMR technique.[4] We achieve an accuracy of 0.1 ppm, which represents an improvement by a factor of 10 relative to an earlier determination.[5] With this new value and the value of the hfs interval for He^3 in the ground $1^2S_{1/2}$ state of He^+, a new value for the hfs anomaly for He^3 is determined.

Section II includes a description of the experimental apparatus. The measurements and results are discussed in Sec. III. The He^3 hfs anomaly is discussed in Sec. IV.

II. EXPERIMENTAL

The basic experimental technique used was that of NMR.[6] The dc magnetic field was produced by a Varian 12-in.-high resolution magnet equipped with field homogenizing coils. Field sweep was provided by driving the magnet power supply sweep input with a triangular voltage.

The measurements were made with gaseous samples at total pressures up to about 60 atm. A high-pressure bulb was built following the design of Garwin and Reich[7] and is shown schematically in Fig. 1. The spherical bulb, fabricated from Araldite 503 epoxy,[8] has a $\frac{1}{4}$-in. radius and 0.025 in. wall thickness. Two rf coils (4 turns of No. 30 copper wire) were wound on the bulb with their polar axes orthogonal. Brass capillary tubing was inserted in the diametrically opposed necks. Feather-edge disks (2 mil brass foil) were soldered to the tubing to prevent leakage around the tubular necks. The whole arrangement was cast in epoxy. Magnetic field modulation coils (1-in.-diam pancakes) were placed on the sample bulb. Destruction of bulbs fabricated in this manner occurred at about 75 atm. The gaseous samples contained high-purity He^3 and commerical-grade O_2 and H_2. The gas mixing was done in a Toepler pumping system and the mixtures were compressed into the sample bulb with a head of mercury backed by a high-pressure N_2 tank. The presence of the H_2 allows a comparison of the helium magnetic moment $\mu_{He^3}(He)$ to the proton moment $\mu_p(H_2)$. The nuclear spin-relaxation time for pure He^3 is long,[7,9] due to the weakness of the nuclear dipole-dipole relaxation mechanism. Hence, oxygen was introduced to reduce the He^3 relaxation time to about 1 sec.

A spectrometer was used which permitted simultaneous observations of the NMR signals from He^3 and protons in H_2 at the same magnetic field site, thus largely eliminating the effects of magnetic field drifts and gradient shifts. A schematic diagram of the spectrometer is shown in Fig. 2. Nuclear sideband detection systems[10] were used to observe the He^3 and proton resonances. The basic system is a simple tank circuit probe driven by an rf signal generator and uses conventional amplification and detection techniques.

† This work was supported in part by the Office of Naval Research and is based on a thesis submitted by W. L. Williams to Yale University in partial fulfillment of the requirements for the Ph.D. degree.

* Present address: Department of Physics, University of Michigan, Ann Arbor, Mich.

[1] H. Collard, R. Hofstadter, E. B. Hughes, A. Johansson, M. R. Yearian, R. B. Day, and R. T. Wagner, Phys. Rev. **138**, B57 (1965); T. A. Griffy and R. J. Oakes, Rev. Mod. Phys. **37**, 402 (1965); R. H. Dalitz and T. W. Thacker, Phys. Rev. Letters **15**, 204 (1965).

[2] L. I. Schiff, Phys. Rev. **133**, B802 (1964).

[3] B. F. Gibson and L. I. Schiff, Phys. Rev. **138**, B26 (1965).

[4] W. L. Williams and V. W. Hughes, Bull. Am. Phys. Soc. **11**, 121 (1966); in *International Nuclear Physics Conference*, edited by R. L. Becker (Academic Press Inc., New York, 1967), p. 1042. A different value for δ_{He^3} was given in this reference as compared to the present paper, principally because a different value for α was used.

[5] H. L. Anderson, Phys. Rev. **76**, 1460 (1949).

[6] A. Abragam, *The Principles of Nuclear Magnetism* (Clarendon Press, Oxford, England, 1961).

[7] R. L. Garwin and H. A. Reich, Phys. Rev. **115**, 1478 (1959).

[8] Ciba Products Co., Fair Lawn, N.J.

[9] G. K. Walters, L. D. Schearer, and F. D. Colegrove, Bull. Am. Phys. Soc. **9**, 11 (1964); R. L. Gamblin and T. R. Carver, *ibid.* **9**, 11 (1964); H. G. Robinson and T. Myint, Appl. Phys. Letters **5**, 116 (1964).

[10] W. A. Anderson, Rev. Sci. Instr. **33**, 1160 (1962).

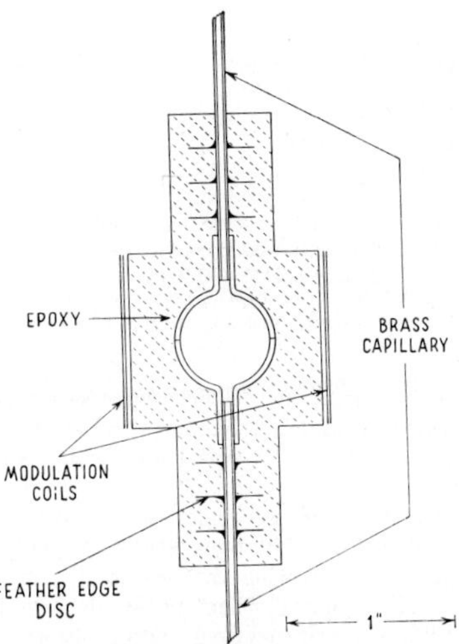

FIG. 1. High-pressure sample bulb.

The magnetic field is modulated at an audio angular frequency ω_m. Under slow passage conditions and with $\gamma H_1 \ll \omega_m$, a spectrum of resonances can be observed at magnetic field values

$$H = H_0 \pm k\omega_m/\gamma = (1/\gamma)(\omega_0 \pm k\omega_m), \quad (1)$$

in which γ is the gyromagnetic ratio of the sample nuclei, H_1 is the amplitude of the rf magnetic field, H_m is the amplitude of the modulation field, and H_0 is the value of the dc magnetic field at resonance when the rf frequency is ω_0. The sideband number is denoted by the integer k. We observe the $k=1$ sidebands. For stronger rf fields the $k=1$ resonances occur at magnetic field values

$$H = H_0 \pm \omega_m/\gamma \mp \gamma H_1^2/2\omega_m, \quad (2)$$

where the correction term of order $(\gamma H_1/\omega_m)^2$ is included. Hence the resonance condition depends on H_1.

The constant driving frequencies for the He^3 and H_2 probes were derived from a Hewlett-Packard 5100 A/5110 A digital-frequency synthesizer (dfs). The dfs has a 0–50-MHz output which can be varied in steps of 0.01 Hz. The rms frequency stability is approximately 1 in 10^{11} at 50 MHz. The H^1 frequency of 50 MHz was obtained with an harmonic generator and crystal filter driven by the dfs standard 1 MHz. The filter has a Q of 19 000 and gives 50-dB rejection of the 49- and 51-MHz harmonics. The He^3 frequency of about 38 MHz was taken directly from the dfs output.

The preamplifiers have noise figures of approximately 3 dB and gains of approximately 20 dB. National HRO-60 and Eddystone 770-R receivers were used as rf amplifiers and detectors. The am outputs were fed to lock-in detectors whose outputs were recorded on a two-pen strip chart recorder. The magnetic field modulation current and lock-in reference signals were obtained from an audio oscillator.

The operation of the spectrometer was checked by measuring the gyromagnetic moment ratio $\gamma_{Li^7}/\gamma_p(H_2O)$. The samples were saturated aqueous solutions of LiCl and $LiNO_3$ in distilled water. The measurement technique used is the same as that described in Sec. III. The results of these measurements were

$$\gamma_{Li^7}(LiNO_3)/\gamma_p(H_2O) = 0.388636251(20),$$

$$\gamma_{Li^7}(LiCl)/\gamma_p(H_2O) = 0.388636264(20),$$

in which 1-standard-deviation errors are indicated. The agreement of these values is taken as meaning that the Li^7 resonances arise from ionized lithium, as previously suggested.[11] The average is

$$\gamma_{Li^7}/\gamma_p(H_2O) = 0.388636258(14),$$

in agreement with previous determinations.[12]

III. EXPERIMENTAL PROCEDURE AND RESULTS

The amplitude of the sideband signals is dependent on H_m and γ. At the beginning of each measurement H_m was adjusted so that the amplitudes of the first sideband resonances from He^3 and protons in H_2 were approximately equal. The rf levels were adjusted to avoid saturation. The dc magnetic field was adjusted until one of the H^1 resonances, e.g., the one at $H_0 + \omega_m/\gamma_p(H_2)$, was in the center of the sweep trace. Here $H_0 = \omega_p/\gamma_p(H_2)$, where ω_p is the driving frequency for the proton resonance. The corresponding upper sideband resonance for He^3 will occur at the same dc magnetic field for the proper driving frequency ω_{He^3}. The separation of the centers of the He^3 and H^1 resonances was measured as a function of ω_{He^3} and the frequency for which the line centers coincide on the sweep trace was determined. A similar measurement was made on the other set of corresponding sidebands $[H = H_0 - \omega_m/\gamma_p(H_2)]$. The effects of resonance shifts due to H_1 [Eq. (2)] were eliminated by taking advantage of the fact that these shifts are symmetric around the center $(H = H_0)$ of the resonance spectrum. Let ω_1 and ω_2 denote, respectively, the He^3 angular frequencies at which corresponding H^1 and He^3 resonance line centers above and below H_0 coincide. From (2) the nuclear magnetic moment ratio is

$$\mu_{He^3}(He)/\mu_p(H_2) = (\omega_1 + \omega_2)/2\omega_p, \quad (3)$$

where the shift terms have cancelled. Further checks for systematic errors were made by measuring the magnetic moment ratio $\mu_{He^3}(He)/\mu_p(H_2)$, with both absorption and dispersion mode signals, with the two

[11] W. C. Dickinson, Phys. Rev. **81**, 717 (1951).

[12] G. Lindstrom, Arkiv Fysik **4**, 1 (1951); G. K. Yagola and E. E. Bogatyryov, Ukr. Fiz. Zh. **7**, 45 (1962).

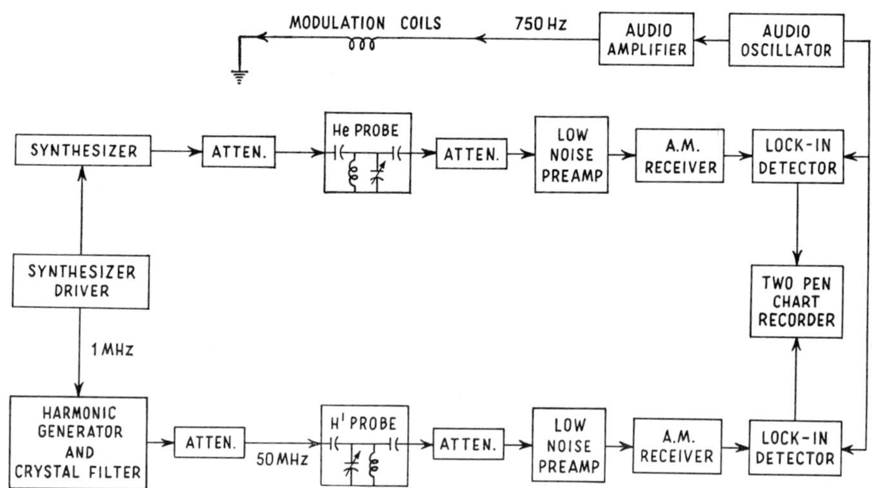

FIG. 2. Schematic diagram of the dual nuclear sideband spectrometer.

detection heads interchanged, with modulation frequencies $\omega_m/2\pi$ of 750 Hz and 1200 Hz, and for three values of the partial pressure ratio $P(H_2):P(O_2):P(\text{He})$. Data were taken with increasing and decreasing magnetic field sweeps.

In a typical run the separation of the resonance centers was varied in 20-Hz steps over a range of approximately 200 Hz. Figure 3 shows a resonance trace. The signal-to-noise ratio is approximately 15. The He^3 and H^1 resonance linewidths are 49 and 76 Hz, respectively. The line centers were coincident for $\omega_{\text{He}^3}/2\pi = 38.088770$ MHz and $\omega_p/2\pi = 50.000000$ MHz.

The observed helium and hydrogen linewidths are given in Table I. Values of the linewidth ratio multiplied by the inverse ratio of the nuclear moments,

$$[\Delta\nu_{1/2}(\text{H}^1)/\Delta\nu_{1/2}(\text{He}^3)][\mu_{\text{He}^3}(\text{He})/\mu_p(\text{H}_2)]$$

are given in the last column of Table I. This ratio is equal to 1 within experimental error, i.e., the linewidths for hydrogen and helium are approximately the same (in magnetic field units). Hence we conclude that the broadening was primarily due to magnetic field inhomogeneity. The average linewidth is 17.5 ± 3.0 mG (1.5 ppm).

The values of the moment ratio $\mu_{\text{He}^3}(\text{He})/\mu_p(\text{H}_2)$ are listed in Table II, where we have indicated the negative sign of μ_{He^3}.[13] They are plotted in Fig. 4 as a function of $P(\text{He}^3)$. The value obtained by Anderson[5] is also shown in Fig. 4.

Analysis of these data for dependence of the measured moment ratio on the partial pressure of He, H_2, and O_2 shows that any pressure shifts are negligible within experimental error. A rough theoretical estimate indicates that no observable pressure shifts would be expected.

The final result is

$$-\mu_{\text{He}^3}(\text{He})/\mu_p(\text{H}_2) = 0.76178685\ (\pm0.1\ \text{ppm}), \quad (4)$$

where the quoted error is 1 standard deviation.

Magnetic shielding corrections must be made to determine the ratio of the magnetic moments of the bare nuclei, μ_{He^3}/μ_p.

$$\mu_{\text{He}^3}/\mu_p = [\mu_{\text{He}^3}(\text{He})/\mu_p(\text{H}_2)][(1+\sigma_{\text{He}})/(1+\sigma_{\text{H}_2})]. \quad (5)$$

The calculated[14] magnetic shielding correction for He^3 is $\sigma_{\text{He}} = 59.935\times10^{-6}$, where relativistic corrections of order α^2Z^2 have been neglected.[15] The shielding correction for H_2 is $\sigma_{\text{H}_2} = (26.43\pm0.60)\times10^{-6}$.[16] Using these values with (4) and (5), we find

$$-\mu_{\text{He}^3}/\mu_p = 0.76181237\pm0.6\ \text{ppm}. \quad (6)$$

IV. DISCUSSION

The hfs anomaly δ is defined in general by the equation[17]

$$\Delta\nu = \Delta\nu_p(1-\delta), \quad (7)$$

[13] G. Weinreich and V. W. Hughes, Phys. Rev. **95**, 1451 (1954).

[14] R. L. Glick, J. Phys. Chem. **65**, 1871 (1961).

[15] V. W. Hughes and W. B. Teutsch, Phys. Rev. **94**, 761 (1954).

[16] T. Myint, D. Kleppner, N. F. Ramsey, and H. G. Robinson, Phys. Rev. Letters **17**, 405 (1966).

[17] P. Kusch and V. W. Hughes, in *Encyclopedia of Physics*, edited by S. Flügge (Springer-Verlag, Berlin, 1959), Vol. 37/1.

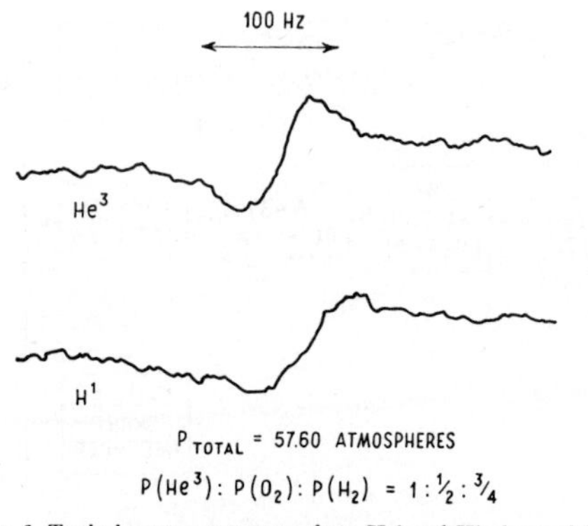

FIG. 3. Typical resonance pattern from He^3 and H^1 observed in the present experiment.

in which $\Delta\nu$ is the hfs interval and $\Delta\nu_p$ is the theoretical value of the hfs interval under the assumption that the nuclear magnetic moment is a point magnetic moment. The theoretical expression for $\Delta\nu$ in He^3 in the ground $1^2S_{1/2}$ state of the He$^+$ ion can be written as[18]

$$\Delta\nu_{\text{theor}}(\text{He}^+) = \left(\frac{128}{3}\alpha^2 c\,\text{Ry}\right)\frac{\mu_{\text{He}^3}}{\mu_0}\left(1-3\frac{m}{M_{\text{He}^3}}\right) \times(1+6\alpha^2+a_e+\epsilon_1+\epsilon_2+\epsilon_3)(1-\delta_{\text{He}}), \quad (8)$$

where

$$a_e=\alpha/2\pi-0.328(\alpha^2/\pi^2),$$

$$\epsilon_1=-2\alpha^2(\tfrac{5}{2}-\ln 2),$$

$$\epsilon_2=-(32\alpha^3/3\pi)(\ln 2\alpha)[\ln 2\alpha-\ln 4+(281/480)],$$

$$\epsilon_3=(4\alpha^3/\pi)(18.4\pm5).$$

Here α=fine-structure constant, c=velocity of light,

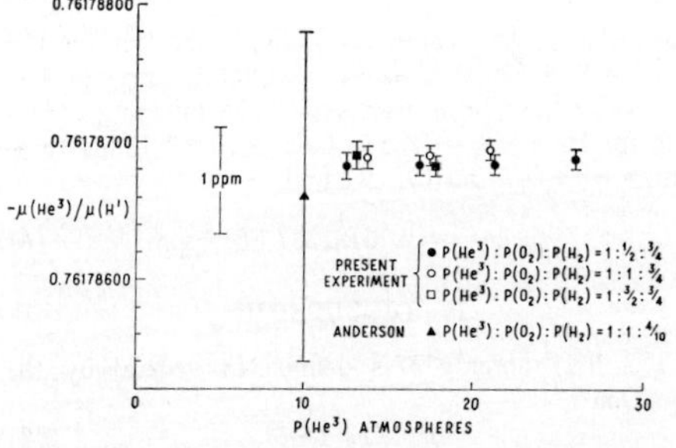

FIG. 4. The ratio $-\mu(\text{He}^3)/\mu(\text{H}^1)$ versus $P(\text{He}^3)$ for various values of the ratio $P(\text{He}^3):P(\text{O}_2):P(\text{H}_2)$. Also shown is Anderson's value Ref. 5.

[18] A. M. Sessler and H. M. Foley, Phys. Rev. **98**, 6 (1955); **110**, 995 (1958); S. J. Brodsky and G. W. Erickson, *ibid.* **148**, 26 (1966).

TABLE I. Linewidths ($\Delta\nu_{1/2}$).

P_{total} (atm)	$\Delta\nu_{1/2}(\text{H}^1)$ (Hz)	$\Delta\nu_{1/2}(\text{He}^3)$ (Hz)	$\frac{\Delta\nu_{1/2}(\text{H}^1)}{\Delta\nu_{1/2}(\text{He}^3)}\;\frac{\mu(\text{He}^3)}{\mu(\text{H}^1)}$
	$P(\text{He}^3):P(\text{O}_2):P(\text{H}_2)=1:\frac{1}{2}:\frac{3}{4}$		
27.4	75±7	51±8	1.1±0.2
37.6	76±6	52±7	1.1±0.2
47.3	73±6	50±6	1.1±0.2
57.6	76±7	49±7	1.2±0.2
	$P(\text{He}^3):P(\text{O}_2):P(\text{H}_2)=1:1:\frac{3}{4}$		
37.6	75±7	50±7	1.1±0.2
47.3	77±8	52±6	1.1±0.2
57.6	72±6	49±6	1.1±0.2
	$P(\text{He}^3):P(\text{O}_2):P(\text{H}_2)=1:\frac{3}{2}:\frac{3}{4}$		
47.3	73±7	54±7	1.1±0.2
57.6	75±6	48±8	1.2±0.2

Ry=rydberg, μ_0=Bohr magneton, m=electron mass, M_{He^3}=He^3 nuclear mass. The relativistic reduced mass correction of order $\alpha m/M_{\text{He}^3}$ is included in δ_{He}. For evaluation of $\Delta\nu_{\text{theor}}$, the factor μ_{He^3}/μ_0 is written in the form

$$\frac{\mu_{\text{He}^3}}{\mu_0}=\frac{\mu_{\text{He}^3}}{\mu_p}\frac{\mu_p}{\mu_e}\frac{\mu_e}{\mu_0}, \quad (9)$$

where μ_e is the spin magnetic moment of the electron. We use μ_{He^3}/μ_p from Eq. (6), and published values for μ_p/μ_e [16] and μ_e/μ_0,[19] to obtain

$$\mu_{\text{He}^3}/\mu_0=1.1587414\times10^{-3}\ (\pm0.9\text{ ppm}). \quad (10)$$

Using Eq. (10), the value of α from the ac Josephson measurements of e/h,[20] and the values of the other

TABLE II. Values of the moment ratio $\mu(\text{He}^3)/\mu(\text{H}^1)$ obtained at various pressures.

$P(\text{He}^3):P(\text{O}_2):P(\text{H}_2)$	$P(\text{He}^3)$ (atm)	$-\mu(\text{He}^3)/\mu(\text{H}^1)$[a]
$1:\frac{1}{2}:\frac{3}{4}$	25.6	0.76178688(9)
$1:\frac{1}{2}:\frac{3}{4}$	21.1	0.76178684(8)
$1:\frac{1}{2}:\frac{3}{4}$	16.7	0.76178684(9)
$1:\frac{1}{2}:\frac{3}{4}$	12.5	0.76178683(9)
$1:1:\frac{3}{4}$	20.9	0.76178694(9)
$1:1:\frac{3}{4}$	17.3	0.76178687(9)
$1:1:\frac{3}{4}$	13.7	0.76178689(10)
$1:\frac{3}{2}:\frac{3}{4}$	17.6	0.76178684(10)
$1:\frac{3}{2}:\frac{3}{4}$	13.0	0.76178690(10)

[a] Uncorrected for diamagnetic shielding.

[19] A. Rich, Phys. Rev. Letters **20**, 967 (1968).
[20] W. H. Parker, B. N. Taylor, and D. N. Langenberg, Phys. Rev. Letters **18**, 287 (1967).

constants as given by Cohen and Dumond,[21] we find from Eq. (8)

$$\Delta\nu_{\rm theor}({\rm He}^+)=(8667.52\pm0.06)(1-\delta_{\rm He})\,{\rm MHz}, \quad (11)$$

in which the error is due principally to the uncertainty in the value of α. Using the experimental value[22]

$$\Delta\nu_{\rm expt}({\rm He}^+)=8665.649905\pm0.00005\ {\rm MHz} \quad (12)$$

with Eq. (11), we find

$$\delta_{\rm He}=(216\pm7)\times10^{-6}. \quad (13)$$

This value agrees with an earlier determination of $\delta_{\rm He}$ based on $\Delta\nu_{\rm expt}({\rm He}^+, 2^2S_{1/2})$, [3] in view of the different values of α that are used.

A relevant quantity which can be obtained independent of the uncertainty in α is

$$\frac{1-\delta_{\rm He}}{1-P}=\frac{\Delta\nu_{\rm expt}({\rm He}^+,\,1^2S_{1/2})}{\Delta\nu_{\rm expt}({\rm H},\,1^2S_{1/2})}\,\frac{\Delta\nu_p({\rm H},\,1^2S_{1/2})}{\Delta\nu_p({\rm He}^+,\,1^2S_{1/2})}, \quad (14)$$

where

$$\frac{\Delta\nu_p({\rm H},\,1^2S_{1/2})}{\Delta\nu_p({\rm He}^+,\,1^2S_{1/2})}=\frac{1}{8}\frac{\mu_p}{\mu_{\rm He^3}}\left(1-3\frac{m}{M_{\rm H^1}}+3\frac{m}{M_{\rm He^3}}\right)$$
$$[(1-\tfrac{9}{2}\alpha^2-\alpha^2(\ln2-\tfrac{5}{2})+8\alpha^3/3\pi$$
$$\times\{4\ln^2 2\alpha-\ln^2\alpha-(\ln4-281/480)(4\ln2\alpha-\ln\alpha)\}$$
$$-3(\alpha^3/\pi)(18.4)]. \quad (15)$$

P includes the $\alpha(m/M_{\rm H^1})$ and nucleon structure corrections for H. With the experimental values for the hfs splittings [2,24] and (6), we have

$$(1-\delta_{\rm He})/(1-P)=1-(182.3\pm0.6)\times10^{-6}. \quad (16)$$

Using the theoretical value[25] for P of $(35\pm3)\times10^{-6}$ with (16), we find

$$\delta_{\rm He}=(217\pm3)\times10^{-6}. \quad (17)$$

The theoretical expression for $\delta_{\rm He}$ is

$$\delta_{\rm He}({\rm theor})=(202\pm3)\times10^{-6}+\delta_{\rm exch}, \quad (18)$$

for a singlet n-p effective range[4,27] $r_{0s}=2.7F$. Here $\delta_{\rm exch}$ represents the contribution of exchange currents to the hfs anomaly. We find, using (17) and (18),

$$\delta_{\rm exch}=(15\pm5)\times10^{-6}. \quad (19)$$

At present there seem to be no theoretical calculations of the contribution of exchange currents.

In view of the satisfactory agreement now of the experimental and theoretical values of P for hydrogen when the new value[20] of α is used and of the existence of accurate experimental values for the hfs anomalies of D, H^3 and He^3, it is important to reconsider the theory of the hfs anomalies for these light nuclei, including the contribution of exchange currents.

ACKNOWLEDGMENTS

The authors would like to thank R. J. Blume and S. A. Lewis for valuable experimental assistance.

[21] E. R. Cohen and J. W. M. Dumond, Rev. Mod. Phys. **37**, 537 (1965). The quantity $M_{\rm He^3}$ is taken from *Handbook of Physics*, edited by E. U. Condon and H. Odishaw (McGraw-Hill Book, Co., New York, 1958), part 9, Chap. 2.

[22] E. N. Fortson, F. G. Major, and H. G. Dehmelt, Phys. Rev. Letters **16**, 221 (1966).

[23] R. Novick and E. D. Commins, Phys. Rev. **111**, 822 (1958).

[24] S. B. Crampton, D. Kleppner, and N. F. Ramsey, Phys. Rev. Letters **11**, 338 (1963).

[25] C. K. Iddings and P. M. Platzman, Phys. Rev. **113**, 192 (1959); C. K. Iddings, *ibid.* **138**, B446 (1965); S. D. Drell and J. D. Sullivan, *ibid.* **154**, 1477 (1967).

[26] D. A. Greenberg and H. M. Foley, Phys. Rev. **120**, 1684 (1960).

[27] Y. C. Tang, E. W. Schmid, and R. C. Herndon, Nucl. Phys. **65**, 203 (1965); G. Breit, K. A. Friedman, and R. E. Seamon, Progr. Theoret. Phys. Suppl. **33-34**, 449 (1965).

VOLUME 33, NUMBER 1 PHYSICAL REVIEW LETTERS 1 JULY 1974

New Experimental Limit on T Invariance in Polarized-Neutron β Decay*

R. I. Steinberg†
Yale University, New Haven, Connecticut 06520, and Institut des Sciences Nucléaires, 38044 Grenoble, France

and

P. Liaud and B. Vignon
Institut des Sciences Nucléaires, 38044 Grenoble, France

and

V. W. Hughes
Yale University, New Haven, Connecticut 06520

(Received 9 May 1974)

We report an improved experimental upper limit for D, the triple-correlation coefficient in polarized-neutron decay. A nonzero value for this coefficient implies a breakdown of T invariance. We find that $D = -(1.1 \pm 1.7) \times 10^{-3}$, consistent with T invariance.

We are reporting a new measurement of D, the triple-correlation coefficient[1] in the β decay of the polarized free neutron. This coefficient is responsible for a term in the decay rate equal to

$$D\vec{P} \cdot (\vec{p}_e \times \vec{p}_{\bar{\nu}})/E_e E_{\bar{\nu}}, \qquad (1)$$

where $\vec{P}$ is the neutron polarization and $\vec{p}_e$, $\vec{p}_{\bar{\nu}}$, E_e, and $E_{\bar{\nu}}$ are the lepton momenta and energy, respectively. Since this quantity is odd under time reversal, a measurement of its coefficient D provides a test of T invariance, provided final-state interactions and momentum-transfer-dependent effects can be neglected. In neutron β decay the only significant final-state interaction is the Coulomb interaction, and the contribution to D from this interaction vanishes in a pure $V-A$ theory.[2] With present measured limits on possible scalar and tensor terms[3] in the effective weak Hamiltonian, this contribution is at most 10^{-3}. The principal momentum-transfer-dependent contribution to D is due to weak magnetism and has been calculated[4] using the conserved-vector-current hypothesis to be 2×10^{-5}. A measurement of D therefore provides a test of time-reversal invariance valid at least to the level of 10^{-3}.

The precision of the best previous measurement[5] of D was severely limited by counting statistics. A polarized-neutron beam intensity of 3×10^7 neutrons/sec yielded a counting rate of only 1/min, which allowed observation of 10^5 decay events. The resulting value of D was -0.01 ± 0.01.

VOLUME 33, NUMBER 1 PHYSICAL REVIEW LETTERS 1 JULY 1974

In the present experiment, a beam intensity of 10^9 neutrons/sec was achieved. Furthermore, since the beam consisted of cold neutrons (mean velocity ≈ 1100 m/sec) rather than the previously used thermal neutrons (2200 m/sec), a further twofold increase in the effective source strength was obtained. In addition, we have improved the detection geometry, so that a counting rate of 1.5 decays/sec was observed. Based upon observation of more than 5×10^6 events, we report the value

$$D = -(1.1 \pm 1.7)\times10^{-3},$$

where the error quoted is 1 standard deviation and is dominated by counting statistics. This value is consistent with T invariance and corresponds to a phase angle φ between the coupling constants g_A and g_V given by

$$\varphi = 180.14 \pm 0.22^\circ.$$

The experiment was performed at the high flux reactor (central flux $= 1.5\times10^{15}$ neutrons/cm^2 sec) of the Institut Laue-Langevin in Grenoble. Figure 1 shows the arrangement of the experiment. The beam was obtained from H14 (A in the figure), one of the five curved guide tubes viewing the liquid-deuterium cold-neutron moderator. The curvature of this guide tube ($\lambda_{\text{cutoff}} = 2.8$ Å) allowed high transmission of cold neutrons but effectively removed nearly all fast neutrons and γ rays originating in the reactor core, which would otherwise have constituted an intense source of background. The flux of the cold-neutron beam leaving H14 was 3×10^9 neutrons/cm^2 sec. The beam was then polarized by a magnetized curved guide tube (B). This instrument has been described previously.[6] The mean polarization of the beam

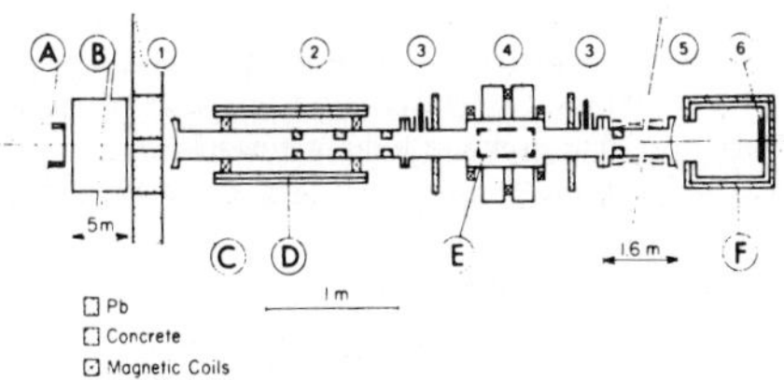

FIG. 1. Vertical section of the experimental setup: A, curved neutron-guide tube H14; B, polarizing guide tube; C, shielding; D, spin flipper; E, detection chamber; F, beam catcher; 1, Al entrance window to vacuum chamber; 2, beam collimators (Li^6F and Pb); 3, gate valves; 4, detectors; 5, LiF; 6, Li^6F.

was measured to be $(70\pm7)\%$ using a second magnetized guide tube as analyzer. The polarization measurement was performed both before and after the experiment as well as at approximately 2-week intervals during the $2\frac{1}{2}$-month period of data collection. No variations in the polarization were observed.

Upon leaving the polarizer, the beam had an intensity of 10^9 neutrons/sec and was roughly 5 cm high and 6 mm wide. The polarization direction, initially vertical, was adiabatically turned into the beam direction by means of the two coils of the spin flipper (D). For simple spin transmission, these two coils generated parallel magnetic fields, while spin flip was accomplished by reversing the current in the first coil, as suggested by Drabkin *et al.* and Hughes and Burgy.[7] With the two coils thus producing opposing magnetic fields, a region in which the combined magnetic field rapidly reversed direction was generated approximately midway between the coils. The neutron beam passed this region nonadiabatically and thereby underwent a spin flip. Depolarization of the beam in this low-field region was prevented by surrounding the entire two-coil spin flipper with a three-layer magnetic shield to exclude stray fields. The spin-flipping efficiency of this apparatus was 97%. During the experiment, the neutron polarization was reversed every second. The 3-G guide field in the decay region (E) was, of course, constant both in direction and magnitude to minimize gain shifts of the photomultiplier tubes. This guide field was maintained parallel to the beam direction within an error of 1°. After the decay region, the beam passed through a drift tube into the beam catcher (F).

A cross section of one section of the decay chamber is shown in Fig. 2. The beam direction was perpendicular to the plane of the paper. The decay chamber consisted of two such sections in series, for a total of eight detectors. Since the momentum of the neutron may be neglected, conservation of linear momentum allows the term (1) to be written

$$D\vec{P}\cdot(\vec{p}_p\times\vec{p}_e)/E_eE_{\bar{\nu}},$$

where $\vec{p}_p$ is the momentum of the recoil proton. The experimental geometry maximized the triple product by arranging the three vectors to be mutually perpendicular. At the same time the symmetrical arrangement greatly reduced systematic errors, as will be made clear below. Decay electrons originating from the beam (1) were detected by means of conventional plastic scintillation de-

VOLUME 33, NUMBER 1 PHYSICAL REVIEW LETTERS 1 JULY 1974

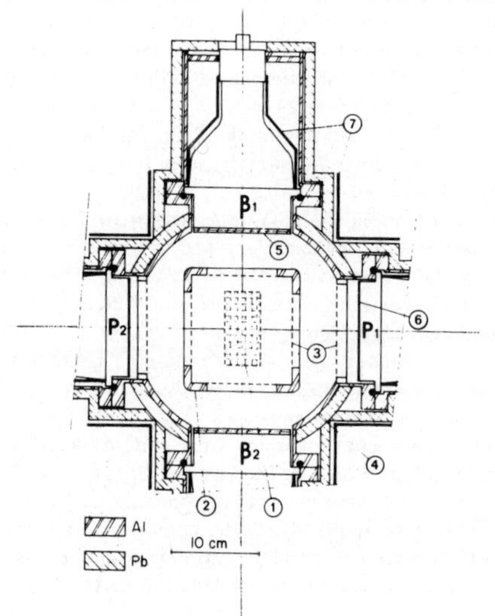

FIG. 2. Cross section of one section of decay chamber: 1, polarized neutron beam; 2, high-voltage box (20 kV); 3, proton-acceleration gap; 4, vacuum-chamber wall; 5, plastic scintillation β detector; 6, vacuum-evaporated 4000-Å layer of NaI(Tl) for proton detection; 7, magnetic shielding for photomultiplier tubes.

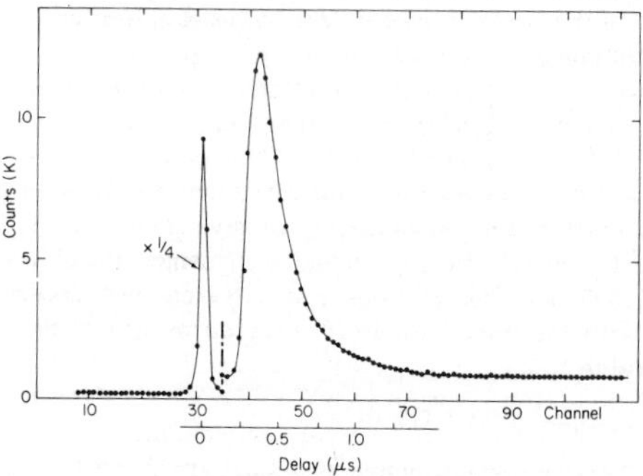

FIG. 3. Time-delay spectrum for coincidence pair $\beta 2p1$ for one of the two spin states. These data represent about two weeks of running time.

tectors (5) biased to accept electron energies between 100 and 500 keV. The recoil protons, after drifting through the field-free region inside the high-voltage box (2), were accelerated to 20 keV in the gap (3) and were counted by scintillation detectors (6) consisting of a vacuum-evaporated 4000-Å layer of NaI(Tl). The extreme thinness of these detectors allowed reduction of background while maintaining sensitivity to the protons. Special handling of the detectors was necessary because of the extremely hygroscopic character of NaI.

The electronics was based upon a single, multiplexed time-to-pulse-height converter which was started by pulses from the β detectors and stopped by pulses from the proton detectors. The sixteen resulting time spectra (four coincidence pairs for each sign of the neutron spin and for each of the two detector sections) were routed into separate regions of the memory of a 4096-channel analyzer. Figure 3 shows a time-delay spectrum for the coincidence pair $\beta 2p1$. The peak at channel 31 corresponding to $t=0$ is caused by background radiation being scattered from one detector into another. The broad peak at 0.4 μsec is due to the recoil protons, while the flat background is caused by accidental coincidences. The 0.4-μsec decay corresponds quite well with calculation of the transit time of the recoil protons from the decay volume through the field-free region and into the proton detector. The number of true coincidences was determined by simple subtraction of the flat background from the integrated recoil-proton peak.

The data are analyzed as follows. Let $\overrightarrow{N}_{\beta i p j}$ and $\overleftarrow{N}_{\beta i p j}$ be the numbers of true coincidences between β detector i and proton detector j for the two directions of the polarization vector. Thus

$$\overleftarrow{N}_{\beta 1 p 1}=c\Omega_{\beta 1}\Omega_{p1}\overleftarrow{e}_{\beta 1}\overleftarrow{e}_{p1}(1+KPD),$$

$$\overrightarrow{N}_{\beta 1 p 1}=c\Omega_{\beta 1}\Omega_{p1}\overrightarrow{e}_{\beta 1}\overrightarrow{e}_{p1}(1-KPD),$$

etc., where c is a constant proportional to the beam intensity, the Ω's are the solid angles subtended by counters $\beta 1$ and $p1$, and the e's are the detection efficiencies, where the possibility of shifts in these efficiencies as a function of the polarization direction has been allowed for. K is an instrumental coefficient, which for our apparatus has been calculated to be 0.45 ± 0.05.

Forming the combination

$$R=\frac{\overleftarrow{N}_{\beta 1 p 1}}{\overrightarrow{N}_{\beta 1 p 1}}\frac{\overrightarrow{N}_{\beta 1 p 2}}{\overleftarrow{N}_{\beta 1 p 2}}\frac{\overrightarrow{N}_{\beta 2 p 1}}{\overleftarrow{N}_{\beta 2 p 1}}\frac{\overleftarrow{N}_{\beta 2 p 2}}{\overrightarrow{N}_{\beta 2 p 2}}=\frac{(1+KPD)^4}{(1-KPD)^4},$$

it will be seen that variations in counter efficiencies, solid angles, and beam intensity are canceled. The symmetrical detector arrangement can also be shown to remove the effect of small misalignments of the polarization axis relative to the beam direction. The value of D is then determined from

$$D=\frac{1}{KP}\frac{R^{1/4}-1}{R^{1/4}+1}.$$

The authors wish to thank the Institut Laue-Langevin for allowing use of its excellent facilities, as well as Professor J. P. Longequeue and Professor J. Yoccoz for their interest in our work. The excellent technical assistance of Mr. C. Barnoux, Mr. J. Pouxe, and others is gratefully acknowledged as well as the help of Dr. V. Sailor and Dr. H. F. Foote, Jr., during the early stages of the experiment. One of us (R.I.S.) would also like to thank the National Science Foundation for a one-year postdoctoral fellowship and the Rutherford High Energy Laboratory for a research associateship.

---

*Work supported in part by the U.S. Atomic Energy Commission under Contract No. AT(11-1)-3075, in part by the Research Corporation, and in part by the Institut National de Physique Nucléaire et de Physique des Particules.

†Present address: Department of Physics and Astronomy, University of Maryland, College Park, Md. 20742.

[1]J. D. Jackson, S. B. Treiman, and H. W. Wyld, Jr., Phys. Rev. 106, 517 (1957).

[2]J. D. Jackson, S. B. Treiman, and H. W. Wyld, Jr., Nucl. Phys. 4, 206 (1957).

[3]H. F. Schopper, *Weak Interactions and Nuclear Beta Decay* (North Holland, Amsterdam, 1966).

[4]C. G. Callan, Jr., and S. B. Treiman, Phys. Rev. 162, 1494 (1967).

[5]B. G. Erozolimsky, L. N. Bondarenko, Yu. A. Mostovoy, B. A. Obinyakov, V. P. Zakharova, and V. A. Titov, Yad. Fiz. 11, 1049 (1970) [Sov. J. Nucl. Phys. 11, 583 (1970)].

[6]K. Berndorfer, Z. Phys. 243, 188 (1971).

[7]G. M. Drabkin, E. I. Zabidarov, Ya. A. Kasman, and A. I. Okorokov, Zh. Eksp. Teor. Fiz. 56, 478 (1969) [Sov. Phys. JETP 29, 261 (1969)]; see also D. J. Hughes and M. T. Burgy, Phys. Rev. 81, 498 (1951).

VOLUME 5, NUMBER 2 PHYSICAL REVIEW LETTERS JULY 15, 1960

FORMATION OF MUONIUM AND OBSERVATION OF ITS LARMOR PRECESSION*

V. W. Hughes, D. W. McColm, and K. Ziock
Gibbs Laboratory, Yale University, New Haven, Connecticut

and

R. Prepost
Nevis Laboratory, Columbia University, New York, New York
(Received June 17, 1960)

After the discovery[1,2] that muons which originate from pion decays are polarized and that positrons which originate from muon decays are emitted with an angular asymmetry with respect to the spin direction of the muons, it was realized[2,3] that it should be possible to observe the atom consisting of a positive muon and an electron (called muonium) and to measure its hyperfine structure. The present Letter reports the formation of muonium in pure argon gas; muonium is observed through its characteristic Larmor precession frequency.

Muonium in its ground $1\ ^2S_{1/2}$ state can be formed when a positive muon is slowed down in matter and captures an electron from an atom in the stopping material. If the muons are polarized, the muonium atoms formed should also have a net polarization.[4] Specifically, if the z axis of quantization is taken along the direction of the momentum of the incident muon beam, then the spin direction of a positive muon will be in the negative z direction to a good approximation.[5] Hence substates $(F, m_F) = (1, -1)$, $(1, 0)$, and $(0, 0)$ will form in the relative amounts 1/2, 1/4, and 1/4, respectively, where F is the quantum number for total atomic angular momentum and m_F is the associated magnetic quantum number. In the states $(1, 0)$ and $(0, 0)$ the muon is unpolarized; the time involved in establishing these unpolarized states subsequent to the electron capture by the muon is determined by the hyperfine structure interaction between the electron and the muon and is of the order of 10^{-10} sec. In the state $(F, m_F) = (1, -1)$ the muon is polarized with its spin along the negative z axis.

Many searches have been made for muonium since 1957. Usually[6,7] the attempt has been to observe the characteristic precession frequency of muonium in the state $(F, m_F) = (1, -1)$ in an external magnetic field H whose direction is perpendicular to the direction of the incoming muon beam (or equivalently to the axis of quantization). The muonium spin angular momentum will precess[8] about H with the frequency $f = \mu H/h$ [μ = magnetic moment of the $(1, -1)$ state of muonium], which is 1.39 Mc/sec-gauss. The published accounts indicate that solid or liquid targets have been used. An experiment has also been done[9] to look for a Zeeman transition between the $F = 1$ magnetic sublevels of muonium induced by a radio-frequency field, when muons are stopped in N_2O gas at 50 atm pressure.

A wide variation has been observed in the amount of depolarization of positive muons stopped in different materials.[6,10,11] The polarization is measured in a precession experiment in which the magnetic field is chosen so as to detect the precession of a free muon; in such an experiment polarized muonium would appear unpolarized because of its high precession frequency. The observed variation in polarization has often been ascribed to varying amounts of muonium formation. However, this explanation is not necessarily correct because one can conceive of other depolarizing mechanisms, e. g., those resulting from chemical reactions of the muon.[12]

The present experiment was designed to search for muonium in pure argon gas at high pressure. The argon gas was contained in a stainless steel cylinder at a pressure of 50 atm and was purified by recirculation over titanium sponge heated to 500°C. First the depolarization of the muons stopped in the gas was measured in a free-muon precession experiment.[1] The value of the ratio of the asymmetry parameter, a, for argon to that for carbon was 0.08 ± 0.15, and thus within the accuracy of the experiment no free polarized muons remain in argon. This result is a necessary condition that all the muons form muonium.

As a specific search for muonium, the characteristic precession frequency of muonium in the state $(F, m_F) = (1, -1)$ was looked for in the manner indicated in Fig. 1. An external magnetic field H between 3 and 5 gauss was applied perpendicular both to the direction of the incoming muon beam and to the direction from the target to the positron telescope, and its value was measured to an accuracy of 1% with an electron reso-

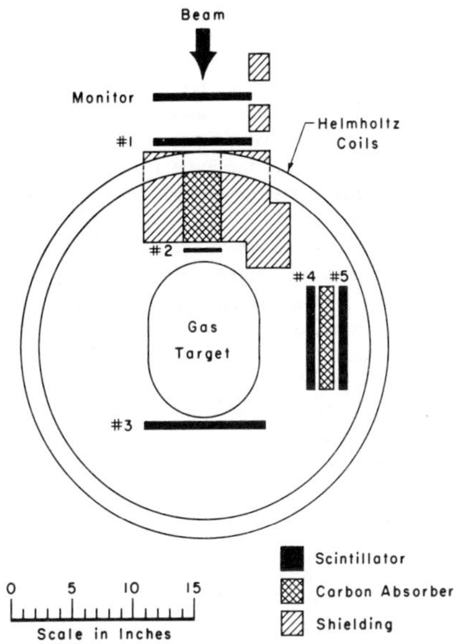

FIG. 1. Experimental arrangement.

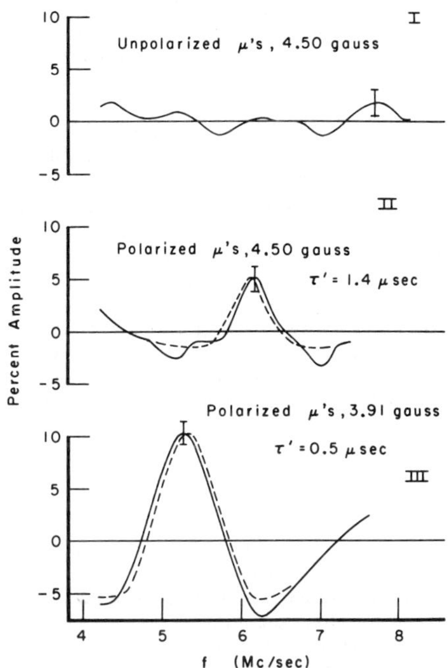

FIG. 2. Frequency analysis.

nance spectrometer using DPH free radicals. The distribution of time intervals between the incoming muon pulses (1-2-$\bar{3}$ coincidences) and those decay positron pulses (4-5 coincidences) which came between 0.2 and 2.0 microseconds after the muon pulse was converted to a spectrum of pulse heights which was fed into a pulse-height analyzer. The characteristic precession of polarized muonium should appear as an oscillation in the pulse-height analyzer data.

The data were assumed to represent the sum of four components, due to (1) polarized muonium, (2) polarized free muons, (3) unpolarized muons, and (4) accidental coincidences. We determined the amplitudes of (2) through (4) by a least squares procedure in which (1), which must be represented by a rapidly varying periodic function, would not contribute. With these computed amplitudes, the contributions of (2) through (4) to the data were then subtracted, leaving numbers y_i containing only the desired oscillation due to polarized muonium. A least squares fit to the y_i was made using the function

$$y_i = e^{-t_i/\tau}\left\{C + Ae^{-t_i/\tau'}\sin[2\pi f(t_i + t_0)]\right\}.$$

Here τ is the muon mean lifetime; τ' is a parameter introduced in order to allow for line broadening, either through field inhomogeneities, drifts in electronics, or depolarization of muonium in collisions; f is the trial value for the precession frequency of the magnetic moment of muonium; t_0 is the time delay between the stopping of the muons and the emission of the earliest positrons used in the data analysis. An IBM-650 computer was used for these computations.

The results of this analysis are shown in Fig. 2 for three sets of data; for two of the cases (II and III) polarized muons were stopped in argon at different magnetic fields and for case I pions were stopped (and hence unpolarized muons were produced). The solid curves were obtained from the analysis of the data and represent the percent amplitude of A compared to the total counting rate. The error bars correspond to an error of one standard deviation in the percent amplitude. The dashed curves are theoretical line shapes centered in each case at the muonium precession frequency predicted from the measured value of the magnetic field. The theoretical lines were computed with the values of τ' and t_0 determined from the data analysis and were normalized to the peak amplitudes of the lines computed from the data.

Resonances are clearly seen in cases II and III at the frequencies which are predicted for muonium precession on the basis of the magnetic field measurements. The observed and predicted resonance frequencies agree within the experimental uncertainty of 0.2 Mc/sec. The observed amplitudes of the resonances are 4 to 5 standard deviations. In case I where unpolarized muons were used, no resonance is observed. In three other cases, not shown here, where polarized muons were used at different magnetic fields, resonances were observed at the precession frequency of muonium; in one additional case with unpolarized muons, no resonance frequency was found. Hence we concluded that muonium is formed in pure argon. Although it is difficult to relate quantitatively the percent amplitude shown in Fig. 2 to the fraction of muons which form muonium, the data do indicate that close to 100% of the muons form muonium in pure argon.

Muonium is of interest principally because it is the simplest system involving a muon and an electron, and hence it offers the greatest promise for a precise study of the interaction between these two particles and thus for a test of the quantum electrodynamic field theory of the muon, electron, and photon system. In particular, it would be of great value to measure the hyperfine structure separation, $\Delta\nu$, in the ground 1 $^2S_{1/2}$ state of muonium. Under the assumption that the muon is a Dirac particle, the theoretical value for $\Delta\nu$ is given by

$$\Delta\nu=\left\{(\tfrac{16}{3})\alpha^2cR_\infty\frac{\mu_\mu}{\mu_0}\right\}\left\{1+\frac{m}{M}\right\}^{-3}\left\{1+\frac{\alpha}{\pi}\right\},$$

in which α = fine structure constant, c = velocity of light, R_∞ = Rydberg constant for infinite mass, m = electron mass, M = muon mass, $\mu_0 = e\hbar/2mc$ = Bohr magneton, $\mu_\mu = e\hbar/2Mc$ = muon magneton. The first bracketed term is the Fermi expression; the second term is a reduced-mass correction; the last term includes the lowest order anomalous magnetic moments of the electron and muon. Use of the known values of the atomic constants[13] gives $\Delta\nu = 4464.0$ Mc/sec. The next order α^2 term has not yet been calculated but is, of course, calculable in principle from a Bethe-Salpeter equation for the bound state of the muon and electron. Comparison of a precise experimental value for $\Delta\nu$ with the theoretical value would provide a critical test of electrodynamics involving the muon and could reveal an anomalous structure of the muon. In view of the abundant formation of muonium that we have found, we are hopeful that a measurement of the hyperfine structure of muonium will be possible and we are preparing to do this experiment.

It is a pleasure to acknowledge encouragement and support from and helpful discussions with Professor L. Lederman and Dr. S. Penman.

---

*This research has been supported in part by the Air Force Office of Scientific Research (with Yale), and also by the Office of Naval Research and the U. S. Atomic Energy Commission (with Columbia).

[1]R. L. Garwin, L. M. Lederman, and M. Weinrich, Phys. Rev. 105, 1415 (1957).

[2]J. I. Friedman and V. L. Telegdi, Phys. Rev. 105, 1681 (1957).

[3]V. W. Hughes, Bull. Am. Phys. Soc. 2, 205 (1957).

[4]G. Breit and V. W. Hughes, Phys. Rev. 106, 1293 (1957).

[5]No direct measurement of the muon spin direction has been made, but measurements of the helicity of the decay positron together with theoretical arguments suggest that the helicity of the positive muon from π decay is negative. If the muon spin were in the $+z$ direction rather than in the $-z$ direction, the labeling of the muonium states would be changed, but none of our conclusions about muonium would be altered. See G. Culligan, S. G. F. Frank, J. R. Holt, J. G. Kluyver, and T. Massam, Nature, 180, 751 (1957); and P. C. Macq, K. M. Crowe, and R. P. Haddock, Phys. Rev. 112, 2061 (1958).

[6]R. A. Swanson, Phys. Rev. 112, 580 (1958).

[7]J. M. Cassels, T. W. O'Keefe, M. Rigby, A. M. Wetherell, and J. R. Wormald, Proc. Phys. Soc. (London) A70, 543 (1957); and J. M. Cassels, Proceedings of the Seventh Annual Rochester Conference on High-Energy Nuclear Physics, 1957 (Interscience Publishers, New York, 1957), Chap. VII, p. 38.

[8]V. Bargmann, L. Michel, and V. L. Telegdi, Phys. Rev. Letters 2, 435 (1959).

[9]D. McColm, V. W. Hughes, A. Lurio, and R. Prepost, Bull. Am. Phys. Soc. 4, 82 (1959).

[10]V. W. Hughes, A. Lurio, D. Malone, L. Lederman, and M. Weinrich, Bull. Am. Phys. Soc. 3, 51 (1958).

[11]M. Weinrich, Ph.D. thesis, Columbia University, 1958 (unpublished).

[12]V. W. Hughes, Phys. Rev. 108, 1106 (1957).

[13]C. M. Sommerfield, Phys. Rev. 107, 328 (1957), and Ann. Phys. 5, 26 (1958). J. W. M. DuMond, Ann. Phys. 7, 365 (1959); R. L. Garwin, D. P. Hutchinson, S. Penman, and G. Shapiro, Phys. Rev. 118, 271 (1960).

VOLUME 8, NUMBER 3 PHYSICAL REVIEW LETTERS FEBRUARY 1, 1962

HYPERFINE STRUCTURE OF MUONIUM*

K. Ziock, V. W. Hughes, R. Prepost, J. Bailey, and W. Cleland
Yale University, New Haven, Connecticut
(Received January 5, 1962)

The discovery of muonium by observation of its characteristic Larmor precession frequency[1] and the rough measurement of its hyperfine structure splitting, $\Delta\nu$, by use of a static magnetic field[2] provided the basic information required to plan a precision measurement of the hfs of muonium in its ground $1\,^2S_{1/2}$ state by a special microwave spectroscopy technique.

If the muon is a particle which obeys the modern Dirac theory and which differs from the electron only in its mass value, then the hfs of muonium can be calculated from the quantum electrodynamic theory of the muon, electron, and photon fields. The result can be expressed as a power series in the small parameters α and (m_e/m_μ), and to terms of order α^2 and $\alpha(m_e/m_\mu)$ is given by[3]

$$\Delta\nu\,(\text{theor})=\left(\frac{16}{3}\alpha^2 cR_\infty\frac{\mu_\mu}{\mu_0}\right)\left(1+\frac{m_e}{m_\mu}\right)^{-3}\left(1+\frac{3}{2}\alpha^2\right)\left(1+\frac{\alpha}{2\pi}-0.328\frac{\alpha^2}{\pi^2}\right)$$

$$\times\left(1+\frac{\alpha}{2\pi}+0.75\frac{\alpha^2}{\pi^2}\right)\left(1-1.81\alpha^2\right)\left(1-\frac{3\alpha}{\pi}\frac{m_e}{m_\mu}\ln\frac{m_\mu}{m_e}\right),$$

in which α = fine structure constant, c = velocity of light, R_∞ = Rydberg constant for infinite mass,

VOLUME 8, NUMBER 3 PHYSICAL REVIEW LETTERS FEBRUARY 1, 1962

μ_μ = muon magneton $(e\hbar/2m_\mu c)$, μ_0 = electron Bohr magneton, m_e = electron mass, and m_μ = muon mass. The first bracketed factor is the Fermi value for the hfs; the second factor is a reduced mass correction; the third factor is the Breit relativistic correction; the fourth and fifth factors are the $g/2$ values for the electron and the muon; the sixth factor is a second order radiative correction; the seventh factor is a relativistic recoil factor. Use of the best modern values of the fundamental atomic constants[4] gives

$$\Delta\nu(\text{theor}) = 4463.13 \pm 0.10 \text{ Mc/sec.}$$

In computing this value we have used

$$(m_\mu/m_e) = (206.76 \pm 0.02),^5 \quad \alpha^{-1} = 137.0391 \pm 0.0006,$$

and[6]

$$\frac{\mu_\mu}{\mu_0} = \left(\frac{\mu_\mu'}{\mu_p}\right)\left(\frac{\mu_p}{\mu_e'}\right) \times \left(1 + \frac{\alpha}{2\pi} - 0.328\frac{\alpha^2}{\pi^2}\right) \Big/ \left(1 + \frac{\alpha}{2\pi} + 0.75\frac{\alpha^2}{\pi^2}\right),$$

in which μ_μ' = muon spin magnetic moment, μ_e' = electron spin magnetic moment, and μ_p = proton magnetic moment. The uncertainty in $\Delta\nu$(theor) is contributed primarily by the uncertainties in the knowledge of α and of (μ_μ'/μ_p).

The effect of a "breakdown of quantum electrodynamics" on the muon magnetic moment has been discussed[7] and any breakdown would also alter the theoretical expression given for $\Delta\nu$. In addition to an alteration in the muon magnetic moment there might also be a structure factor for the muon which could be observed in a measurement of $\Delta\nu$ but not in a measurement of the magnetic moment, as for the case of the proton for which a knowledge of the structure as well as the magnetic moment is needed to determine the value of $\Delta\nu$ for hydrogen.[8]

The present experiment involves the observation of an induced microwave transition between the two hfs magnetic substates of muonium designated by their strong-field quantum numbers $(m_J, m_\mu) = (\frac{1}{2}, \frac{1}{2}) \leftrightarrow (\frac{1}{2}, -\frac{1}{2})$ (m_J and m_μ are the magnetic quantum numbers of the electron and the muon). The transition is observed under approximately strong-field conditions for which the transition frequency is roughly $\Delta\nu/2$. Use of the Breit-Rabi formula[9] allows an exact calculation of $\Delta\nu$ from the observed resonance condition of microwave frequency and static magnetic field. Parity nonconservation in the decay of π mesons[10] results in polarized muons and hence in the initial formation of muonium in only the $m_\mu = +\frac{1}{2}$ states. Parity nonconservation in the decay of the muons[10] allows the determination of the muon spin state through the nonisotropic angular distribution of the decay positrons, and hence the observation of an induced transition between different muon spin states.

The muons are obtained from the Columbia University Nevis synchrocyclotron. The experimental arrangement is shown in Fig. 1 and, apart from the microwave cavity, is similar to that of our previous experiment.[2] The microwave cavity operates in the TM_{110} mode with the axial direction coincident with that of the static magnetic field, and it is fed by a 1-kilowatt klystron amplifier. It is a thin-walled, high-Q cavity filled with highly purified argon gas at a pressure of about 55 atm and is contained in a stainless steel pressure tank. The stopping of a muon in the gas target is indicated by a $12\bar{3}$ coincident count. A decay positron is indicated by a $34\bar{2}$ coincident count and is registered as an "event" if it occurs in the time interval between 0.1 and 3.3 μsec subsequent to the $12\bar{3}$ count. The number of events and the number of $12\bar{3}$ counts are measured, with the microwave field off and with the microwave field on, as a function of the static magnetic field. The ratio R is then computed,

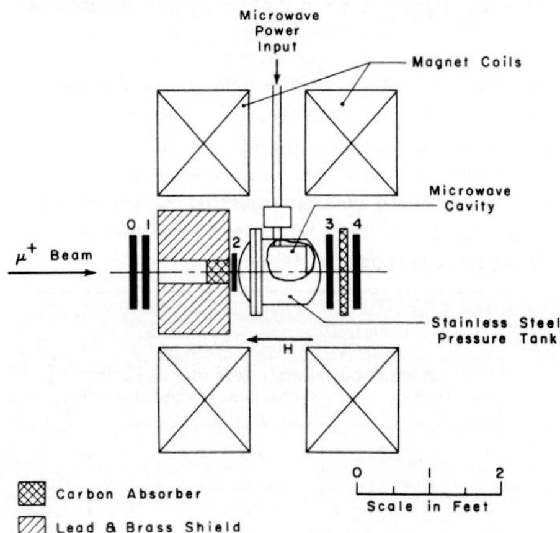

FIG. 1. Experimental arrangement. 0, 1, 2, 3, and 4 are plastic scintillation counters.

where

$$R = \frac{(\text{events}/12\bar{3})_{\text{microwaves on}}}{(\text{events}/12\bar{3})_{\text{microwaves off}}}.$$

If the microwave field induces a transition between the two hfs magnetic substates $(m_J, m_\mu) = (\frac{1}{2}, \frac{1}{2}) \rightarrow (\frac{1}{2}, -\frac{1}{2})$, the angular distribution of the decay positrons will be changed from $C(1+A\cos\theta)d\Omega$ to $C(1-A\cos\theta)d\Omega$, in which C is a constant, $A \simeq \frac{1}{3}$, and θ is the angle between the direction of the static magnetic field and the direction of emission of the positron. Hence the transition should be observed as an increase of R to a value greater than 1.

An observed curve is shown in Fig. 2. Measured values of the quantity R are plotted vs values of the magnetic field for a fixed microwave frequency; the solid curve has a Lorentzian form whose constants are chosen by use of the Yale IBM-709 computer to give a least-squares fit to the experimental data. The linewidth of 120 gauss is due primarily to the high microwave power level used. Another resonance curve was obtained with about $\frac{1}{4}$ the power level and it has a linewidth of 44 gauss. A resonance curve was also obtained at a reduced argon pressure of 35 atm in order to test for a dependence of $\Delta\nu$ on pressure. Within our experimental accuracy no pressure shift was observed, which is consistent with the known pressure shift for hydrogen in argon.[11] It was observed that the resonance curve disappeared when 200 parts per million of air was added to the pure argon. We believe that this probably indicates the occurrence of chemical reactions of oxygen or nitrogen molecules with muonium, which is similar chemically to atomic hydrogen.

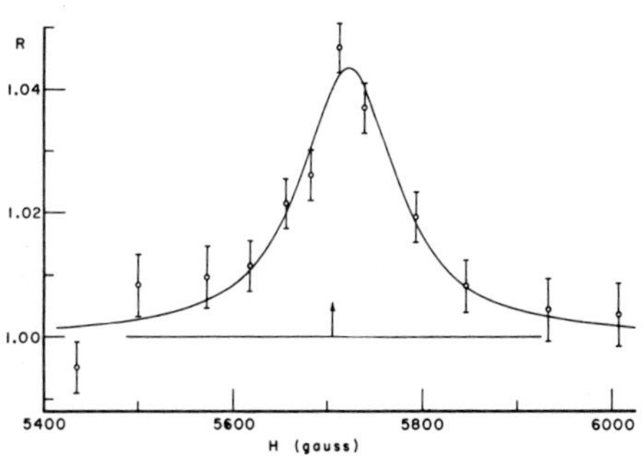

FIG. 2. Experimental values of R vs static magnetic field for a microwave frequency of 1850.08 Mc/sec, and with an argon pressure of 55 atm. The solid curve is a least-squares fit Lorentzian curve with $H_{\text{center}} = 5725$ gauss and with width = 120 gauss. Error flags are ±(one sample standard deviation). The arrow indicates the theoretical line center, ignoring the pressure shift.

On the basis of four resonance curves we determine the hfs splitting of muonium in its ground $1\,^2S_{1/2}$ state to be

$$\Delta\nu(\text{expt}) = 4461.3 \pm 2.2 \text{ Mc/sec}.$$

The error quoted is due primarily to the inhomogeneity of the magnetic field, but it also includes an estimated upper limit of the unknown pressure shift. The experimental value agrees with the theoretical value. Experiments are being planned with a more homogeneous magnetic field in order to obtain higher accuracy.

---

*This research was supported in part by the Air Force Office of Scientific Research (with Yale University) and also by the Office of Naval Research and the U. S. Atomic Energy Commission (with Columbia University).

[1]V. W. Hughes, D. W. McColm, K. Ziock, and R. Prepost, Phys. Rev. Letters 5, 63 (1960).

[2]R. Prepost, V. W. Hughes, and K. Ziock, Phys. Rev. Letters 6, 19 (1961).

[3]R. Karplus and A. Klein, Phys. Rev. 85, 972 (1952); N. M. Kroll and F. Pollock, Phys. Rev. 86, 876 (1952); R. Arnowitt, Phys. Rev. 92, 1002 (1953).

[4]E. R. Cohen, K. M. Crowe, and J. W. M. DuMond, Fundamental Constants of Physics (Interscience Publishers, Inc., New York, 1957); J. W. M. DuMond, Ann. Phys. (New York) 7, 365 (1959).

[5]J. Lathrop, R. A. Lundy, V. L. Telegdi, R. Winston, and D. D. Yovanovitch, Nuovo cimento 17, 109 (1960); J. Lathrop, R. A. Lundy, S. Penman, V. L. Telegdi, R. Winston, D. D. Yovanovitch, and A. J. Bearden, Nuovo cimento 17, 114 (1960); S. Devons, G. Gidal, L. M. Lederman, and G. Shapiro, Phys. Rev. Letters 5, 330 (1960).

[6]D. P. Hutchinson, J. Menes, G. Shapiro, A. M. Patlach, and S. Penman, Phys. Rev. Letters 7, 129 (1961); S. H. Koenig, A. G. Prodell, and P. Kusch, Phys. Rev. 88, 191 (1952); N. F. Ramsey, in Experimental Nuclear Physics, edited by E. Segrè (John Wiley & Sons, New York, 1953), Vol. I, p. 430.

[7]G. Charpak, F. J. M. Farley, R. L. Garwin, T. Muller, J. C. Sens, V. L. Telegdi, and A. Zichichi, Phys. Rev. Letters 6, 128 (1961).

[8]A. C. Zemach, Phys. Rev. 104, 1771 (1956).

[9]P. Kusch and V. W. Hughes, Encyclopedia of Physics, edited by S. Flügge (Springer-Verlag, Berlin, 1959), Vol. 37, Part 1.

[10]R. L. Garwin, L. M. Lederman, and M. Weinrich, Phys. Rev. 105, 1415 (1957).

[11]L. W. Anderson, F. M. Pipkin, and J. C. Baird, Jr., Phys. Rev. 120, 1279 (1960); Phys. Rev. 122, 1962 (1961).

Reprinted from THE JOURNAL OF CHEMICAL PHYSICS, Vol. 44, No. 11, 4354–4355, 1 June 1966
Printed in U. S. A.

Muonium Chemistry*

R. M. MOBLEY, J. M. BAILEY,† W. E. CLELAND,†
V. W. HUGHES, AND J. E. ROTHBERG
Gibbs Laboratory, Yale University, New Haven, Connecticut
(Received 17 March 1966)

MUONIUM (M) is the atom consisting of a positive muon and an electron, and it will behave as a light isotope of hydrogen with regard to its atomic interactions and chemical reactions since the muon mass is 207 times the electron mass and since the muon mean lifetime of 2.2 μsec is long compared to electron atomic orbital times. The present Letter reports on two methods for studying muonium chemistry and on initial results for the interactions of muonium with a number of molecules.[1]

The methods utilize the powerful techniques of elementary particle physics and rely upon parity nonconservation in the weak interactions responsible for the production and decay of the muon. From the decay of positive pions, polarized positive muons are produced, and in the decay of positive muons into positrons and neutrinos the positrons are emitted preferentially in the direction of the muon spin. Hence collisions which produce changes in the muon spin direction can be studied. Polarized muons are stopped in pure argon gas to form muonium atoms[2] and effects due to the admixture of small fractional amounts of various molecules as impurities are observed.

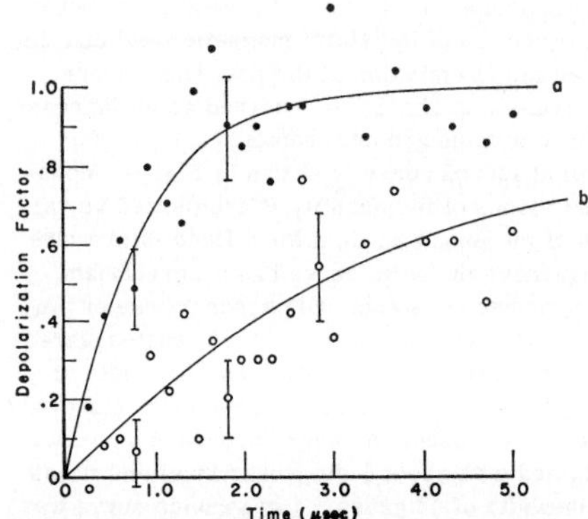

FIG. 2. Depolarization factor, $1-e^{-\lambda t}$, vs time, where λ is the fitted depolarization rate, under the assumption that the muon polarization varies as $P=P_0e^{-\lambda t}$. The partial pressures of NO are 0.37 mm of Hg for Curve a and 0.13 mm of Hg for Curve b.

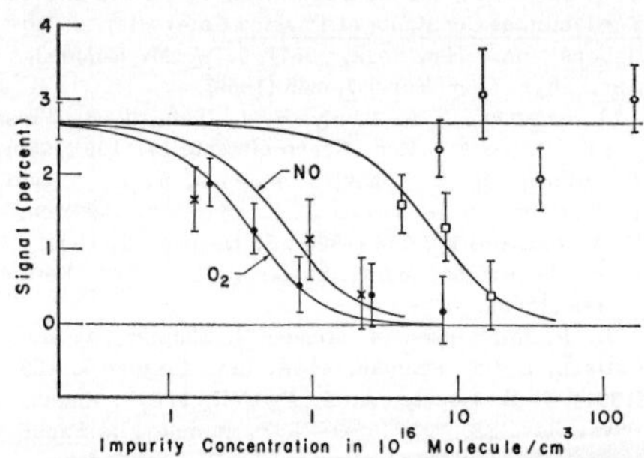

FIG. 1. Signal at resonance of the transition between muonium hfs states $(m_J=\frac{1}{2},\ m_\mu=\frac{1}{2})\rightleftarrows(m_J=\frac{1}{2},\ m_\mu=-\frac{1}{2})$ at 5200 G vs impurity concentration (m_J and m_μ are the electron and muon magnetic quantum numbers). The fitted curves are of the form $S=S_0\{[(1/\tau)+\gamma_c]^2+4|b|^2\}^{-1}$, in which $|b|^2$ is a measured constant proportional to the rf power, τ is the muon mean lifetime, and $\gamma_c=n\bar{v}_{rel}\sigma_1$, where σ_1 is the fitted cross section, n is the impurity concentration, and $\bar{v}_{rel}$ is the mean relative velocity of muonium and impurity atom. ●, O_2; □, C_2H_4; ×, NO; ○, H_2.

The first method involves the use of a magnetic resonance transition between two hyperfine-structure magnetic substates of muonium.[3] The intensity of the resonance signal is studied as a function of the impurity content with the results shown in Fig. 1. Decrease in the signal implies collisions which remove muonium from the resonant states. These data are analyzed to yield an effective cross section σ_1 for such a collision and the results are given in Table I. For the paramag-

TABLE I. Muonium-molecule cross sections.

Gas	σ_1 (10^{-16} cm^2)	σ_2 (10^{-16} cm^2)	σ_1/σ_2
NO	3.2±1.5	0.27±0.08	12±7
O_2	5.4±2.5	0.31±0.08	17±9
C_2H_4	0.29±0.16	0.024±0.006	12±7

netic molecules NO and O_2 an electron spin exchange collision which transfers muonium from one hfs magnetic substate to another is probably the reaction mechanism.[4] For C_2H_4, which is an unsaturated hydrocarbon, a muonium-containing molecule may be formed. No reaction is observed with H_2, which is consistent with the facts that H_2 is not paramagnetic and that the chemical reaction $M+H_2 \rightarrow MH+H$ is forbidden on energetic grounds for thermal muonium due to the high vibrational energy of MH.

The second method involves the measurement of the polarization of the muons as a function of time and of impurity concentration by use of a precision digital time analyzer following the scintillation counters for the positrons.[5] Such data are shown in Fig. 2 for NO. These data are analyzed to yield an effective cross section σ_2 for depolarizing collisions (see Table I). The cross sections σ_1 are much larger than σ_2. For an electron spin exchange reaction occurring in a strong magnetic field this difference is due to the fact that the most probable transitions are between two hfs substates with different directions of the electron spin but the same direction of the muon spin, and this type of collision results in a decrease in the resonance signal but not in depolarization, which occurs only in transitions between two hfs substates with different directions of the muon spin. The magnetic field was about 5200 G and hence of an intermediate strong-field character so that the relative effective cross sections for depolarization and resonance signal quenching should be about 1 to 12 based on a calculation of the intermediate field character of the eigenstates, in agreement with the results of Table I.

Further data are being obtained which will include the effect of magnetic field on the effective depolarizing collision rate and studies of other molecules.

* This research has been supported in part by the U.S. Air Force Office of Scientific Research (with Yale University) and by the U.S. Office of Naval Research (with Columbia University).

† Present address: CERN, Geneva, Switzerland.

[1] R. M. Mobley, J. M. Bailey, W. E. Cleland, V. W. Hughes, and J. E. Rothberg, Proc. Intern. Conf. Phys. Electronic At. Collisions 4th Univ. Laval, Quebec, 1965, 194 (1965); Bull. Am. Phys. Soc. **10**, 80 (1965).

[2] V. W. Hughes, D. W. McColm, K. Ziock, and R. Prepost, Phys. Rev. Letters **5**, 63 (1960).

[3] W. E. Cleland, J. M. Bailey, M. Eckhause, V. W. Hughes, R. M. Mobley, R. Prepost, and J. E. Rothberg, Phys. Rev. Letters **13**, 202 (1964).

[4] H. C. Berg, Phys. Rev. **137**, A1621 (1965).

[5] S. L. Meyer, E. W. Anderson, E. Bleser, L. M. Lederman, J. L. Rosen, J. Rothberg, and I.-T. Wang, Phys. Rev. **132**, 2693 (1963).

Reprinted from

THE JOURNAL OF CHEMICAL PHYSICS VOLUME 47, NUMBER 8 15 OCTOBER 1967

Letters to the Editor

Muonium Chemistry II*

R. M. MOBLEY, J. J. AMATO, V. W. HUGHES, J. E. ROTHBERG, AND P. A. THOMPSON

Gibbs Laboratory, Yale University, New Haven, Connecticut

(Received 24 July 1967)

Initial studies of the interactions of muonium (M= μ^+e^-) with atoms and molecules were reported in a previous Letter to *The Journal of Chemical Physics*.[1,2] Two related experimental methods were described for studying the interactions of muonium with molecules admixed as small fractional impurities in an argon buffer gas. The first method involved the observation of the intensity of a resonance transition as a function of impurity concentration, and yielded a signal-quenching cross section σ_1. The second method involved the observation of the polarization of the muons as a function of time and of impurity concentration, and yielded a depolarization cross section σ_2. The present Letter reports on much more extensive data on σ_2 as a function of static magnetic field and for additional molecules, and on the results of the analysis of all data obtained. Analysis of the data on σ_2 versus magnetic field establishes that the reaction mechanism for collisions of M with the paramagnetic molecules NO and O_2 is an electron-spin exchange process.

The muon polarization P can be determined from the angular distribution of the decay positions and is assumed to vary with time t as

$$P(t)=P_0\exp(-\lambda_2 t). \tag{1}$$

P_0 is the initial polarization of the muons which form muonium in the hyperfine structure levels of its ground state, and it depends on the external magnetic field because the hfs energy eigenstates of muonium are field dependent.[2] The quantity $\lambda_2=n\bar{v}\sigma_2$ is the depolarization rate due to collisions, where n is the number of interacting molecules per cm^3 and $\bar{v}$ is the mean relative velocity of the muonium atoms and the molecules. Measured values of λ_2/n as a function of the value of a static magnetic field H antiparallel to the direction of the incoming muon beam are shown as the points with error bars in Fig. 1, for the molecules NO, O_2, C_2H_4, and NO_2.

If the collision depolarization mechanism is an electron-spin exchange process,[3] as is expected for a paramagnetic molecule, then the theory of the muon depolarization can be given in terms of a density matrix formulation of the equations for the populations of the four hfs levels. The validity of Eq. (1) is established and the dependence of λ_2 on H is found to be

$$\lambda_2=\lambda_{20}/(1+x^2)^{1/2}, \tag{2}$$

in which λ_{20} is the depolarization rate for $H=0$ and $x=H(\text{gauss})/1585$. The theory treats the paramagnetic molecule as having either electronic spin $S=\frac{1}{2}$ (NO) or electronic spin $S=1$ (O_2) and neglects molecular fine structure; for the case $S=\frac{1}{2}$, $\lambda_{20}=n\bar{v}\sigma_{SE}$ and for

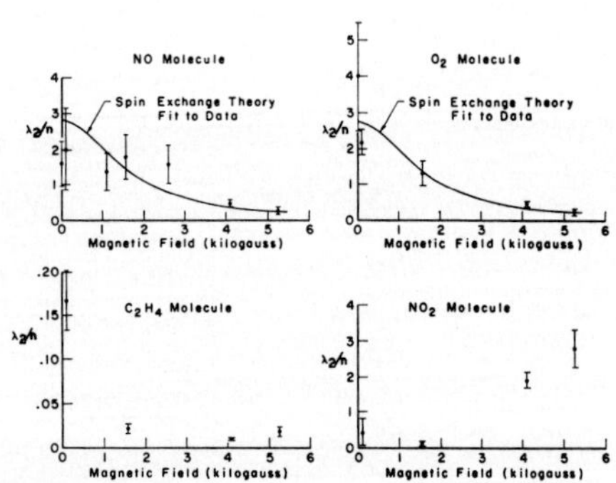

FIG. 1. Observed values of λ_2/n versus H for the molecules NO, O_2, C_2H_4, and NO_2 in units of 10^{-16} (μsec·molecules/cm^3)$^{-1}$. The error bars indicate one standard deviation statistical counting errors. The solid curves for NO and O_2 are theoretical curves based on an electron-spin-exchange collision process.

TABLE I. Muonium—molecule cross sections.

Molecule	Postulated interaction	σ at 5250 G (unit = 10^{-16} cm^2)	σ_{SE} (unit = 10^{-16} cm^2)
NO_2	$NO_2+M\rightarrow NO+OM$	≥ 23	...
O_2	Spin exchange	5.4 ± 2.5	5.9 ± 0.6
NO	Spin exchange	3.2 ± 1.5	7.1 ± 1.0
C_2H_4	$C_2H_4+M\rightarrow C_2H_4M$	0.29 ± 0.16	...
H_2, N_2, SF_6	...	≤ 0.01	...

$S=1$, $\lambda_{20}=(32/27)\bar{n}v\sigma_{SE}$ in which σ_{SE} is the electron-spin-exchange cross section. The solid curves in Fig. 1 for NO and O_2 are fits of Eq. (2) to the experimental points; these curves agree well with the experimental points and thus confirm that the depolarization mechanism is a spin-exchange process and determine the values of σ_{SE} given in Table I. The spin-exchange cross section for M and NO and for M and O_2 are about $\frac{1}{3}$ the corresponding cross sections[4] for H and NO [$(25\pm2.5)\times10^{-16}$ cm^2] and for H and O_2 [$(21\pm2.1)\times10^{-16}$ cm^2], which may be due to the fact that if M and H have the same kinetic energy, fewer partial waves are important for the muonium collisions.[5]

For the nonparamagnetic molecules C_2H_4 and NO_2, the experimental points (Fig. 1) cannot be fit by curves of the form of Eq. (2). Postulated reaction mechanisms are indicated in Table I. The strong dependence of the depolarization rate on H for C_2H_4 may indicate that the internal magnetic field on the muon in C_2H_4M is small (less than about 100 G). For NO_2 the magnetic-field dependence is not understood. In Table I, values of signal-quenching cross sections σ_1 are given for all the molecules studied by this method. These values were observed at a field of 5250 G, but they are not expected to vary much with H. The comparison of values of σ_1 and σ_{SE} has been discussed.[1,2]

Other molecules, including C_2H_6, Cl_2, CH_3Cl, CO_2, and H_2O were studied using the depolarization method. With large concentration of impurity gases, we obtained the following depolarization cross sections at 100 G: C_2H_6[$(2\pm1)\times10^{-18}$ cm^2], CH_3Cl [$(2\pm0.5)\times10^{-19}$ cm^2], Cl_2[1.1+0.9)$\times10^{-17}$ cm^2], and CO_2 [$(4\pm2)\times10^{-19}$ cm^2]. Since the concentrations were greater than 1000 ppm, these results may be due to reactions with nonthermal muonium, since of the order of 10^3 elastic collisions in argon are required for thermalization.

* This research has been supported in part by the U.S. Air Force Office of Scientific Research (with Yale University) and by the U.S. Office of Naval Research (with Columbia University).

1 R. M. Mobley, J. M. Bailey, W. E. Cleland, V. W. Hughes, and J. E. Rothberg, J. Chem. Phys. **44**, 4354 (1966); R. M. Mobley, J. J. Amato, V. W. Hughes, J. E. Rothberg, and P. A. Thompson, Bull. Am. Phys. Soc. **12**, 104 (1967).

2 V. W. Hughes, Ann. Rev. Nucl. Sci. **16**, 445 (1966).

3 E. M. Purcell and G. B. Field, Astrophys J. **124**, 542 (1956); P. L. Bender, Phys. Rev. **132**, 2154 (1963).

4 H. C. Berg, Phys. Rev. **137**, A1621 (1965).

5 A. E. Glassgold and S. A. Lebedeff, Ann. Phys. (N.Y.) **28**, 181 (1964).

PHYSICAL REVIEW LETTERS

VOLUME 34 9 JUNE 1975 NUMBER 23

Formation of the Muonic Helium Atom, $\alpha\mu^-e^-$, and Observation of Its Larmor Precession*

P. A. Souder, D. E. Casperson, T. W. Crane, V. W. Hughes, D. C. Lu, H. Orth,† H. W. Reist,‡ and M. H. Yam
Gibbs Laboratory, Physics Department, Yale University, New Haven, Connecticut 06520

and

G. zu Putlitz
University of Heidelberg, Heidelberg, West Germany

(Received 9 April 1975)

The muonic helium atom, $\alpha\mu^-e^-$, has been formed by stopping polarized negative muons in He gas at 14 atm with a Xe admixture of 2%, and observed through its characteristic Larmor precession frequency of 1.4 MHz/G. In addition a nonzero residual polarization of $P=0.06\pm0.01$ for μ^- stopped in pure He gas has been measured for the first time, which corresponds to a depolarization factor of 18 ± 3.

The muonic helium atom $\alpha\mu^-e^-$ is the simple basic atom in which one of the two electrons in a normal helium atom is replaced by a negative muon. In its structure the muonic helium atom is similar to hydrogen with the relatively small muonic helium "nucleus" $(\alpha\mu^-)^+$ corresponding to the proton.[1,2] It is the simplest prototype of an electronic atom with a muonic nucleus[3] and provides an interesting and potentially useful system for study of a very different type of atomic structure, and also for study of the μ^--e^- interaction and for the precise determination of the magnetic moment and mass of the negative muon.

When a negative muon is stopped in He gas, it is captured by a He atom in an Auger process, and then as a result of further Auger and radiative processes will form $(\alpha\mu^-)^+$ in its ground $1S$ state. In a collision with He at thermal energy $(\alpha\mu^-)^+$ cannot capture an electron to form $\alpha\mu^-e^-$ because of the 11.0 eV difference in binding energy of an electron in He and in $\alpha\mu^-e^-$.[4] Hence to form $\alpha\mu^-e^-$ an electron donor with an ionization potential less than that of $\alpha\mu^-e^-$ is needed; xenon is chosen and the reaction will be[2,5]

$$(\alpha\mu^-)^+ + \mathrm{Xe} \to \alpha\mu^-e^- + \mathrm{Xe}^+ . \qquad (1)$$

The residual polarization of μ^- in $\alpha\mu^-$ $(1S)$ is an essential factor for our experiment. The theory[6] of the depolarization of a negative muon in its capture and cascade to the $1S$ state due to the μ^- spin-orbit interaction predicts a residual polarization $P=0.17$, when the initial polarization is 1 and a spin-0 nucleus is involved. For C and many other materials, this value of residual polarization has been observed,[7] but in experiments done thus far in liquid helium[8] and in gaseous helium at 50 atm[9] the residual polarizations observed were about $P=0.01\pm0.02$ and $P=0.035\pm0.024$, respectively. If the $\alpha\mu^-e^-$ atom were formed by Reaction (1), the resulting μ^- polarization in a weak external magnetic field would be $\frac{1}{2}$ that of μ^- in $\alpha\mu^-$ because of the hfs interac-

VOLUME 34, NUMBER 23 PHYSICAL REVIEW LETTERS 9 JUNE 1975

tion in the atom.[10]

The method of our experiment is the classic and usual one for studying the Larmor precession of muonium[11] (1.4 MHz/G) or of free muons[12] (13.6 kHz/G). Our experiment was done at the Space Radiation Effects Laboratory with a 100-MeV/c μ^- beam of polarization 0.65 from their muon channel. A diagram of the apparatus is shown in Fig. 1; all the counters were plastic scintillation counters. The He gas used was produced[13] with an impurity concentration of less than 1 ppm, and during our experiment was circulated over Ti at 750°C to maintain its purity. The Xe used had an impurity content of less than 25 ppm.[14] Three pairs of mutually orthogonal Helmholtz pair coils nulled out the residual laboratory magnetic field, and an additional Helmholtz pair produced a field B of up to 67 G transverse to the spin direction of the incoming muons. The resultant magnetic field was homogeneous to about ± 3% over the gas target, and its stability was better than 0.3% as measured with a Rb optical-pumping magnetometer.[15]

A stopping muon (μ_S) was defined by the coincidence-anticoincidence count $S_1S_2S_3S_4\bar{E}_1\bar{E}_2\bar{E}_3$, and with an incident muon beam S_1 of 3×10^5 sec^{-1}, the μ_S rate was about 600 sec^{-1} with 14 atm of He in the target, and with the target evacuated about 150 sec^{-1} or 25% of the full-target rate. Decay electrons were detected as $e_F = E_1E_2\bar{S}_1\bar{S}_2\bar{S}_3$ or $e_D = E_3E_4$. The time intervals between e_F or e_D and μ_S were obtained with time-to-amplitude converters and pulse-height analyzers.

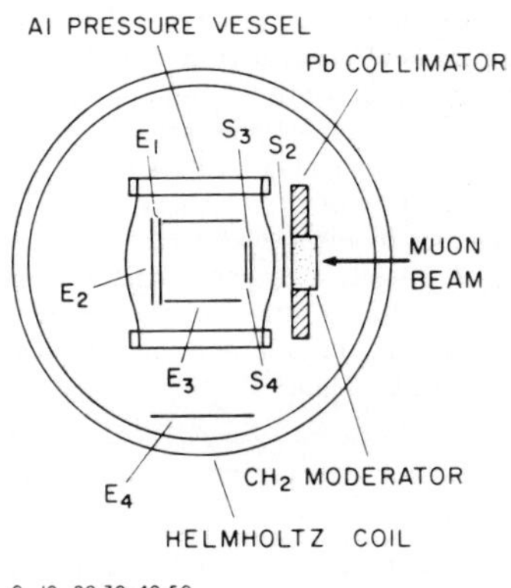

FIG. 1. Vertical cross section of the apparatus. All counters are disks except for E_3 which is a cylinder and E_4 which is square. Not shown is the upstream counter S_1. The counters S_1, S_2, and S_3 were 0.16 cm thick, and S_4 was 0.012 cm thick. The E counters were 0.6 cm thick.

In order to study the residual polarization of μ^- by observation of a free-muon precession signal, data were obtained with pure He at pressures of 7 and 14 atm and in an evacuated target, with $B = 67$ and 4 G. In addition data were taken with Xe admixtures of 0.2 and 1.2%. The time-distribution data were fitted by the equation[16]

$$N(t) = N_0 \exp(-t/\tau) \times [1 + A \exp(-t/\tau_D)\cos(2\pi f t + \varphi)] + B, \quad (2)$$

where $N(t)$ is the observed number of events at time t (an event is an e_F or e_D count occurring at a time interval t after μ_S); N_0 is a normalization constant; τ is the muon lifetime; τ_D is the depolarization time; A is the precession amplitude; f is the precession frequency; φ is the initial phase; and B is a constant background term.

A summary of the results of the data analysis using e_F for the amplitude A_μ at the Larmor frequency for free-muon precession of 0.91 MHz is given in Table I. In pure He at pressures of both 7 and 14 atm a statistically significant signal A_μ was observed; it corresponds to a residual polarization $P \simeq 0.06$, when normalized to an incident μ^- beam with polarization 1. The smaller value of A_μ observed with the target evacuated is consistent with the residual μ^- polarization reported for polystyrene (scintillator); in view of the relatively small number of events associated with target-empty, or wall, μ_S (20% of the full-target event rate) and the large width of the μ^- stopping distribution in grams per square centimeter as compared to the stopping power of the He, the A_μ observed in pure He cannot be due to wall stops. The addition of 1.2% Xe reduced the

TABLE I. Results of data analysis for A_μ at $f = 0.91$ MHz.

Data group	Helium pressure (atm)	Xenon (%)	B (G)	Number of μ_S (10^7)	A_μ (%)
1	14	0	67	3.1	1.24 ± 0.17
2	7	0	67	1.9	1.39 ± 0.24
3	0	0	67	0.9	0.72 ± 0.46
4	14	0.2	67	1.9	1.24 ± 0.22
5	14	1.2	67	2.3	0.25 ± 0.22
6	14	0	4	0.9	−0.34 ± 0.29
7	14	1.2	4	3.0	−0.39 ± 0.18

VOLUME 34, NUMBER 23 PHYSICAL REVIEW LETTERS 9 JUNE 1975

residual polarization significantly, whereas 0.2% Xe did not. The signal at 0.91 MHz also vanished when the magnetic field was reduced to 4 G.

The observed residual polarization in pure He is smaller by about a factor of 3 compared to the value expected on the basis of the conventional theory of depolarization.[6] The cause of this additional depolarization has not yet been established but is probably associated with collisional Stark mixing of different L levels of $\alpha\mu^-$ during its cascade from high-n states to the $1S$ state.[17] The reduction in A_μ due to the admixture of Xe we interpret as due to the formation of $\alpha\mu^-e^-$ by Reaction (1).

In view of this residual polarization of μ^- in He and its quenching by addition of Xe, data were taken with μ^- to search for the characteristic muonic-helium-atom Larmor precession (the same frequency as that of muonium) in several magnetic fields—3.10, 3.42, 3.73, and 4.64 G—with the stopping gas of He + 2% Xe. An equal amount of data was obtained at each field. Data were also taken under these same conditions with μ^+ stopping in the target to form muonium.[16]

All of the data taken at the four magnetic field values were combined by calculating the amplitude $A(\gamma)$ as a function of the gyromagnetic ratio $\gamma = f/B$, and the result for the e_D spectra is shown in Fig. 2(a). A clear maximum is obtained at $\gamma = 1.4$ MHz/G, which is the characteristic Larmor precession frequency for $\alpha\mu^-e^-$. Its value is $A_{\alpha\mu^-e^-} = (0.53 \pm 0.09)\%$. The width of the resonance is due principally to the inhomogeneity of the magnetic field. Figure 2(b) shows $A(\gamma)$ for a μ^+ run in which the muonium Larmor precession is apparent. The two curves are very similar as is to be expected. The amplitude $A_{\alpha\mu^-e^-}$ is about $\frac{1}{2}$ that of A_μ for μ^- in pure He (Table I), which is also the expected value.

The observation of $\alpha\mu^-e^-$ with the same residual polarization as $\alpha\mu^-$ implies that the $\alpha\mu^-e^-$ atoms are being formed through Reaction (1) in a time short compared to their Larmor precession period. Hence the cross section σ for Reaction (1) at thermal energies is $\sigma \gtrsim 1.5 \times 10^{-17}$ cm^2. This value is consistent with our present knowledge of this cross section.[18]

The formation of polarized muonic helium atoms, $\alpha\mu^-e^-$, should make possible precision measurements of its hyperfine-structure interval $\Delta\nu$ and Zeeman effect similar to those of muonium.[10] The approximate theoretical value for $\Delta\nu$ has been given[1] as $\Delta\nu = 4494.1$ MHz. This value differs from $\Delta\nu$ for muonium[19] ($\Delta\nu_M = 4463.32$ MHz) principally because of the different reduced-mass factors and because of the structure of the $(\alpha\mu^-)^+$ nucleus as compared to the structureless μ^+. Such precision measurements of the muonic helium atom should provide a test of the theory of this atom, and in addition yield precise values for the magnetic moment and mass of the negative muon, for comparison with those of the positive muon as a test of CPT invariance.[20]

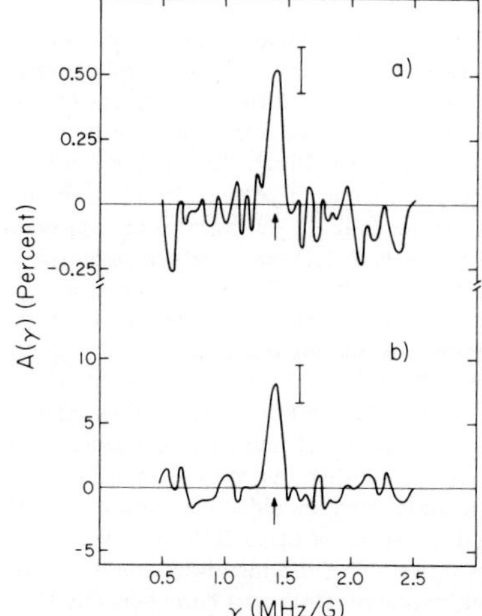

FIG. 2. Observed Larmor precession amplitudes $A(\gamma)$ versus gyromagnetic ratio $\gamma = f/B$: (a) μ^- stopped in He + 2% Xe, and forming $\alpha\mu^-e^-$ (3×10^8 μ_S); (b) μ^+ stopped in He + 2% Xe, and forming μ^+e^- (1.8×10^6 μ_S). The arrows indicate the expected gyromagnetic ratio $\gamma = 1.4$ MHz/G.

We are happy to acknowledge enthusiastic support from R. Siegel, C. Hansen, and H. Sowers of the Space Radiation Effects Laboratory, and helpful discussions of atomic cross sections with J. Bayfield and W. Maier.

---

*Research (Yale Report No. COO-3075-106) supported by the U. S. Energy Research and Development Administration under Contract No. AT(11-1)-3075 and by the Space Radiation Effects Laboratory which is supported by the National Science Foundation, the National Aeronautics and Space Administration, and the Commonwealth of Virginia.

†Also of University of Heidelberg, Heidelberg, West Germany.

‡Also of University of Bern, Bern, Switzerland.

[1]V. W. Hughes and S. Penman, Bull. Amer. Phys. Soc. 4, 80 (1959); K. N. Huang *et al.*, Bull. Amer. Phys. Soc. 18, 1503 (1973); K. N. Huang, Ph.D. thesis, Yale University, 1974 (unpublished).

[2]K. N. Huang *et al.*, in *Fifth International Conference on High Energy Physics and Nuclear Structure*, edited by G. Tibell (North-Holland, Amsterdam, 1973), p. 312.

[3]E. W. Otten, Z. Phys. 225, 393 (1969); V. G. Varlamov *et al.*, Pis'ma Zh. Eksp. Teor. Fiz. 17, 186 (1973) [JETP Lett. 17, 132 (1973)].

[4]V. W. Hughes *et al.*, Bull. Amer. Phys. Soc. 5, 75 (1960).

[5]M. Camani, Helv. Phys. Acta 46, 47 (1973).

[6]R. A. Mann and M. E. Rose, Phys. Rev. 121, 293 (1961); I. M. Shmushkevich, Nucl. Phys. 11, 419 (1959).

[7]A. E. Ignatenko, Nucl. Phys. 23, 75 (1961).

[8]D. C. Buckle *et al.*, Phys. Rev. Lett. 20, 705 (1968).

[9]P. Souder *et al.*, in *Proceedings of the Fourth International Conference on Atomic Physics, Heidelberg, Germany, 1974. Abstracts of Contributed Papers*, edited by J. Kowalski and H. G. Weber (Heidelberg Univ. Press, Heidelberg, Germany, 1974), p. 32; R. Stambaugh, Ph.D. thesis, Yale University, 1974 (unpublished).

[10]V. W. Hughes, Annu. Rev. Nucl. Sci. 16, 445 (1966).

[11]V. W. Hughes *et al.*, Phys. Rev. Lett. 5, 63 (1960), and Phys. Rev. A 1, 595 (1970).

[12]R. L. Garwin, L. M. Lederman, and M. Weinrich, Phys. Rev. 105, 1415 (1957).

[13]Matheson-Grade helium. Matheson Gas Products, East Rutherford, N. J.

[14]Research-Grade xenon. Air Products and Chemicals, Inc., Emmaus, Pa.

[15]W. Farr and E. W. Otten, Appl. Phys. 3, 367 (1974).

[16]R. D. Stambaugh *et al.*, Phys. Rev. Lett. 33, 568 (1974).

[17]A. Placci *et al.*, Nuovo Cimento 1A, 445 (1971).

[18]H. H. Fleischmann and R. A. Young, Phys. Rev. Lett. 19, 941 (1967); W. B. Meier II, Phys. Rev. A 5, 1256 (1972).

[19]P. A. Thompson *et al.*, Phys. Rev. A 8, 86 (1973).

[20]T. D. Lee and C. S. Wu, Annu. Rev. Nucl. Sci. 15, 381 (1965).

Proceedings of the International Conference on Sector-Focused Cyclotrons and Meson Factories, CERN, April 23–26, 1963, eds. F. T. Howard and N. Vogt-Nilsen.

A VERY HIGH-INTENSITY PROTON LINEAR ACCELERATOR AS A MESON FACTORY

E.R. Beringer, W.A. Blanpied, R.L. Gluckstern, V.W. Hughes,
H.B. Knowles, S. Ohnuma and G.W. Wheeler
Yale University, New Haven
(Presented by G.W. Wheeler)

Characteristics of the Linac as a Meson Factory

A "meson factory" is a complete research installation built around a high intensity proton accelerator of maximum energy below 1 GeV, and which is characterized principally by the unusually high intensity of the secondary particle beams.

There is much significant and interesting information not yet known in many fields of physics in the energy range below 1 GeV. The primary fields are : 1) particle physics involving nucleons, pions, muons, and neutrinos and their strong, electromagnetic and weak interactions, and 2) nuclear structure. High intensity beams, such as provided by a meson factory, are required to obtain much of the new information because they make possible : 1) studies of processes with small cross-sections or probabilities such as neutrino-induced events, rare decay modes, or triple scattering from thin targets, 2) precise experiments with high energy resolution, high purity beams, and thin targets, 3) high counting rates.

A proton linear accelerator has the following outstanding characteristics as the accelerator for a meson factory : 1) very high intensity primary proton beam (10^3 times that of existing synchro-cyclotrons and 10 times that of proposed sector-focused cyclotrons), 2) external proton beam with full intensity of the internal beam and with excellent geometrical properties, 3) energy variable in small steps, 4) acceleration of polarized protons, 5) reasonably high duty cycle, 6) relative freedom of the accelerator from problems of radioactivity and radiation damage, because of the ease of beam extraction.

The characteristics of the proton linear accelerator being designed at Yale University are shown in Table I.

The secondary particle beams which can be derived from this proton linac will be a factor of 10^3 to 10^4 times more intense than the secondary beams derived from the Berkeley 184 inch synchro-cyclotron, which is at present the most intense source for pion and muon beams in the energy range under consideration[1]. Table II shows some typical useful secondary beams associated with the proton linac. Particularly noteworthy are the beams of various types of neutrinos, which are obtained from the decays of pions and muons at rest or in flight, and which have adequate intensities for various experiments[2]. Fig. 1 and 2 show neutrino spectra which can be produced by 750 MeV protons. Table III[3] lists some reactions which might be studied.

Table I

Characteristics of the Proton Linear Accelerator

I. Beam Energy

a) Maximum energy : 750 MeV

b) Energy variable in steps of 7 to 10 MeV from 200 to 750 MeV

c) Energy spread : approximately 0.3%

II. Beam Intensity

a) Average current : 1 mA or 6×10^{15} protons/s

b) Peak pulse current : 20 mA

c) Pulse length and rate : 2 ms; 25 pulses/s

d) Beam duty cycle : 5%

e) Within each pulse the beam will be bunched into 4×10^5 packets; each packet has a duration of approximately 0.07 ns and adjacent packets are spaced by 5 ns.

f) Beam power, average : 750 kW

g) Beam quality, area in transverse phase space : $4.6\,\pi \times 10^{-4}$ cm rad

III. Physical Characteristics

a) Total length : 2000 ft

b) Total peak power : 85 MW

c) Injector : 750 kV Cockcroft-Walton generator

d) Drift tube accelerator at 200 MHz, 0.75 to 200 MeV. The first cavity about 5 m long and followed by six cavities about 25 m long.

e) Iris-loaded guide at 800 MHz, 200 MeV to 750 MeV. 59 cavities each 7.5 m long

f) Transverse focusing by magnetic quadrupoles

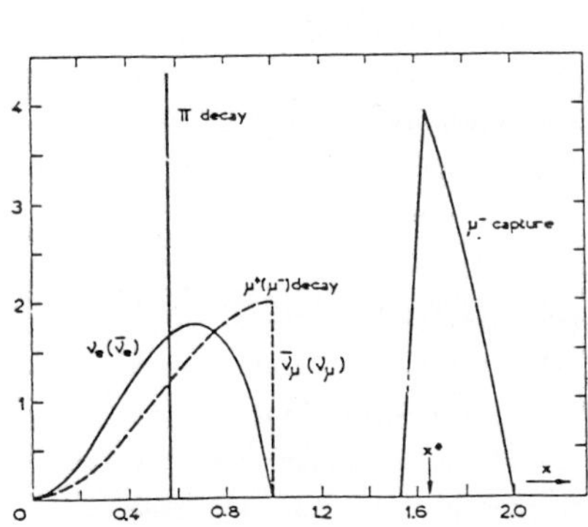

Fig. 1 Neutrino energy spectra available from 750 MeV stopped protons. The ν_μ are obtained from stopped negative muons. x = energy in units of one half the muon rest energy (U. Überall, preprints).

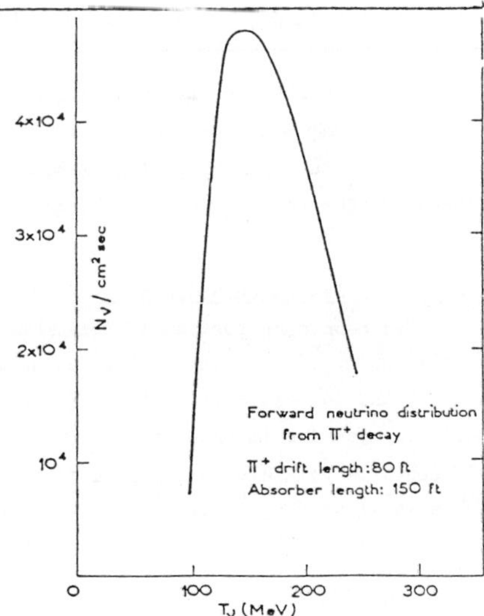

Fig. 2 Neutrino energy spectra available from decay of pions in flight, produced by 750 MeV protons.

Table II

Examples of Secondary Beams Derived from Proton Linac

Particle	Energy	Δ p/p	Intensity	Remarks
π^+ π^-	300 MeV 300 MeV	± 1% ± 1%	1×10^8/s 1.5×10^7/s	Solid angle : 10^{-4}sr Primary target : 12 g/cm^2, 30° meson collection
π^+ π^-	stopped stopped	-- --	4×10^8/s 7×10^7/s	Stopping target 25 cm^2, 6 g/cm^2
μ^+ μ^-	200 MeV 200 MeV	± 1% ± 1%	1×10^8/s 2×10^7/s	Use of quadrupole lens channel. Beams have maximum area of 80 cm^2 and maximum angular divergence of 0.0870 rad
μ^+ μ^-	stopped stopped	-- --	5×10^7/s 6×10^6/s	Stopping target 25 cm^2, 6 g/cm^2
ν_μ	30 MeV	Monoenergetic	1×10^7/cm^2 s	Derived from stopped protons and π^+ decay at rest
$\bar{\nu}_\mu + \nu_e$	90 MeV (max. total)	--	1×10^7/cm^2 s	Derived from stopped protons and μ^+ decay at rest
ν_μ	125 to 240 MeV	--	1×10^6/cm^2 s	Derived from π^+ decay in flight
n	730 MeV	Δ E/E = 2%	1×10^8/cm^2 s	

An unusual facility is also provided by the high-energy, high-intensity neutron secondary beam which has a time definition of 10^{-10} s as determined by the RF fine structure of the primary proton beam, and hence is suitable for experiments involving time-of-flight techniques. The beams of pions and muons have extremely high intensity as well as high purity, small momentum spread, and small angular divergence.

Accelerator Design Considerations

The remaining portion of this paper deals with specific considerations relating to the design of the linac. The features of very high intensity, excellent beam quality and low radio-activation of the structure require careful attention to the particle dynamics in order to achieve a minimum loss of particles during acceleration. It should be possible to design a linac in which no particles are lost at energies above about 25 MeV. However, a loss of 0.1% of the particles above 25 MeV would be

Table III

Some Energetically Possible Reactions with Neutrinos

Produced by a 750 MeV Proton Beam

(1) $\bar{\nu}_e + p \rightarrow e^+ + n$ (Cowan-Reines experiments)

(2) $\nu_e + n \rightarrow e^- + p$ (Davis Cl^{37} experiment)

(3) $\nu_\mu + n \rightarrow \mu^- + p$ (Columbia-Brookhaven experiment)

(4) $\bar{\nu}_\mu + p \rightarrow \mu^+ + n$

(5) $\nu_\mu + n \rightarrow e^- + p$ (probably forbidden)

(6) $\bar{\nu}_\mu + p \rightarrow e^+ + n$ (probably forbidden)

(7) $\bar{\nu}_\mu + p \rightarrow n + e^+ + \pi^0$

(8) $\bar{\nu}_\mu + p \rightarrow \Lambda^0 + e^+$

tolerable. Several features of the design are being closely examined : types of accelerating structures, injection into the linac, control of phase and amplitude of the field in each cavity, transition in type of structure, effects of beam loading and transverse focusing systems. For economic reasons, attention must be given to minimizing the cost of the accelerator.

Types of Structures. Existing proton linacs are of the Alvarez type, employing from one to three drift-tube loaded, standing wave cavities. The Alvarez structure is characterized by a serious decrease in the shunt impedance as the particle velocity increases. Electron linacs operate with $\beta = 1$ over their entire length except for a very short buncher section. The iris-loaded waveguide operating in a travelling or standing wave mode has proved practical for this purpose. The shunt impedance of the iris-loaded waveguide decreases with decreasing particle velocity. For both types of structure in the region of $\beta = 1/2$, the shunt impedance has decreased so much and the power loss per unit energy gain has become so large that the structure becomes economically undesirable. Extensive work has been done to determine the shapes for drift tubes and irises which will lead to the highest possible values of the shunt impedance[4]. In the region of $\beta = 1/2$, savings in RF power by 50% are indicated as compared to the early types of structures.

A number of studies have been attempted to find alternate accelerating structures which would have a higher shunt impedance in the region of $\beta = 1/2$. Some recent results at Harwell may yield a new and useful structure. However, at present we feel that a transition from drift tube structure directly to iris-loaded waveguide must be made near $\beta = 1/2$. The exact point for the transition is determined by detailed economic consideration. For a particular type of structure the shunt impedance increases with increasing frequency. The drift-tube section will operate at a frequency of 200 MHz which is the highest frequency consistent with the requirements of fabrication. A value of 800 MHz has been chosen tentatively for the iris-loaded waveguide for reasons discussed below.

Injection. The design of the buncher and first drift-tube cavity should incorporate two important features : High capture efficiency to ensure the high intensity beam and rapid concentration of the particles near the synchronous phase and energy for a minimum loss of particles.. It is certain that some particles will not be captured in the phase stable region and these will ultimately be lost to the drift tubes. Those particles which are far from the phase stable region will quickly be lost before gaining enough energy to cause activation of the structure. However, other particles which are just outside the stable region may gain considerable amounts of energy (25 MeV or more) before being lost. Several methods are being studied for increasing the fraction of beam captured and for losing the unstable portion of the beam before it can gain enough energy to cause trouble.

Control. Once particles have been trapped within the phase stable region of the "perfect" accelerator, there can be no further loss of particles from the beam. In a practical accelerator there will be several forms of errors which may induce the phase oscillations to grow rather than damp, resulting in loss of particles. These errors include : Incorrect amplitude of the electric field in a cavity with respect to adjacent cavities and incorrect longitudinal position of the drift tubes or irises within a cavity. In addition, the interaction of the beam with a cavity must be considered. The phase and amplitude of the electric field must be servo-controlled to an accuracy such that the residual errors will not stimulate the phase oscillation to a serious degree. Similarly, care must be taken in the fabrication of the cavities.

A computer code has been set up to examine the effects of these residual errors on the beam. The present program considers only the longitudinal motion but the transverse motion is being added. The program permits tracing of a particle bunch through the accelerator with any distribution of phase and amplitude errors. A particle is considered lost if it moves outside of the phase stable region. Preliminary results indicate that if the phase shift between cavities is held to 1 degree and the amplitude of the cavity fields to 1 percent, no particles will be lost from the bunch above about 25 MeV when the drift-tube section is operated at 200 MHz and the iris-loaded section at 800 MHz.

The control of the phases and amplitudes of the fields in the cavities will be accomplished by servo-systems which are capable of maintaining the required tolerance and making corrections within the time of one RF pulse. When the accelerator is set into operation, the initial adjustment of the servo-systems must be made by direct observation of the beam. The computer programs mentioned above will be used to investigate methods for this adjustment.

Transition. If the drift tube-section of tha accelerator is carefully adjustod to minimize the errors which cause the particle bunch to grow in longitudinal phase space, the bunch leaving this section will have a phase spread of about 10° measured with respect to the 200 MHz wave. The phase spread of the bunch on entering the iris-loaded waveguide will be increased by the ratio of the frequencies of the iris-loaded section to drift-tube section. Residual errors in the iris-loaded section will offset the normal phase damping. Hence, it is believed that the initial phase spread of 40° resulting from the choice of 800 MHz for the iris-loaded section is reasonably conservative. It would seem dangerous to use 1200 MHz with the resulting initial spread of 60°.

Beam Loading. A preliminary analysis of beam loading effects in a standing wave linac has been made. For the standing wave linac, the resonant modes form a complete set of functions for the discussion of all transient and steady state phenomena. In the steady-state condition found during the body of a long pulse, the major effects will be a decrease in the accelerating field in the cavity and a shift in phase of the RF relative to some standard in the power source. Detailed examination indicates that the bunch will seek a new synchronous phase of lower value (as a result of the amplitude decrease) and will perform phase oscillations about this new value. It will be possible to increase the accelerating field with increasing beam current by means of a field amplitude servo in order to maintain a phase stable region of adequate size. The amplitude and phase of the longitudinal beam oscillation will depend on the initial conditions and on the variation of the parameters from cavity to cavity. It can be shown that a truly periodic beam pulse will excite higher modes in the cavity only if the beam frequency harmonics coincide (accidentally) with the cavity harmonics. This coincidence will have to be within the line width of the cavity harmonics, which is of the order of 1 part in 10^4 for typical cavities. The likelihood of such a coincidence is therefore small, and may be removed, should it occur, by shifting the cavity harmonics. A section of waveguide is being fabricated to study the distribution of the cavity harmonics and related matters.

However, with a beam pulse of varying magnitude (e.g. during beam build-up), the discrete beam harmonic frequencies broaden according to the way in which the beam current changes. This increases the possibility of overlap with the cavity harmonics and presumably corresponds to the serious transient effects[5)] which lead to beam blow up in travelling wave electron linacs. Nevertheless, any accidental condition of resonance between the beam frequency and the cavity harmonics in one cavity is not likely to be duplicated in other cavities since the geometry is different from cavity to cavity.

Transverse Focusing. Quadrupole magnets will be used to supply the necessary transverse focusing throughout the accelerator. Magnets will be mounted in every drift tube in the first one or two cavities. Succeeding cavities will have magnets in every second or third drift tube. For this iris-loaded section, doublet pairs will be mounted between cavities. Calculations show that this configuration will contain the beam with reasonable fields in the magnets. In addition, the system is sufficiently flexible so that the low energy beam (200 MeV) may be carried the entire length of the accelerator by suitably readjusting the magnets in the iris-loaded section.

Cost Minimization. In addition to the technical considerations mentioned above, a serious effort is needed to produce the most economical design which is consistent with the high performance capabilities of this accelerator. It is known that the product of the total peak RF power and the length of a linac is a constant for a given energy gain and type of structure. For a given duty cycle, there will be a particular combination of length and power which leads to the minimum cost. A longer accelerator with lower peak power is required for cost minimization when a larger duty cycle is used. For the linac with 5 percent duty cycle, this minimization procedure gives a rate of energy gain of about 1.28 MeV/m for the drift-tube section and about 1.23 MeV/m for the iris-loaded section.

References

1. R.P. Haddock, Nucl. Instr. and Meth. 18-19 p 387 (1962).
2. W.A. Blanpied and V.W. Hughes, "Neutrino Beams", Conference on Advances in Meson and Nuclear Research Below One BeV. November 12, 13, 1962, Gatlinburg, Tennessee.
3. Y. Yomaguchi, Progr. Theor. Phys. (Kyoto), 6, 1117 (1960).
4. R.L. Gluckstern, "Proceedings of the Conference on High Energy Accelerators", New York City, September 1961, pp 129-141.
5. J.E. Leiss and R.A. Schrack, "Transient and Beam Loading Phenomena in Linear Electron Accelerators", National Bureau of Standards, October 30, 1962.

DISCUSSION

HUGHES : Since the duty factor question with respect to linacs has been raised a number of times, I should like to add a few comments. The macroscopic duty factor of the proposed linac is 5%. There is also a microscopic duty factor within each linac pulse due to RF structure. The beam appears in 0.07 ns bursts spaced 5 ns apart. However, for purposes of background comparisons, this factor 5/0.07 = 71 should not be divided into the macroscopic duty factor because the background from decaying particles (e.g. π-μ-e) is spread out by decay lifetimes and does not show the RF structure. For some experiments the RF structure is even an advantage because it allows time-of-flight analysis with very short time resolution.

SMITH : Is the 800 Mc/s for the travelling wave part of the accelerator locked in phase to the 200 Mc/s of the cavity section?

WHEELER : Yes, it must be.

LANGEVIN : What is the estimated cost of this accelerator?

WHEELER : For the purpose of comparison the accelerator proper, not including the experiment area, or the buildings, costs $ 16,000,000. This is without the use of the Harwell-type of structure. It must be pointed out that any improvements that are made in structure alone would lower the cost of this device.

VOLUME 49, NUMBER 14 PHYSICAL REVIEW LETTERS 4 OCTOBER 1982

Higher Precision Measurement of the hfs Interval of Muonium and of the Muon Magnetic Moment

F. G. Mariam, W. Beer,[(a)] P. R. Bolton, P. O. Egan, C. J. Gardner, V. W. Hughes, D. C. Lu, and P. A. Souder
Yale University, New Haven, Connecticut 06520

and

H. Orth and J. Vetter[(b)]
Physikalisches Institut der Universität Heidelberg, D-6900 Heidelberg, Federal Republic of Germany

and

U. Moser
University of Bern, Bern, Switzerland

and

G. zu Putlitz
Physikalisches Institut der Universität Heidelberg, D-6900 Heidelberg, Federal Republic of Germany, and Gesellschaft für Schwerionenforschung, D-6100 Darmstadt, Federal Republic of Germany

(Received 23 July 1982)

New higher precision measurements of the hyperfine Zeeman transitions in the ground state of muonium have been performed with use of the high-stopping-density surface μ^+ beam at the Clinton P. Anderson Meson Physics Facility. The results are $\Delta\nu = 4\,463\,302.88(16)$ kHz (0.036 ppm) and $\mu_\mu/\mu_p = 3.183\,346\,1(11)$ (0.36 ppm). The current theoretical value of $\Delta\nu$ agrees well with experiment within the 0.77-ppm error of $\Delta\nu_{\rm theor}$, which is due principally to inaccuracy in evaluation of the nonrecoil radiative correction term. The most precise current value of m_μ/m_e is obtained from our value of μ_μ/μ_p.

PACS numbers: 36.10.Dr, 12.20.Fv, 14.60.Ef

Muonium (μ^+e^-) is the hydrogenlike atom consisting of a positive muon and an electron. It provides an ideal simple system for determining the properties of the muon and for measuring the muon-electron interaction. With muonium we can sensitively test the theory of quantum electrodynamics for the two-body bound state and can search for effects of weak, strong, or unknown interactions on the electron-muon bound state.[1]

The present paper reports new measurements[2] of the hyperfine structure interval $\Delta\nu$ in the ground state of μ^+e^- and of the magnetic moment of the positive muon, μ_μ, with substantially improved precision. These new experimental values, taken together with recent calculations[3,4] of relativistic recoil and radiative contributions to $\Delta\nu$, provide a considerably more sensitive comparison of theory and experiment for $\Delta\nu$.

The muonium Zeeman transitions $(M_J, M_\mu) = (\frac{1}{2}, \frac{1}{2}) \leftrightarrow (\frac{1}{2}, -\frac{1}{2})$, designated ν_{12}, and $(-\frac{1}{2}, -\frac{1}{2}) \leftrightarrow (-\frac{1}{2}, +\frac{1}{2})$, designated ν_{34}, are observed at strong magnetic field by the microwave magnetic resonance method.[1,5,6] Our increased precision is due principally to the use of the high-intensity, low-momentum "surface" muon beam[7] of the Los Alamos 800-MeV proton linear accelerator [Clinton P. Anderson Meson Physics Facility (LAMPF)]. Measurement of the two transitions ν_{12} and ν_{34} allows the determination of both $\Delta\nu$ and μ_μ, which appear in the relevant Hamiltonian term for μ^+e^-:

$$\mathcal{H} = h\Delta\nu\vec{I}_\mu\cdot\vec{J} - \mu_B{}^\mu g_\mu'\vec{I}_\mu\cdot\vec{H} + \mu_B{}^e g_J\vec{J}\cdot\vec{H}, \qquad (1)$$

in which $\vec{I}_\mu$ is the muon spin operator, $\vec{J}$ is the electron total angular momentum operator, g_μ' (g_J) is the muon (electron) gyromagnetic ratio in muonium, $\vec{H}$ is the external static magnetic field, and $\mu_B{}^\mu$ ($\mu_B{}^e$) is the muon (electron) Bohr magneton. The quantities g_μ' and g_J are related[8] to the free muon and electron g values g_μ and g_e by

$$g_\mu' = g_\mu\left(1 - \frac{\alpha^2}{3} + \frac{\alpha^2}{2}\frac{m_e}{m_\mu}\right)$$

and

$$g_J = g_e\left(1 - \frac{\alpha^2}{3} + \frac{\alpha^2}{2}\frac{m_e}{m_\mu} + \frac{\alpha^3}{4\pi}\right).$$

A diagram of the experimental apparatus[5] is shown in Fig. 1. The LAMPF stopped-muon channel was tuned to accept and transmit μ^+ originating from the decay of π^+ stopped near the surface

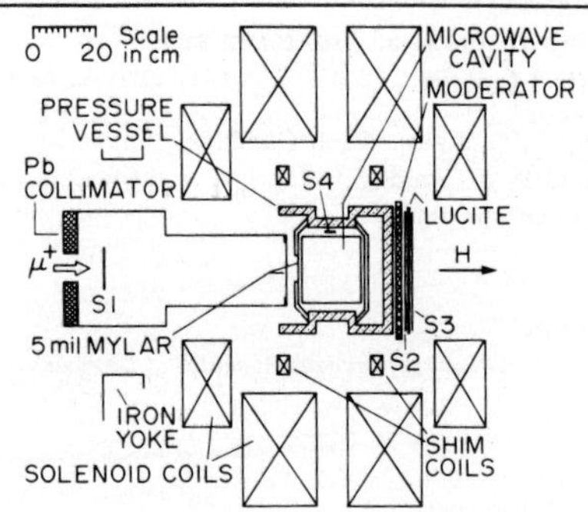

FIG. 1. Schematic diagram of the experimental setup. S1, S2, S3, and S4 are plastic scintillation counters of thicknesses 0.005, 0.25, 0.25, and 0.25 in., respectively. Counter S4 was used to indicate maximum μ^+ stopping rate in center of microwave cavity.

of the pion production target in the primary proton beam. With an 800-MeV proton beam of 300 μA average, the μ^+ beam had an instantaneous flux of 3×10^7 s^{-1} (2×10^6 s^{-1} average) after collimation. The muon momenta were between 25 and 28 MeV/c, and the longitudinal polarization was close to 1. The ratio of μ^+ to e^+ in the beam was about 6/1. The μ^+ flux was monitored with a thin (0.12 mm) plastic scintillator S1 by integrating the anode current over the 600-μs proton beam pulse. The target vessel was filled with Kr gas at pressures of $\frac{1}{2}$ or 1 atm, and about half of the incident μ^+ stopped in the gas. Scintillation counters S2 and S3 detected e^+ from μ^+ decays, and the plastic and aluminum moderator downstream from the cavity helped reduce the background from the e^+ contamination in the beam. A central element of the experiment was the high-precision solenoid electromagnet[9] which provided the magnetic field of 13.6 kG, which was homogeneous over the volume of the microwave cavity to about 3 ppm rms and had a long term stability of better than 0.3 ppm. The microwave cavity was resonant in the TM_{110} mode at 1.918 GHz (ν_{12} at 13.6 kG) and in the TM_{210} mode at 2.545 GHz (ν_{34}). It was excited with an input power of ~20 W switched on and off at a repetition period of 160 ms, alternating between modes for successive microwave-on periods.

Typical resonance curves are shown in Fig. 2 and were observed by varying the magnetic field H in small steps with fixed microwave frequency and power. The signal at each value of H is de-

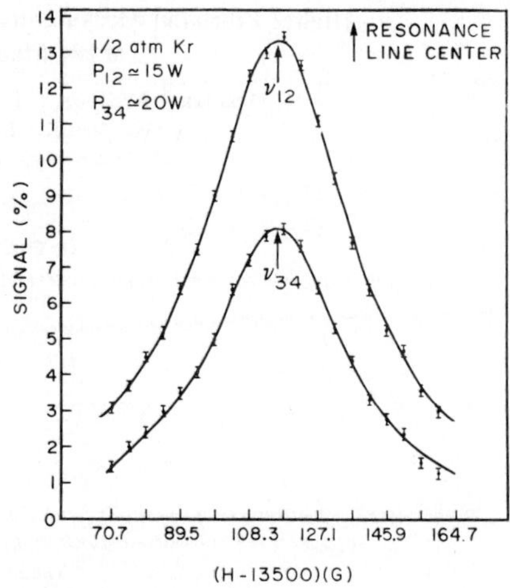

FIG. 2. Typical resonance lines with theoretical line shapes (solid lines) fitted to the data points. Data points were taken alternately on opposite sides of the line center. Data-taking time for each pair of resonance lines was less than 2 h.

fined by $S=\{(S2\cdot S3/S1)_{\rm rf\,on}/(S2\cdot S3/S1)_{\rm rf\,off}\}-1$. The theoretical line shape, which includes as free parameters essentially a resonance line center, a linewidth, and a height, is fitted to the experimental points and determines the resonance magnetic field value. The theoretical line shape incorporates the measured distribution of the magnetic field H, the measured μ^+ stopping distribution, the microwave power distribution over the cavity, and the solid angle for detection of an e^+ from μ^+ decay. A total of 184 resonance lines obtained in about 600 h of data taking (102 at 0.5 atm and 54 at 1 atm in the present experiment, and 18 at 1.7 atm and 10 at 5.2 atm from a previous experiment[5]) were analyzed. The resulting transition frequencies, after adjustment to correspond to a magnetic field H of 13 616.0 G and correction for a small measured quadratic density shift,[10] were extrapolated to zero gas density with use of the linear density dependence $\nu(D) = \nu(0)(1+aD)$. Using[5] the Hamiltonian of Eq. (1), we then obtain

$$\Delta\nu = 4\,463\,302.88(16)\text{ kHz } (0.036\text{ ppm}); \qquad (2)$$
$$\mu_\mu/\mu_p = 3.183\,346\,1(11)\ (0.36\text{ ppm}),$$

in which the one-standard-deviation total error

including systematic and random errors is given. Table I lists the sources of error.[11]

The value of $\Delta\nu$ given in Eq. (2) agrees with our earlier measurement[5] of $\Delta\nu$, but the error is less by a factor of 3.3. Our value of μ_μ/μ_p agrees with the most recent muon spin rotation measurement[16] done in liquid bromine which gave $\mu_\mu/\mu_p = 3.183\,344\,1(17)$ (0.53 ppm).

The hfs and g_J fractional density shifts in Kr can be obtained from the density-shift results given above

$$(1/\Delta\nu)(\partial\Delta\nu/\partial D) = -10.57(4)\times10^{-9}\ \text{Torr}^{-1}\ (0\,^\circ\text{C},\text{Kr});$$

$$(1/g_J)(\partial g_J/\partial D) = -1.83(32)\times10^{-9}\ \text{Torr}^{-1}\ (0\,^\circ\text{C},\text{Kr}).$$

Use of our value of μ_μ/μ_p in Eq. (2) together with that of Ref. 16, and of experimental values of g_{μ^+} (Ref. 17) and of $\mu_p/\mu_B{}^e$ (Ref. 18) determines the most precise value for m_μ/m_e, as follows:

$$m_\mu/m_e = (g_\mu/2)(\mu_p/\mu_\mu)(\mu_B{}^e/\mu_p) = 206.768\,259(62)\ (0.30\ \text{ppm}).$$

The current theoretical value for $\Delta\nu$ is given by[1,3,4]

$$\Delta\nu = \tfrac{16}{3}\alpha^2 cR_\infty(\mu_\mu/\mu_p)(\mu_p/\mu_B{}^e)(1+m_e/m_\mu)^{-3}[1+\tfrac{3}{2}\alpha^2+a_e+\epsilon_1+\epsilon_2+\epsilon_3-\delta_\mu{}'], \qquad (3)$$

in which the bracketed term includes the radiative and relativistic corrections to the leading Fermi term, where

$$\epsilon_1 = \alpha^2(\ln 2 - \tfrac{5}{2});\quad \epsilon_2 = -(8\alpha^3/3\pi)\ln\alpha[\ln\alpha - \ln 4 + \tfrac{281}{480}];\quad \epsilon_3 = (\alpha^3/\pi)(18.4\pm5);$$

$$\delta_\mu{}' = \frac{3\alpha}{\pi}\frac{m_R}{m_\mu - m_e}\ln m_\mu/m_e + \alpha^2\frac{m_R}{m_\mu+m_e}[2\ln\alpha + 8\ln 2 - 3\tfrac{11}{18}] + (\alpha/\pi)^2 m_e/m_\mu$$
$$\times[2\ln^2(m_\mu/m_e) - \tfrac{31}{12}\ln(m_\mu/m_e) + (\tfrac{28}{9}+\pi^2/3 - 1.9)],$$

where $m_R = m_e m_\mu/(m_e+m_\mu)$. The term $\frac{3}{2}\alpha^2$ is a relativistic correction; the terms a_e, ϵ_1, ϵ_2, and ϵ_3 are nonrecoil radiative corrections; the term $\delta_\mu{}'$ is relativistic recoil correction, where the first two terms involve recoil only and the third term is a QED radiative recoil contribution including the small hadronic vacuum polarization term. The following values of the fundamental atomic constants are used[18]: $R_\infty = 1.097\,373\,152\,1(11)\times10^5\ \text{cm}^{-1}$ (0.001 ppm)[19]; $c = 2.997\,924\,580(12)\times10^{10}\ \text{cm s}^{-1}$ (0.004 ppm); $\alpha^{-1} = 137.035\,963(15)$ (0.11 ppm)[20]; $\mu_p/\mu_B{}^e = 1.521\,032\,209(16)\times10^{-3}$ (0.01 ppm); $\mu_\mu/\mu_p = 3.182\,345\,47(95)$ (0.30 ppm); $m_\mu/m_e = 206.768\,259(62)$ (0.30 ppm); $a_e = (g_e-2)/2 = 1\,159\,652\,200(40)\times10^{-12}$.[21] Hence we ob-

TABLE I. Sources of error in $\Delta\nu$ and μ_μ/μ_p.

Source	$\delta\Delta\nu$ (kHz)	$\delta(\mu_\mu/\mu_p)$ (ppm)
1. Statistical error (e^+ counts)	0.000 ± 0.073	0.000 ± 0.196
2. Random error associated with μ^+ beam monitor	0.000 ± 0.031	0.000 ± 0.084
3. Muon stopping distribution and detector solid angle distribution	0.000 ± 0.008	0.000 ± 0.119
4. Magnetic field measurement (±0.31 ppm)	0.000 ± 0.000	0.000 ± 0.093
5. Microwave power averaging	$+0.021\pm0.035$	-0.102 ± 0.104
6. Gas density	0.000 ± 0.065	0.000 ± 0.001
7. Temperature dependence of a[a]	-0.073 ± 0.073	-0.013 ± 0.013
8. Quadratic density shift	0.000 ± 0.041	0.000 ± 0.005
9. Field-dependent line-shape systematics[b]	$+0.037\pm0.083$	$+0.356\pm0.217$
10. Bloch-Siegert term and nonresonant states[c]	$+0.004\pm0.000$	-0.005 ± 0.000
11. Small approximations in line fitting	0.000 ± 0.000	0.000 ± 0.047
Total correction and one-standard-deviation error	-0.011 ± 0.160	$+0.236\pm0.359$

[a]Data of Ref. 5 and of this paper were taken at two temperatures differing by 2.5 °C. Corrections to the data of Ref. 5 were made for the dependence of a on temperature based on experimental (Ref. 12) and theoretical (Ref. 13) information on hydrogen density shifts.

[b]Based on measurements of broadened resonance lines at high microwave power.

[c]Calculation with Refs. 14 and 15.

tain

$$\Delta\nu_{\text{theor}} = 4\,463\,303.7(1.7)(3.0)\text{ kHz} \quad (0.77\text{ ppm}), \quad (4)$$

in which the 1.7-kHz uncertainty comes from combining a 1.3-kHz uncertainty from μ_μ/μ_p with a 1.0-kHz uncertainty from α. The 3.0-kHz theoretical uncertainty is due to ϵ_3.

Comparison of the experimental and theoretical values of $\Delta\nu$ in Eqs. (2) and (4) gives $\Delta\nu_{\text{expt}} - \Delta\nu_{\text{theor}} = -0.8 \pm 3.4$ kHz where the dominant error comes from the theoretical value. The agreement is excellent and provides an important test of the validity of muon electrodynamics.

An alternative approach is to equate $\Delta\nu_{\text{expt}}$ of Eq. (2) to $\Delta\nu_{\text{theor}}$ of Eq. (3) and hence determine α. The result $\alpha^{-1} = 137.035\,974(50)$ (0.37 ppm) is in good agreement with the value of α obtained from the ac Josephson effect.[20]

The standard electroweak theory predicts[22] an axial-vector–axial-vector coupling contribution to $\Delta\nu$ of +0.07 kHz or a fractional contribution of 1.6×10^{-8}. This is about $\frac{1}{2}$ the experimental error in $\Delta\nu$, but about $\frac{1}{50}$ the present theoretical error in $\Delta\nu$. Recent high-energy colliding-beam experiments[23] have measured the charge asymmetry in $e^+e^- \to \mu^+\mu^-$, which is believed due to this weak neutral-current coupling between leptons. As yet there has been no measurement of a weak-interaction-energy contribution in an atom, and it would be of interest to measure this term in muonium where low momentum transfer is involved.

The contribution of a conjectured Hamiltonian term coupling muonium to antimuonium[24] with a Fermi coupling strength G_F would modify the hfs energy levels by about 0.26 kHz. However, because of the small component of antimuonium in the muonium wave function under our experimental conditions with muonium formed in a gas, the effect of such a coupling on our measurements would be negligible.[25]

Improvement in the precision of measurement of $\Delta\nu$ and μ_μ/μ_p by at least a factor of 5 should be possible with the use of a pulsed muon beam which would permit resonance line-narrowing techniques. Improvement in our knowledge of $\Delta\nu_{\text{theor}}$ requires most urgently an improved calculation of the nonrecoil radiative correction term ϵ_3.

This work was supported in part by the U. S. Department of Energy under Contract No. DE-AC02-76ER03074, NATO under Grant No. 1589, the Schweizer National Fonds, and the Max Kade Foundation. One of us (V.W.H.) was the recipient of a Senior Scientist Award from the Alexander von Humboldt Foundation.

[(a)]Present address: Laboratorium für Hochenergiephysik der ETH Zurich, c/o SIN, CH-5234 Villigen, Switzerland.

[(b)]Present address: Siemens AG, Erlangen, Federal Republic of Germany.

[1]V. W. Hughes and T. Kinoshita, in *Muon Physics*, edited by V. W. Hughes and C. S. Wu (Academic, New York, 1977), Vol. 1, p. 12; V. W. Hughes, Annu. Rev. Nucl. Sci. 16, 445 (1966).

[2]F. G. Mariam *et al.*, Bull. Am. Phys. Soc. 27, 480 (1982).

[3]G. T. Bodwin *et al.*, Phys. Rev. Lett. 48, 1799 (1982).

[4]E. A. Terray and D. R. Yennie, Phys. Rev. Lett. 48, 1803 (1982).

[5]D. E. Casperson *et al.*, Phys. Rev. Lett. 38, 956, 1504 (1977).

[6]J. M. Bailey *et al.*, Phys. Rev. A 3, 871 (1971); W. E. Cleland *et al.*, Phys. Rev. A 5, 2338 (1972).

[7]H. W. Reist *et al.*, Nucl. Instrum. Methods 153, 61 (1978).

[8]H. Grotch and R. A. Hegstrom, Phys. Rev. A 4, 59 (1971).

[9]R. D. Stambaugh, Ph.D. thesis, Yale University, 1974 (unpublished).

[10]D. E. Casperson *et al.*, Phys. Lett. 59B, 397 (1975).

[11]F. G. Mariam, Ph.D. thesis, Yale University, 1981 (unpublished).

[12]C. L. Morgan and E. S. Ensberg, Phys. Rev. A 7, 1494 (1973).

[13]B. K. Rao *et al.*, Phys. Rev. A 2, 1411 (1970).

[14]J. H. Shirley, Phys. Rev. 138, B979 (1965).

[15]H. Salwen, Phys. Rev. 99, 1274 (1955).

[16]E. Klempt *et al.*, Phys. Rev. D 25, 652 (1982).

[17]J. Bailey *et al.*, Nucl. Phys. B150, 1 (1979).

[18]E. R. Cohen and B. N. Taylor, J. Phys. Chem. Ref. Data 2, 663 (1973).

[19]S. R. Amin *et al.*, Phys. Rev. Lett. 47, 1234 (1981).

[20]E. R. Williams and P. T. Olsen, Phys. Rev. Lett. 42, 1575 (1979).

[21]H. Dehmelt, *Atomic Physics 7*, 337 (1981).

[22]M. A. B. Bég and G. Feinberg, Phys. Rev. Lett. 33, 606 (1974), and 35, 130 (1975).

[23]W. Bartel *et al.*, Phys. Lett. 108B, 140 (1982); R. Brandelik *et al.*, Phys. Lett. 110B, 173 (1982); B. Adeva *et al.*, Phys. Rev. Lett. 48, 1701 (1982).

[24]B. Pontecorvo, Zh. Eksp. Teor. Fiz. 33, 549 (1957) [Sov. Phys. JETP 6, 429 (1958)]; G. Feinberg and S. Weinberg, Phys. Rev. Lett. 6, 381 (1961); E. Derman, Phys. Rev. D 19, 317 (1979); A. Halprin, Phys. Rev. Lett. 48, 1313 (1982); P. Némethy and V. W. Hughes, Comments Nucl. Part. Phys. 10, 147 (1981).

[25]D. L. Morgan, Ph.D. thesis, Yale University, 1965 (unpublished).

VOLUME 52, NUMBER 11 PHYSICAL REVIEW LETTERS 12 MARCH 1984

Formation of Muonium in the 2*S* State and Observation of the Lamb Shift Transition

A. Badertscher, S. Dhawan, P. O. Egan, V. W. Hughes, D. C. Lu, M. W. Ritter, and K. A. Woodle
Gibbs Laboratory, Physics Department, Yale University, New Haven, Connecticut 06520

and

M. Gladisch, H. Orth, and G. zu Putlitz
Physikalisches Institut der Universitat Heidelberg, D-6900 Heidelberg, Federal Republic of Germany

and

M. Eckhause and J. Kane
College of William and Mary, Williamsburg, Virginia 23185

and

F. G. Mariam
Los Alamos National Laboratory, Los Alamos, New Mexico 87545

and

J. Reidy
University of Mississippi, University, Mississippi 38677

(Received 2 November 1983; revised manuscript received 19 December 1983)

Muonium in the $2S$ state has been produced by the beam-foil method with a μ^+ beam of 10 MeV/c. The metastable $2S$ state of muonium has been detected by a static electric field quenching method, and the transition $2S_{1/2}$, $F=1 \rightarrow 2P_{1/2}$, $F=1$ induced by a radio-frequency electric field has been observed with an event rate of about 4/h.

PACS numbers: 36.10.Dr, 12.20.Fv

Muonium (M) is an atom consisting of a positive muon and an electron and is an ideal system to test quantum electrodynamics (QED).[1] High precision measurements of the hyperfine structure interval $\Delta\nu$ and of the Zeeman effect in its ground state have provided very sensitive tests of QED, as well as precise values of the fine structure constant α and the ratio of the muon and proton magnetic moments.[2,3] In these experiments muonium was formed by stopping μ^+ in gaseous targets.

Another important QED test would be a precise measurement of the Lamb shift in muonium in the $n=2$ state, since it would be free of the effects of proton structure which complicate the interpretation of the Lamb shift in hydrogen.[4] For this measurement muonium in the $2S$ state must be obtained in vacuum to avoid collisional quenching of M($2S$) atoms. With a method similar to beam neutralization in proton-beam-foil spectroscopy[5] we have shown that muonium in its ground state is formed by passing a μ^+ beam through a thin foil in vacuum.[6] From proton data[7] we expect that metastable M($2S$) atoms will be formed as well, and that M($2S$)/M($1S$) will be about 0.1. Higher excited states will also be formed but their formation probabilities are lower and most of these excited state atoms will decay rapidly. This paper reports on further developments for producing muonium in vacuum, the detection of M($2S$) by a static electric field quenching method, and the observation of the Lamb shift transition $2S \rightarrow 2P$ induced by an rf electric field of about 1140 MHz.

Our experiment was done at the Stopped Muon Channel at the Los Alamos Clinton P. Anderson Meson Physics Facility (LAMPF). In our previous experiment to obtain muonium in vacuum,[6] the surface muon beam[8] of 28 MeV/c was degraded so that about one half of the muons were stopped in the production foil. Monte Carlo calculations of the energy spectrum of μ^+ emerging from the foil showed that this spectrum is rather flat and extends up to a kinetic energy of 2.5 MeV. Thus, only a small fraction of about 10^{-3} of the muons passing through the foil could capture an electron due to

VOLUME 52, NUMBER 11 PHYSICAL REVIEW LETTERS 12 MARCH 1984

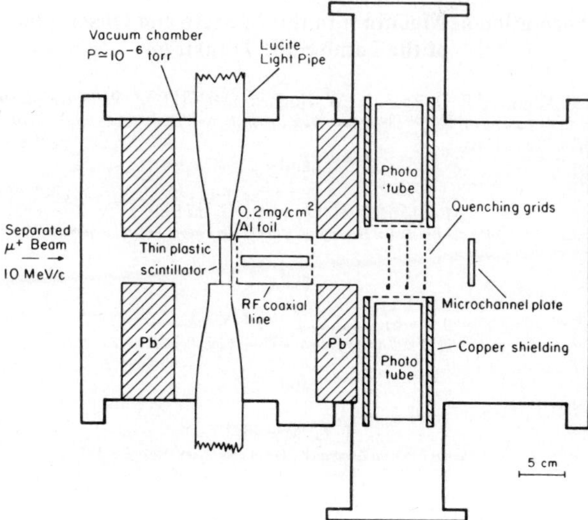

FIG. 1. Schematic diagram of the apparatus.

the rapid decrease of the capture cross section with increasing kinetic energy.[9,10] Our calculations also showed that the absolute yield of muonium atoms is roughly independent of the incident muon beam momentum p whereas the muon flux was measured to be proportional to $p^{3.5}$. Hence for lower beam momenta the signal-to-noise ratio for any muonium experiment should be improved. Our calculations indicate that muonium atoms emerge approximately isotropically from the foil and that the velocity distribution of M(1S) or M(2S) is determined by the charge capture cross section and is not dependent on p.

Figure 1 shows a diagram of our apparatus. The 20-μm-thick muon scintillation counter (NE 102A) served as a degrader, and muonium was formed in a thin Al foil (0.2 mg/cm^2) at ground potential immediately downstream of the scintillator. With a parasitic apparatus mounted downstream of the apparatus shown in Fig. 1 (with microchannel plate removed) and similar to the apparatus of Ref. 6, we optimized the beam momentum for maximal M(1S) yield per incoming muon. In the search for M(2S) we were limited to accept an average muon rate of 70 kHz at a duty factor of 9% because of pile-up considerations. Figure 2 shows that at 9.75 MeV/c the M(1S) flux is about 4% of the muon flux measured with the scintillation counter. The muonium flux from the downstream side of the foils is obtained from the measured rate with the parasitic apparatus assuming isotropic distribution of M(1S) from the foil. This "subsurface" muon beam was obtained by tuning the muon channel to transport muons of about 10 MeV/c; an electrostatic separator was used in our beam line to suppress the very high positron contamination ($e^+/\mu^+ \simeq 10^3$ at 5 MeV/c).

Figure 3 gives the energy level diagram for

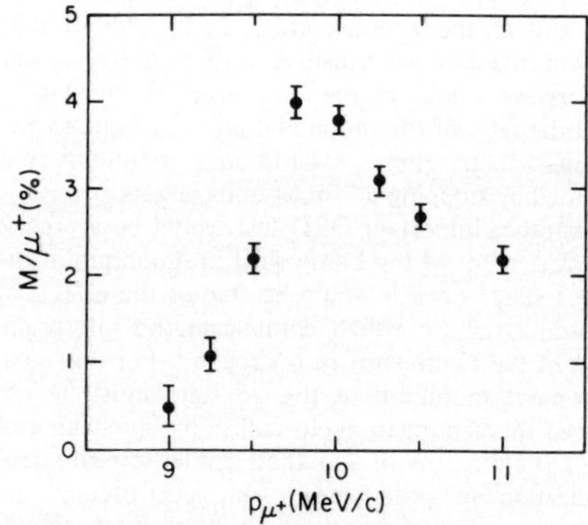

FIG. 2. Muonium yield at the foil as a function of the μ^+ beam momentum.

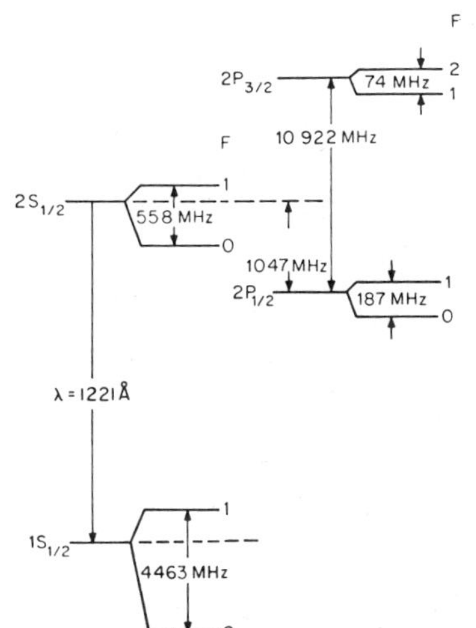

FIG. 3. Energy level diagram of the $n=1$ and $n=2$ states of muonium.

muonium in its $n=1$ and $n=2$ states including fine structure, Lamb shift, and hyperfine structure. The value of the Lamb shift interval of $2S_{1/2}$ to $2P_{1/2}$ for muonium has been calculated from QED.[11] The $2S$ atomic state is metastable with the mean lifetime of $\frac{1}{7}$ s for two-photon decay to the $1S$ state; however, the mean life of 2.2 μs for μ^+ decay determines the lifetime of M($2S$). In muonium the $2P$ state decays to the $1S$ state with a mean lifetime of 1.6 ns through an electric dipole transition with emission of a Lyman α photon of 1221 Å. M($2S$) is quenched in an external static electric field[12] through admixture of the $2P$ state; a Lyman α photon is emitted, and the mean lifetime is about 2 ns in a field of 600 V/cm.

Our first approach to observation of M($2S$) was to detect the static electric field quenching of M($2S$). Our apparatus (Fig. 1) includes four photomultiplier tubes for detection of Lyman α uv radiation. These are the Hamamatsu type R2050 photomultipliers with a MgF_2 window of 5 cm diameter and a quantum efficiency of 10% at 1221 Å including transmission through the window. The total solid angle of the four tubes is 30% of 4π. In addition, there is a microchannel plate just downstream of the phototubes to detect M($1S$) atoms and μ^+; in our first run the active plate diameter was 4 cm and in the second, 7.5 cm. Three quenching grids located at the axial position of the phototubes produced an axial static electric field of 600 V/cm. The center grid was at -1200 V and the two outer grids were grounded.

We searched for M($2S$) by modulating the static electric field between on (600 V/cm) and off and looking for a difference in the event rate due to Lyman α photons emitted during the field-on phase. An event required a triple coincidence between an incoming muon, a delayed Lyman α photon in one of the four uv phototubes, and a further delayed M($1S$) atom detected in the microchannel plate. The distance between the scintillator and the center plane of the uv phototubes is 20 cm and that between the scintillator and the microchannel plate 27.5 cm; the time gates were set to accept M($2S$) atoms with kinetic energies between 4 and 26 keV. The phototube pulse had to occur between 30 and 80 ns after an incoming muon, and the microchannel plate pulse between 10 and 40 ns after a phototube pulse was detected in the first gate. Combining about two days of data taking from our two runs, we observed 174 events with the quenching field on and 72 events with the field off. Assigning the difference between grid on and grid off to events due to Lyman α photons from quenched M($2S$) gives a Lyman α event rate induced by the static electric field of (4.7 ± 0.7)/h. Within the uncertainties of our detection efficiencies this value agrees with the expected rate of M($2S$). For the normalized difference signal (S) between quenching field on and field off defined with the events (E) normalized to the number of incoming muons, $S=[(E/\mu)_{\text{on}}-(E/\mu)_{\text{off}}]/[(E/\mu)_{\text{on}}+(E/\mu)_{\text{off}}]$, we obtained $S=(41.5\pm5.8)\%$. For delayed coincidences with delay times outside the expected flight times of M($2S$) atoms, the signal is consistent with zero, namely $S=(-1.7\pm6.4)\%$. Formation and static electric field quenching of M($2S$) has also been reported from an experiment at TRIUMF.[13] However, the M($2S$) formation rate reported by them is in disagreement with ours, being about a factor of 10 higher when the different beam momenta are taken into account.

Our next step in establishing the formation of M($2S$) atoms was to search for the Lamb shift transition $2S_{1/2}$, $F=1 \rightarrow 2P_{1/2}$, $F=1$ driven with an rf electric field. In Fig. 4 the solid curve shows the theoretical non-power-broadened resonance lines for two transitions between the hyperfine sublevels of the $2S_{1/2}$ and $2P_{1/2}$ states. The natural linewidth of 100 MHz (full width at half maximum) is due to the lifetime of the $2P_{1/2}$ state. Our rf interaction region (Fig. 1) consisted of a coaxial line. It was 8.3 cm long and had tungsten meshes with a

VOLUME 52, NUMBER 11 PHYSICAL REVIEW LETTERS 12 MARCH 1984

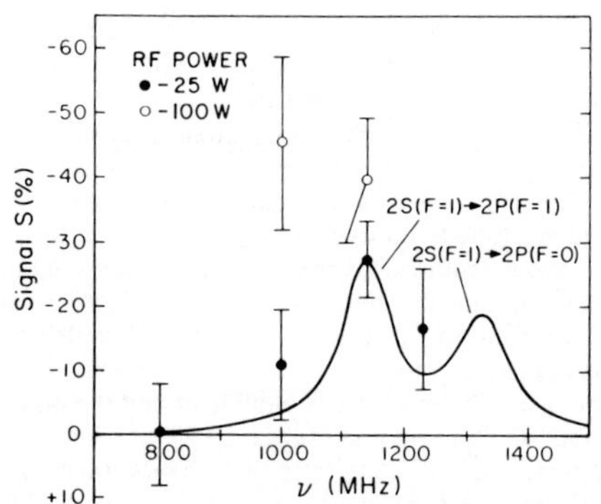

FIG. 4. Muonium $2S_{1/2} \to 2P_{1/2}$ resonance lines. The solid curve shows the theoretical non-power-broadened resonance line shape. The data points for 100- and 25-W rf input power are shown.

transmission of 80% each at the entrance and exit planes to allow the muon beam to pass through. The diameter of the inner conductor was 1 cm and the inner diameter of the outer conductor was 5.1 cm, which resulted in a variation of a factor of 25 in the power density between the inner and the outer radii.

Initially we searched for the $2S_{1/2}$, $F=1 \to 2P_{1/2}$, $F=1$ transition at 1140 MHz with the maximum available microwave input power of 100 W (70 W transmitted power) where large microwave power broadening of the line is expected. The microwave power was modulated between on and off with a frequency of 1 Hz. The static electric field in the quenching grids was held constant at 600 V/cm. With microwave power on, M(2S) atoms should be quenched in the rf region, while with power off, quenching will take place only in the region of the static field quenching grids which is viewed by the phototubes. Event counts were defined as above and our signal should be the reduction in event counts due to the microwave power. The signal S is defined as the normalized difference between rf power on and power off. A large signal of the expected sign was found: $S=(-40\pm10)\%$.

Our six data points are shown in Fig. 4 and were obtained in six days of data taking in two runs. After obtaining a second point with 100 W at 1000 MHz which is off the center of the resonance, we took the four data points shown with 25 W input power, which is calculated to be approximately the optimum power for studying the resonance line. Again no signals were observed with delayed coincidences outside the expected times of flight of muonium atoms. The solid curve shows the theoretical non-power-broadened resonance line shape with its height normalized to the data point at 1140 MHz with a microwave input power of 25 W. The four data points are clearly consistent with the theoretical curve and indicate that a more correct theoretical curve would include some power broadening. The magnitude of the signal at the peak of 1140 MHz, $(-28\pm6)\%$, agrees well with that expected from the measured static electric field signal, namely $S_{\text{rf}}=-(0.6)S_{\text{static}}=-25\%$. We note that the large S value observed at 1000 MHz with 100 W input power is expected due to power broadening and provides further evidence of our observation of the $2S(F=1)\to 2P(F=1)$ transition. Our resonance curve with 25 W input power is in agreement with the predicted value of the Lamb shift.

We conclude from our data that we have formed muonium in the 2S state with the expected rate and that we have observed the Lamb shift transition. We are proceeding to make measurements of the $2S_{1/2} \to 2P_{1/2}$ transitions to determine the Lamb shift and hfs intervals.

This work was supported in part by Department of Energy Contract No. DE-AC02-76ER03075, by NATO Research Grant 1589, German Ministry for Science and Technology (BMFT), and National Science Foundation Grant No. PHY8215754. We are happy to acknowledge the helpful support of Dr. L. Rosen, Dr. L. Agnew, Dr. R. Werbeck, and the LAMPF staff. We thank R. Amittay for his effective work on the microwave system.

[1]V. W. Hughes and T. Kinoshita, in *Muon Physics I*, edited by V. W. Hughes and C. S. Wu (Academic, New York, 1977), p. 12.

[2]F. G. Mariam *et al.*, Phys. Rev. Lett. 49, 993 (1982).

[3]J. R. Sapirstein *et al.*, Phys. Rev. Lett. 51, 982 (1983); J. R. Sapirstein, Phys. Rev. Lett. 51, 985 (1983).

[4]S. R. Lundeen and F. M. Pipkin, Phys. Rev. Lett. 46, 232 (1981).

[5]H. G. Berry, Rep. Prog. Phys. 40, 155 (1977).

[6]P. R. Bolton *et al.*, Phys. Rev. Lett. 47, 1441 (1981).

[7]G. Gabrielse, Phys. Rev. A 23, 775 (1981).

[8]H.-W. Reist *et al.*, Nucl. Instrum. Methods 153, 61 (1978).

[9]H. Tawara and A. Russek, Rev. Mod. Phys. 45, 178 (1973).

[10]A. Chateau-Thierry *et al.*, Nucl. Instrum. Methods 132, 553 (1976).

[11]D. A. Owen, Phys. Lett. 44B, 199 (1973).

[12]W. E. Lamb, Jr., and R. C. Retherford, Phys. Rev. 79, 549 (1950).

[13]C. J. Oram *et al.*, J. Phys. B 14, L789 (1981).

VOLUME 66, NUMBER 21 PHYSICAL REVIEW LETTERS 27 MAY 1991

New Search for the Spontaneous Conversion of Muonium to Antimuonium

B. E. Matthias,[(1),(a)] H. E. Ahn,[(1)] A. Badertscher,[(5)] F. Chmely,[(1)] M. Eckhause,[(3)] V. W. Hughes,[(1)] K. P. Jungmann,[(2)] J. R. Kane,[(3)] S. H. Kettell,[(1),(b)] Y. Kuang,[(3)] H.-J. Mundinger,[(1),(2),(c)] B. Ni,[(1),(d)] H. Orth,[(4)] G. zu Putlitz,[(2)] H. R. Schaefer,[(1)] M. T. Witkowski,[(3)] and K. A. Woodle[(1),(6)]

[(1)]*Yale University, New Haven, Connecticut 06520*
[(2)]*Physikalisches Institut der Universität Heidelberg, D-6900 Heidelberg, Germany*
[(3)]*College of William and Mary, Williamsburg, Virginia 23185*
[(4)]*Gesellschaft für Schwerionenforschung, D-6100 Darmstadt 11, Germany*
[(5)]*Paul-Scherrer-Institut, CH-5234 Villigen, Switzerland*
[(6)]*Brookhaven National Laboratory, Upton, New York 11973*

(Received 7 March 1991)

To search for spontaneous conversion of muonium to antimuonium with very low background, a new signature was implemented that required the coincident detection of the decay products of the antimuonium atom, the energetic e^- and the atomic e^+. No conversion events were seen, which sets an improved upper limit of 6.5×10^{-7} (90% C.L.) on the conversion probability. The corresponding limit on the coupling constant, using a $V-A$ form for the conversion Hamiltonian, is $G_{M\bar{M}}<0.16G_F$ (90% C.L.), where G_F is the Fermi coupling constant.

PACS numbers: 13.10.+q, 13.35.+s, 14.60.−z, 36.10.Dr

There has been continuing interest in the possibility of conversion of muonium ($M\equiv\mu^+e^-$) to antimuonium ($\bar{M}\equiv\mu^-e^+$) from both the experimental[1,2] and theoretical sides.[3-7] As different neutrino flavors were unknown at the time, Pontecorvo suggested[8] in 1957 that $M\to\bar{M}$ might proceed through an intermediate state of two neutrinos in analogy to the K^0-$\bar{K}^0$ coupling via intermediate pions. The minimal standard model[9] forbids the $M\to\bar{M}$ coupling as it violates the separate, additive conservation of muon and electron number. Thus, searching for a mixing of M and $\bar{M}$ is a probe for physics beyond the standard model.

Traditionally, the results of searches for M to $\bar{M}$ conversion have been stated as upper limits on the coupling constant $G_{M\bar{M}}$, as it appears in the $V-A$ Hamiltonian density,[10]

$$\mathcal{H}_{M\bar{M}}=\frac{G_{M\bar{M}}}{\sqrt{2}}\bar{\mu}\gamma^\lambda(1-\gamma_5)e\bar{\mu}\gamma_\lambda(1-\gamma_5)e+\text{H.c.} \quad (1)$$

This form is motivated by an analogy to the effective Hamiltonian densities for observed weak processes.

In the absence of external electromagnetic fields, the ground-state hyperfine levels of M and $\bar{M}$ are degenerate in energy. The nonvanishing matrix elements of the Hamiltonian are diagonal in the angular momentum quantum numbers. For the ground state, one obtains[10]

$$\langle\bar{M}(F,m_F)|H_{M\bar{M}}|M(F,m_F)\rangle=(1.07\times10^{-12}\,\text{eV})\frac{G_{M\bar{M}}}{G_F}, \quad (2)$$

where F and m_F are the total angular momentum and its z component, respectively. Given an initial state of pure M, the coupling $\mathcal{H}_{M\bar{M}}$ leads to the development with time of an $\bar{M}$ component in the wave function of the system. The probability of decay from the $\bar{M}$ state is given by[10] $P_{\bar{M}}=(2.57\times10^{-5})(G_{M\bar{M}}/G_F)^2$.

Since the incident μ^+ beam used for M formation is fully polarized in the backward direction and the electron that is captured is unpolarized, M is formed equally in the $m_F=0$ (unpolarized) and in the $m_F=-1$ (polarized) substates, where the quantization axis has been chosen to lie along the beam axis. Magnetic fields shift the energy levels of M and $\bar{M}$ oppositely since the magnetic moments of the component leptons of $\bar{M}$ are reversed in sign from those of M. This removal of the degeneracy in energy of M and $\bar{M}$ causes a reduction of the conversion probability between polarized levels by a factor of 2 at 26 mG and between unpolarized levels by a factor of 2 at 1.6 kG. In general, diagonalization of the $n=1$ hyperfine states of the coupled $M\leftrightarrow\bar{M}$ system[11] yields magnetic-field-dependent eigenstates and eigenenergies with admixtures of M and $\bar{M}$ in each of the eight new states. Since the different interactions of M and $\bar{M}$ with atoms of a host medium remove their degeneracy in matter,[12] a sensitive search for the $M\to\bar{M}$ conversion requires that M be produced in vacuum.

Signatures used in previous searches for $M\to\bar{M}$ relied on radiations induced by μ^- interaction with the material of the detection medium. The most sensitive of these was recently completed by a group at TRIUMF with the result[1] of $P_{\bar{M}}<2.1\times10^{-6}$ and $G_{M\bar{M}}<0.29G_F$ at 90% confidence. The experiment described here is the first to seek the detection of *both* charged products of the $\bar{M}$ atom breakup; namely, the energetic e^- and the atomic e^+.

This new search[13] for the M to $\bar{M}$ conversion was performed at the Los Alamos Clinton P. Anderson Meson

Physics Facility (LAMPF). The apparatus (Fig. 1) included a section in which thermal muonium was formed in a SiO_2 powder target and a detector which consisted of a multiwire proportional chamber (MWPC) spectrometer to detect high-energy e^+ or e^- from μ^+ or μ^- and a second spectrometer for the coincident detection of a low-energy atomic e^- or e^+ from M or $\overline{M}$.

A well collimated 20-MeV/c subsurface μ^+ beam[14] from the Stopped Muon Channel (SMC) with a momentum bite of $\Delta p/p \approx 10\%$ and a duty factor of 6.4% was used. The incoming muons were counted and moderated in a 150-μm plastic scintillator and then stopped in a fine SiO_2 powder target of 9 mg/cm^2 thickness[15] where they formed M that diffused into the vacuum region downstream of the target at thermal velocities. The detector could observe either M or $\overline{M}$ decays from the vacuum.

After passing through a 100-μm Al vacuum window, the decay positrons or electrons, whose energies range up to 52.83 MeV, were observed in an array of four MWPCs placed on an axis at a right angle to the beam line. A wide-gap C-yoke dipole magnet with a central field of 522 G was placed between the second and third MWPCs to deflect decay e^+ and e^- in opposite directions. Two layers of plastic scintillator behind MWPC4 provided the timing signal on a candidate track and a cylindrical NaI(Tl) crystal measured the energy of a particle in the spectrometer.

The atomic e^- or atomic e^+, remaining in the vacuum with a mean kinetic energy of 13.5 eV after a M or $\overline{M}$ atom decay, was electrostatically collected, focused, and accelerated to 5.7 keV with a system of eleven electrodes arranged in three stages. Either e^- or e^+ at this energy were then charge and momentum selected in an iron-free bending magnet with a central field of about 15 G and transported and focused with a solenoidal field of 11 G onto a 75-mm-diam chevron pair microchannel plate detector (MCP). Additional coils allowed control over all components of the magnetic field in the target region to counteract effects of the C-magnet fringe field on the e^+ and e^- trajectories, and axial coils (S1, S2, and S3) enhanced the transport efficiency of the system.

The production of thermal muonium in our apparatus was studied by the established technique,[16,17] which involved use of a low-intensity μ^+ rate, observation of the decay e^+ track with the MWPC system, and measurement of the time of flight between a pulse in the muon beam counter and a decay e^+ track in the spectrometer. The formation fraction observed was $(5.02 \pm 0.06)\%$ M per stopped μ^+, with $(56 \pm 2)\%$ of the incident μ^+ stopping in the powder. Hence, an overall fraction of $(2.8 \pm 0.1)\%$ of the incoming μ^+ captured an electron from the powder to form M that diffused into the vacuum region downstream of the target at thermal velocities. In the vacuum region, M was found to move from the powder surface with a mean velocity of 0.7 cm/μs, corresponding to a temperature of 300 K.

The first observation of thermal M in vacuum by the coincident detection of its decay e^+ and the atomic e^- served to verify the experimental method and to calibrate the acceptances of the apparatus. When detecting thermal M in vacuum in this way, the polarities of the C magnet, the bending magnet, the solenoid, the steering coils, and the potentials in the electrostatic optics were set appropriately. The time-of-flight (TOF) spectrum taken with M decays started by an event in the MWPC spectrometer and stopped by a count in the MCP is shown in Fig. 2. This spectrum required that the decay origin of the M atom lie in the vacuum downstream of

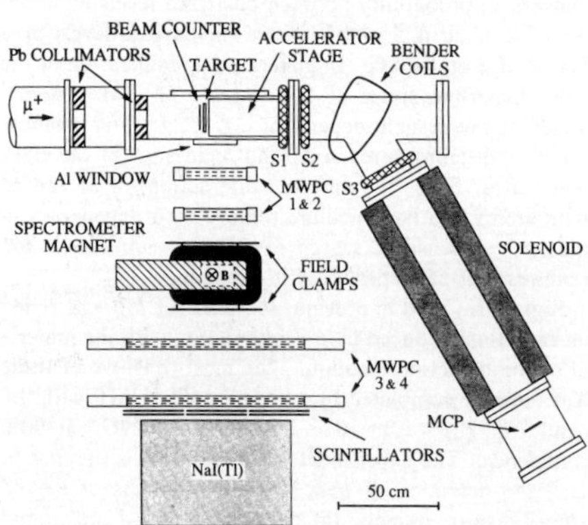

FIG. 1. Schematic view of the apparatus. Beam counter and target were mounted at 50° with respect to the incident muon beam.

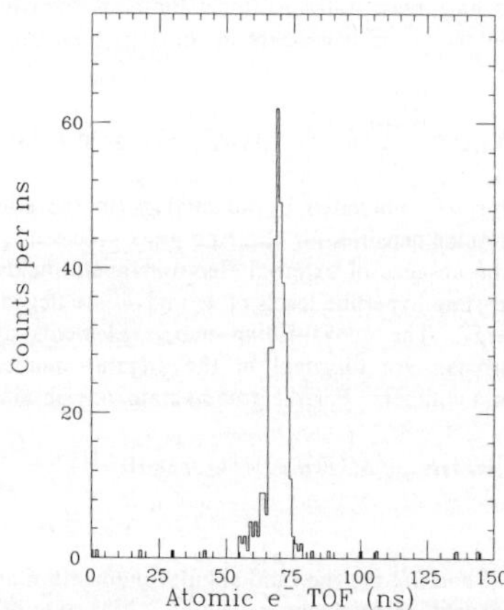

FIG. 2. Sample spectrum of the time of flight between decay e^+ and atomic e^- from M-atom decays in vacuum.

the target. The peak is therefore due to atomic e^- transported and detected after a M-atom decay. The transport acceptance and the MCP efficiency combined to allow detection of $(15.5 \pm 0.8)\%$ of the atomic electrons from thermal M decays in the vacuum. The acceptance of the high-energy spectrometer, including the solid angle, the detection efficiencies of the MWPCs and the plastic scintillators, and the efficiency of track reconstruction was $(2.50 \pm 0.02) \times 10^{-3}$.

To search for $\overline{M}$ decays, the electrostatic potentials and current directions in the high-energy and low-energy spectrometers were reversed in polarity and the full intensity of the incident μ^+ beam of $10^6\ \mathrm{s}^{-1}$ was used. To reduce the trigger rate and hence dead time for the data-acquisition system, the wires of all four planes detecting the position component perpendicular to the magnetic field were coarsely grouped to preselect e^- in the trigger by their curvature in the field. This preselection was 66% efficient in accepting e^- tracks and rejected 98.9% of the e^+ tracks. The remaining e^+ events were used as a monitor of M formation during the search for $\overline{M}$. To test the transport properties for low-energy e^+, the SiO_2 target was replaced with a venetian-blind arrangement of W foils that moderated the β^+ from a ^{22}Na source to energies of about 1 eV. The slow e^+ were detected by the MCP and were characterized by the same tuning behavior in each element of the system as the secondary e^- from e^+ impact on the foils.

We took approximately 270 h of data with a total of 9.8×10^{11} incident μ^+. To look for $\overline{M}$ decay candidates, it was required that a track in the MWPC spectrometer be successfully fitted, that its curvature indicate a negatively charged particle, that its fitted momentum lie above 22.5 MeV/c, that the track project back through the vacuum window, and that there be a count in the 75-ns window of the TOF spectrum determined from the M coincidence signal.

These events are displayed in a histogram of the decay position projected onto the axis perpendicular to the target plane (see Fig. 3). To find the most probable number of $\overline{M}$ events, a maximum-likelihood analysis of this distribution was carried out. A pure $\overline{M}$ signal was represented by a calculated reference distribution, while the characteristic distribution for the sum of all background processes was obtained from events that satisfy all $\overline{M}$ cuts except the TOF condition. The most probable number of $\overline{M}$ events was found to be zero, with a 90%-confidence-level upper limit of 2 counts. Thus, all events passing the cuts designed to emphasize a conversion signal were found to be most likely due to background. The total number of muonium atoms which could produce an observable conversion was determined from a maximum-likelihood analysis of the decay origin distribution along the axis perpendicular to the target plane. The result including an estimate for the total error is $(6.17 \pm 0.28) \times 10^6$, where the effect of the uncertainty on our final result is negligible.

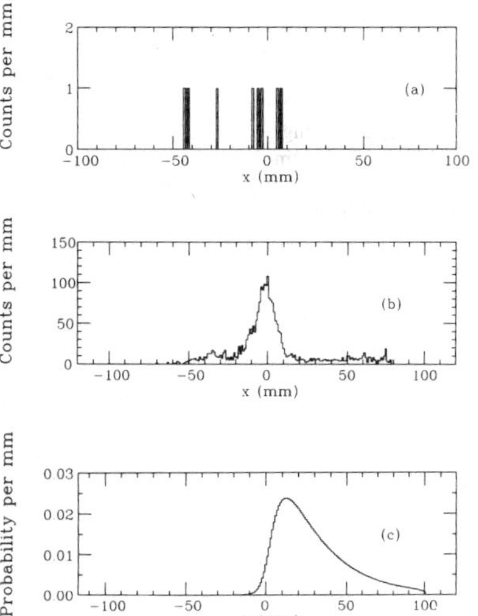

FIG. 3. Decay position distributions projected onto the target normal of (a) all $\overline{M}$ candidates, (b) sample of measured knock-on e^- background, and (c) calculated pure $\overline{M}$ signal. Positive values of x correspond to the downstream side of the target.

Since the magnetic field in the muonium cloud was about 10 G, a conversion would have been suppressed by a factor of 2. The resulting upper limit on the probability for the M to $\overline{M}$ conversion per atom is $P_{\overline{M}} < 6.5 \times 10^{-7}$ (90% C.L.). This gives an upper limit of

$$G_{M\overline{M}} < 0.16 G_F \quad (90\%\ \mathrm{C.L.}) \tag{3}$$

on the effective coupling constant of Eq. (1).

Background events are due to accidental coincidences of random counts in the MCP ($10^3\ \mathrm{s}^{-1}$ instantaneous) with knock-on electrons produced by decay e^+ in the MWPC spectrometer material before the C magnet (about 2×10^{-5} observed knock-on e^- above 22.5 MeV/c per e^+ reconstructed as originating in the vacuum region). We estimate this background at about 10^{-13} event per incident μ^+. The suppression of background on the MCP correlated to the incident beam is achieved by the TOF cut; further, 89% of the knock-on e^- are rejected by the momentum cut, while maintaining an acceptance of 87% for decay e^-.

The $M \rightarrow \overline{M}$ conversion is one of many processes which are strictly forbidden in the present standard model of particle physics. Speculative theories based on many different specific models[18] which violate the standard model can allow for these rare decays, but general criteria for comparing the sensitivity of different rare-decay searches are not known. The M to $\overline{M}$ conversion as well as the muon decay mode $\mu^+ \to e^+ \bar{\nu}_e \nu_\mu$ both violate additive conservation of muon and electron number, but they are allowed by the multiplicative conservation laws for muon and electron parity,[10] according to which the quantities $\prod(-1)^{L_\mu}$ and $\prod(-1)^{L_e}$ are conserved. However, other rare muon decay processes, such as $\mu \to e\gamma$ and $\mu \to 3e$, violate both the additive and multiplicative conservation laws. Hence, $M \to \overline{M}$ conversion and the forbidden mode of μ^+ decay are of particular interest to test the applicability of a multiplicative rather than an additive law. A search for the decay $\mu^+ \to e^+ \bar{\nu}_e \nu_\mu$ in an experiment[19] at LAMPF established the limit $R < 0.1$ (90% C.L.) relative to the normal decay mode, thus favoring the additive law. Our experiment establishes a limit for $M \to \overline{M}$ and hence also supports the additive law.

A particularly interesting model which would allow $M \to \overline{M}$ conversion at the present level of experimental sensitivity is the left-right symmetric model of Mohapatra and Senjanović[20] with an additional Higgs-boson triplet that violates lepton number conservation in its couplings. The doubly charged member of this triplet Δ^{++} mediates the M to $\overline{M}$ conversion at the tree level.[7] Within this model, the forbidden decay $\mu^+ \to e^+ \bar{\nu}_e \nu_\mu$ is also allowed as a first-order weak interaction through the singly charged Higgs boson Δ^+. However, a direct comparison of $M \to \overline{M}$ conversion and the forbidden μ^+ decay cannot be made because the masses of the Δ^{++} and the Δ^+ are not known.

We acknowledge the support of the entire LAMPF staff and we thank P. Herczeg for discussions of theoretical developments and B. Dieterli and E. B. Hughes for providing the NaI(Tl) detector hardware. This work was supported in part by the U.S. DOE, the Bundesministerium für Forschung und Technologie, and the NSF.

[(a)]Present address: Physikalisches Institut der Universität Heidelberg, D-6900 Heidelberg, Germany.

[(b)]Present address: Physics Department, Temple University, Philadelphia, PA 19122.

[(c)]Present address: Leybold AG, D-5000 Köln 51, Germany.

[(d)]Present address: Indiana University Cyclotron Facility, Indiana University, Bloomington, IN 47405.

[1]T. M. Huber *et al.*, Phys. Rev. D **41**, 2709 (1990).

[2]B. Ni *et al.*, Phys. Rev. Lett. **59**, 2716 (1987); G. A. Beer *et al.*, Phys. Rev. Lett. **57**, 671 (1986); G. M. Marshall *et al.*, Phys. Rev. D **25**, 1174 (1982); W. C. Barber *et al.*, Phys. Rev. Lett. **22**, 902 (1969); J. J. Amato *et al.*, Phys. Rev. Lett. **21**, 1709 (1968).

[3]See P. Herczeg, in *Rare Decay Symposium*, edited by D. Bryman *et al.* (World Scientific, Singapore, 1989).

[4]M. L. Swartz, Phys. Rev. D **40**, 1521 (1989).

[5]D. Chang and W.-Y. Keung, Phys. Rev. Lett. **62**, 2583 (1989).

[6]R. N. Mohapatra, in *Proceedings of the Eighth Workshop on Grand Unification*, edited by K. Wali (World Scientific, Singapore, 1988), p. 200.

[7]A. Halprin, Phys. Rev. Lett. **48**, 1313 (1982).

[8]B. Pontecorvo, Zh. Eksp. Teor. Fiz. **33**, 549 (1957) [Sov. Phys. JETP **6**, 429 (1958)].

[9]A. Salam, in *Elementary Particle Theory: Relativistic Groups and Analyticity*, edited by N. Svartholm, Proceedings of the Eighth Nobel Symposium (Almqvist & Wiksell, Stockholm, 1968), p. 367; S. Weinberg, Phys. Rev. Lett. **19**, 1264 (1967); A. Salam and J. C. Ward, Phys. Lett. **13**, 168 (1964); S. L. Glashow, Nucl. Phys. **22**, 579 (1961).

[10]G. Feinberg and S. Weinberg, Phys. Rev. **123**, 1439 (1961); Phys. Rev. Lett. **6**, 381 (1961).

[11]B. E. Matthias, Ph.D. thesis, Yale University, 1991 (unpublished).

[12]D. L. Morgan and V. W. Hughes, Phys. Rev. A **7**, 1811 (1973).

[13]LAMPF Proposal No. 1073, 1987, H. R. Schaefer and V. W. Hughes, spokesmen.

[14]A. Badertscher *et al.*, Nucl. Instrum. Methods Phys. Res., Sect. A **238**, 200 (1985).

[15]Cab-O-Sil fumed silica, grades M-5 and PTG (bulk density of 0.032 g/cm^3, typical grain diameters of 7 nm). Cabot Corporation, Tuscola, IL 61953.

[16]K. A. Woodle *et al.*, Z. Phys. D **9**, 59 (1988).

[17]A. C. Janissen *et al.*, Phys. Rev. A **42**, 161 (1990).

[18]For a review, see, for example, J. D. Vergados, Phys. Rep. **133**, 1 (1986).

[19]S. E. Willis *et al.*, Phys. Rev. Lett. **44**, 522 (1980).

[20]R. N. Mohapatra and G. Senjanović, Phys. Rev. D **23**, 165 (1981); Phys. Rev. Lett. **44**, 912 (1980).

Volume 37, Number 19 PHYSICAL REVIEW LETTERS 8 November 1976

Deep Inelastic Scattering of Polarized Electrons by Polarized Protons*

M. J. Alguard, W. W. Ash, G. Baum, J. E. Clendenin, P. S. Cooper, D. H. Coward, R. D. Ehrlich, A. Etkin, V. W. Hughes, H. Kobayakawa, K. Kondo, M. S. Lubell, R. H. Miller, D. A. Palmer, W. Raith, N. Sasao, K. P. Schüler, D. J. Sherden, C. K. Sinclair, and P. A. Souder

University of Bielefeld, Bielefeld, West Germany, and City University of New York, New York, New York 10031, and Nagoya University, Nagoya, Japan, and Stanford Linear Accelerator Center, Stanford, California 94305, and University of Tsukuba, Ibaraki, Japan, and Yale University, New Haven, Connecticut 06520

(Received 5 August 1976)

We report measurements of the asymmetry in deep inelastic scattering of longitudinally polarized electrons by longitudinally polarized protons. The antiparallel-parallel asymmetries are positive and large in agreement with predictions of quark-parton models of the proton. A limit is obtained on parity nonconservation in the scattering of longitudinally polarized electrons by unpolarized nucleons.

Experimental and theoretical studies of deep inelastic electron scattering from protons and neutrons have led in the past eight years to the important discovery of scaling and to the quark-parton model of nucleon structure.[1] Deep inelastic muon[2] and neutrino[3] scattering have confirmed these general ideas.[4]

For deep inelastic electron-proton scattering,

VOLUME 37, NUMBER 19 PHYSICAL REVIEW LETTERS 8 NOVEMBER 1976

accurate data have been obtained on the differential cross section $d^2\sigma/d\Omega dE'$ over a wide range of the energy loss, ν, of the electron and the square of the four-momentum transfer, q^2, to the proton. The two spin-averaged proton structure functions $W_1(\nu,q^2)$ and $W_2(\nu,q^2)$ have been determined from these data. Important, independent information is contained in two additional spin-dependent proton structure functions whose determination requires the measurement of spin correlation asymmetries.[5]

In this Letter we report the first results of an experiment done at the Stanford Linear Accelerator Center (SLAC) to measure the asymmetry, A, in the deep inelastic scattering of longitudinally polarized electrons by longitudinally polarized protons, where A is given by

$$A=[d\sigma(\uparrow\downarrow)-d\sigma(\uparrow\uparrow)]/[d\sigma(\uparrow\downarrow)+d\sigma(\uparrow\uparrow)], \tag{1}$$

with $d\sigma$ denoting the differential cross section $d^2\sigma(E,E',\theta)/d\Omega dE'$ for electrons of incident (scattered) energy E (E') and laboratory scattering angle θ, and the arrows denoting the antiparallel and parallel spin configurations.

If the scattering is described by the one-photon-exchange approximation, then for unpolarized electrons the virtual photons are linearly polarized, whereas for polarized electrons the photons are elliptically polarized. The differential cross section for the scattering of longitudinally polarized electrons by longitudinally polarized protons is

$$\frac{d^2\sigma}{d\Omega dE'}=\left(\frac{d\sigma}{d\Omega}\right)_{\rm M}\left(\frac{1}{\epsilon(1+\nu^2/Q^2)}\right)W_1\{1+\epsilon R\pm(1-\epsilon^2)^{1/2}\cos\psi A_1\pm[2\epsilon(1-\epsilon)]^{1/2}\sin\psi A_2\}, \tag{2}$$

in which $(d\sigma/d\Omega)_{\rm M}$ is the Mott differential cross section, $\epsilon=[1+2(1+\nu^2/Q^2)\tan^2\frac{1}{2}\theta]^{-1}$, $Q^2=-q^2$, $R=\sigma_L/\sigma_T$ is the ratio of the cross sections for absorption of longitudinal and transverse virtual photons, and ψ is the angle between the directions of the virtual photon momentum and the proton spin. The + (−) signs in Eq. (2) refer to the antiparallel (parallel) spin configurations.

The spin-dependent terms A_1 and A_2 are two new measurable quantities which can be expressed in terms of two spin-dependent structure functions.[5,6] Equivalently, they can be expressed in terms of the total absorption cross sections of circularly polarized photons on polarized protons as

$$A_1=(\sigma_{1/2}-\sigma_{3/2})/(\sigma_{1/2}+\sigma_{3/2}),\quad A_2=2\sigma_{TL}/(\sigma_{1/2}+\sigma_{3/2}), \tag{3}$$

where $\sigma_{1/2}$ ($\sigma_{3/2}$) is the total absorption cross section when the z component (z is the direction of the virtual photon momentum) of angular momentum of the virtual photon plus proton is $\frac{1}{2}$ ($\frac{3}{2}$), and σ_{TL}, which may be negative, is a term which arises from the interference between transverse and longitudinal photon-nucleon amplitudes. It should be noted that $\sigma_{1/2}$ and $\sigma_{3/2}$ are related to σ_T by $\sigma_{1/2}+\sigma_{3/2}=2\sigma_T$.

For the case of protons polarized along the incident beam direction, the asymmetry A of Eq. (1) is

$$A=D(A_1+\eta A_2), \tag{4}$$

where

$$D=(E-E'\epsilon)/E(1+\epsilon R)=(1-\epsilon^2)^{1/2}\cos\psi/(1+\epsilon R), \tag{5}$$

and

$$\eta=\epsilon(Q^2)^{1/2}/(E-E'\epsilon)=[2\epsilon/(1+\epsilon)]^{1/2}\tan\psi\simeq\tan\psi. \tag{6}$$

The quantity D can be regarded as a kinematic depolarization factor of the virtual photon and is ~0.3 for our kinematic points. Positivity limits imposed on A_1 and A_2 are[7]

$$|A_1|\leq 1,\quad |A_2|\leq\sqrt{R}. \tag{7}$$

In this experiment we determine the combination $A_1+\eta A_2$ by dividing the measured electron-proton asymmetry A by the depolarization factor D. Although we do not separately determine A_1 and A_2, our result is dominated by A_1 because the kinematic factor η is small.

On the basis of a high-energy sum rule derived with the algebra of currents for a quark model, it has been predicted[8] that A_1 has a positive value greater than 0.2 over a large region of the deep inelastic continuum. Scaling relations are predicted for the spin-dependent proton structure functions, and hence also for A_1[9]:

$$A_1(\nu,Q^2)\to A_1(\omega)\text{ as }\nu,Q^2\to\infty,\text{ with }\omega\text{ held constant} \tag{8}$$

($\omega=2M\nu/Q^2$, M is the proton mass). Specific models of proton structure make widely varying predictions for A_1. The simplest quark-parton

TABLE I. Results of asymmetry measurements.

E (GeV)	θ (deg)	Q^2 $[(\mathrm{GeV}/c)^2]$	W[a] (GeV)	ω	Δ (%)	A[b]	D[c]	$A_1+\eta A_2$[b]	$\|\eta A_2\|$
9.711	9.000	1.680	2.059	3	0.44 ± 0.11	0.191 ± 0.057 (0.044)	0.284	0.67 ± 0.20 (0.16)	<0.146
12.948	9.000	2.735	2.519	3	0.50 ± 0.17	0.215 ± 0.089 (0.080)	0.352	0.61 ± 0.25 (0.23)	<0.109
9.711	9.000	1.418	2.560	3	0.28 ± 0.11	0.141 ± 0.058 (0.051)	0.412	0.34 ± 0.14 (0.12)	<0.087

[a] W is the missing mass of undetected hadron system.
[b] The total errors are the statistical counting errors added in quadrature to the systematic errors in P_e, P_p, and F; the numbers in parentheses are the 1-standard-deviation counting errors.
[c] D is obtained from Eq. (5) using $R=0.14$.

model predicts that $A_1=\frac{5}{9}$, and more elaborate models also predict large positive values for $A_1(\omega)$.[5,10]

The method of measuring the experimental asymmetry, Δ, for deep inelastic electron-proton scattering was the same as that described for elastic scattering in the preceding Letter.[11] For the inelastic case, the scattered electron counting rate was lower (0.02 to 0.06 electrons per pulse).

TABLE II. False asymmetries.[a]

Combination of miniruns	Average asymmetry[b] (%)	$\chi^2(0)$ per degree of freedom
	$\omega=3$, $Q^2=1.680$	
$\frac{(1234)-(5678)}{(1234)+(5678)}$	0.04 ± 0.11	18/34
$\frac{(1357)-(2468)}{(1357)+(2468)}$	-0.04 ± 0.11	38/34
$\frac{(2367)-(1458)}{(2367)+(1458)}$	$+0.14\pm0.11$	27/34
	$\omega=3$, $Q^2=2.735$	
$\frac{(1234)-(5678)}{(1234)+(5678)}$	-0.30 ± 0.17	33/30
$\frac{(1357)-(2468)}{(1357)+(2468)}$	-0.03 ± 0.17	26/30
$\frac{(2367)-(1458)}{(2367)+(1458)}$	$+0.24\pm0.017$	40/30
	$\omega^2=5$, $Q^2=1.418$	
$\frac{(1234)-(5678)}{(1234)+(5678)}$	-0.12 ± 0.11	34/35
$\frac{(1357)-(2468)}{(1357)+(2468)}$	-0.10 ± 0.11	34/35
$\frac{(2367)-(1458)}{(2367)+(1458)}$	-0.03 ± 0.11	30/35

[a] See preceding Letter (Ref. 11) for definitions of false asymmetries.
[b] Irrespective of sign of target polarization.

Background due to misidentified pions was again negligible.

The antiparallel-parallel asymmetry Δ was measured for three deep inelastic kinematic points and the results are given in Table I. Several false asymmetries were also measured and are listed in Table II, together with the χ^2 values for the agreement with zero of the measured false asymmetries for the indicated degrees of freedom (number of individual runs). No statistically significant false asymmetries were found.

The asymmetry A of Eq. (1) is related to Δ by

$$\Delta=P_eP_pFA. \tag{9}$$

The electron polarization, P_e, was 0.51 ± 0.06, and the average target polarization, P_p, measured for each kinematic point, was $\simeq 0.40$ with 10% uncertainty. The quantity F is the fraction of detected electrons scattered from free protons. This is taken as the ratio of the number of free protons to the total number of nucleons in the target, including measured contributions from helium and other background sources. A small correction for the difference in scattering cross sections of neutrons and protons was also included. The value for F, determined for each point, was $\simeq 0.11$ with a 10% uncertainty.

The measured values of A are listed in Table I. The uncertainties are dominated by counting statistics. No radiative corrections have yet been made. Also listed are the quantities D (evaluated using $R=0.14$),[1] $A/D=A_1+\eta A_2$, and upper limits for $|\eta A_2|$ (taking $A_2=\sqrt{R}$). From Table I it is seen that A/D is dominated by A_1. Furthermore, parton theories predict[12] that the interference term A_2 will be considerably smaller than its positivity limit $\sqrt{R}$. It is therefore valid to compare our measured value of A/D to theoretical predictions for A_1 as shown in Fig. 1.

With the explicit assumption that $A/D=A_1$, our

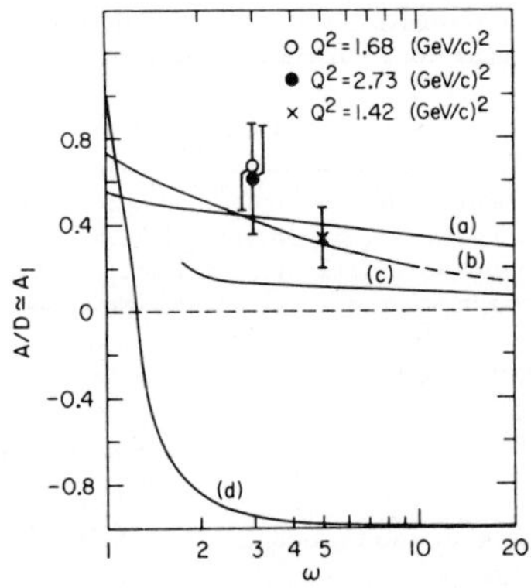

FIG. 1. Experimental values of $A/D \approx A_1$ and theoretical predictions of the virtual-photon–proton asymmetry A_1 versus ω. Theoretical curves a, b, c, and d are obtained from Refs. 5, 10, 13, and 14, respectively. For curve c the quark model with symmetry breaking is used: The model does not give values for A_1 in the range $1<\omega<2$, but rather gives $A_1(1)=1$. For curve d the quantity μ^2/m_p^2 in the theory is taken equal to 0.12.

values of A_1 are indeed positive and large in accord with early theoretical expectations from sum rules.[8] The two values for $\omega=3$ agree within their errors, which is consistent with the expectation that A_1 satisfies the scaling relation, given by Eq. (8). Our data are consistent with the predictions of the quark-parton models shown as curves a[5] and b[10] in Fig. 1, but disagree strongly with the resonance model[13] (curve c) and the bare-nucleon–bare-meson model[14] (curve d). We note that the theoretical curves are all given for the scaling limit.

Data from this experiment can also be used to place a limit on parity nonconservation in the scattering of longitudinally polarized electrons from unpolarized nucleons, i.e., an interaction term of the form $\vec{\sigma}_e\cdot\vec{p}_e$ in which $\vec{\sigma}_e$ is the electron spin and $\vec{p}_e$ is the electron incident momentum. If we define Δ^+ (Δ^-) as the asymmetry for protons polarized along (against) the beam direction and if the magnitude of P_p is the same for both cases, then we can define an asymmetry, $\Delta_{\rm PNC}$, associated with parity nonconservation by[15]

$$\Delta_{\rm PNC}=(\Delta^+-\Delta^-)/2\equiv rP_e, \qquad (10)$$

in which $r=(d\sigma^- - d\sigma^+)/(d\sigma^- + d\sigma^+)$ is the asymmetry for electron polarization $P_e=1$, and the minus and plus superscripts refer to the electron beam helicity. From the deep inelastic scattering data summarized in Table I for Q^2 between 1.4 and 2.7 $(\text{GeV}/c)^2$, we find that r is consistent with zero. For the combined data we have an upper limit of $r<5\times10^{-3}$ with a 95% confidence level. For the elastic scattering data reported in the preceding Letter,[11] again r is consistent with zero and its upper limit is less than 3×10^{-3} with a 95% confidence level. The gauge theories of weak and electromagnetic interactions, which contain parity nonconservation, predict[16,17] considerably smaller values of $r\simeq(10^{-5}$ to $10^{-4})Q^2/M^2$.

We are happy to acknowledge helpful and stimulating discussions with J. D. Bjorken, F. Gilman, and J. Kuti.

---

*Research supported in part by the U. S. Energy Research and Development Administration under Contract No. E(11-1)-3075 (Yale) and Contract No. E(04-3)-515 (Stanford Linear Accelerator Center), the German Federal Ministry of Research and Technology and the University of Bielefeld, the Japan Society for the Promotion of Science, and the National Science Foundation.

[1]R. E. Taylor, in *Proceedings of the International Symposium on Lepton and Photon Interactions at High Energies, Stanford, California, 1975,* edited by W. T. Kirk (Stanford Linear Accelerator Center, Stanford, Calif., 1975), p. 679. See also references therein.

[2]L. Mo, in *Proceedings of the International Symposium on Lepton and Photon Interactions at High Energies, Stanford, California, 1975,* edited by W. T. Kirk (Stanford Linear Accelerator Center, Stanford, Calif., 1975), p. 651.

[3]D. H. Perkins, in *Proceedings of the International Symposium on Lepton and Photon Interactions at High Energies, Stanford, California, 1975,* edited by W. T. Kirk (Stanford Linear Accelerator Center, Stanford, Calif., 1975), p. 571.

[4]C. H. Llewellyn-Smith, in *Proceedings of the International Symposium on Lepton and Photon Interactions at High Energies, Stanford, California, 1975,* edited by W. T. Kirk (Stanford Linear Accelerator Center, Stanford, Calif., 1975), p. 709. See also references therein.

[5]J. Kuti and V. F. Weisskopf, Phys. Rev. D 4, 3418 (1971).

[6]F. Gilman, Phys. Rep. 4, 95 (1972); F. Gilman, SLAC Report No. SLAC-167, 1973 (unpublished), Vol. I, p. 71.

[7]M. G. Doncel and E. de Rafael, Nuovo Cimento 4A, 363 (1971).

[8]J. D. Bjorken, Phys. Rev. D 1, 1376 (1970).

[9]L. Galfi *et al.*, Phys. Lett. 31B, 465 (1970).

[10]F. Close, Nucl. Phys. B80, 269 (1974), and references therein.

[11]M. J. Alguard *et al.*, preceding Letter [Phys. Rev.

Lett. 37, 1258 (1976)].

[12]J. D. Bjorken and F. Gilman, private communication.

[13]G. Domokos *et al.*, Phys. Rev. D 3, 1191 (1971).

[14]S. D. Drell and T. D. Lee, Phys. Rev. D 5, 1738 (1972).

[15]In the actual analysis, target polarization differences were included. Since $P_p FA \simeq 0.01$ is small, these differences have little effect.

[16]S. M. Berman and J. R. Primack, Phys. Rev. D 9, 2171 (1974).

[17]G. Feinberg, Phys. Rev. D 12, 3575 (1975).

Volume 77B, number 3 PHYSICS LETTERS 14 August 1978

PARITY NON-CONSERVATION IN INELASTIC ELECTRON SCATTERING ☆

C.Y. PRESCOTT, W.B. ATWOOD, R.L.A. COTTRELL, H. DeSTAEBLER, Edward L. GARWIN, A. GONIDEC [1], R.H. MILLER, L.S. ROCHESTER, T. SATO [2], D.J. SHERDEN, C.K. SINCLAIR, S. STEIN and R.E. TAYLOR
Stanford Linear Accelerator Center, Stanford University, Stanford, CA 94305, USA

J.E. CLENDENIN, V.W. HUGHES, N. SASAO [3] and K.P. SCHÜLER
Yale University, New Haven, CT 06520, USA

M.G. BORGHINI
CERN, Geneva, Switzerland

K. LÜBELSMEYER
Technische Hochschule Aachen, Aachen, West Germany

and

W. JENTSCHKE
II. Institut für Experimentalphysik, Universität Hamburg, Hamburg, West Germany

Received 14 July 1978

We have measured parity violating asymmetries in the inelastic scattering of longitudinally polarized electrons from deuterium and hydrogen. For deuterium near $Q^2 = 1.6\ (\mathrm{GeV}/c)^2$ the asymmetry is $(-9.5 \times 10^{-5})Q^2$ with statistical and systematic uncertainties each about 10%.

We have observed a parity non-conserving asymmetry in the inelastic scattering of longitudinally polarized electrons from an unpolarized deuterium target. In this experiment a polarized electron beam of energy between 16.2 and 22.2 GeV was incident upon a liquid deuterium target. Inelastically scattered electrons from the reaction

$$\mathrm{e(polarized)} + \mathrm{d} \rightarrow \mathrm{e}' + \mathrm{X}, \tag{1}$$

were momentum analyzed in a magnetic spectrometer at 4° and detected in a counter system instrumented to measure the electron flux, rather than to count individual scattered electrons. The momentum transfer, Q^2, to the recoiling hadronic system varied between 1 and 1.9 $(\mathrm{GeV}/c)^2$ (see table 1).

Parity violating effects may arise from the interference between the weak and electromagnetic amplitudes. Calculations of the expected effects in deep inelastic experiments have been reported by several authors [1–7], and asymmetries at the level of $10^{-4}\,Q^2$ are predicted for the kinematics of our experiment. Previous experiments with muons [8] and electrons [9,10] have not achieved sufficient accuracy to observe such small effects. This same interference of amplitudes may also give rise to measurable effects in

☆ Work supported by the Dept. of Energy.
[1] Permanent address: Annecy (LAPP), 74019 Annecy-le-Vieux, France.
[2] Permanent address: National Laboratory for High Energy Physics, Tsukuba, Japan.
[3] Present address: Department of Physics, Kyoto University, Kyoto, Japan.

Volume 77B, number 3 PHYSICS LETTERS 14 August 1978

Table 1
Kinematic conditions at which data were taken. The average Q^2 and y values were calculated for the shower counter using a Monte Carlo program.

Beam energy E_0 (GeV)	$g-2$ precession angle θ_{prec} (rad)	Spectrometer setting E' (GeV)	Kinematic quantities averaged over spectrometer	
			Q^2 $(\mathrm{GeV}/c)^2$	y
16.18	5.0π	12.5	1.05	0.18
17.80	5.5π	13.5	1.25	0.19
19.42	6.0π	14.5	1.46	0.21
22.20	6.9π	17.0	1.91	0.21

atomic spectra; experiments on transitions in the spectrum of bismuth have already been reported [11–13].

Of crucial importance to this experiment was the development of an intense source of longitudinally polarized electrons. The source consisted of a gallium arsenide crystal mounted in a structure similar to a regular SLAC gun with the GaAs replacing the usual thermionic cathode. The polarized electrons were produced by optical pumping with circularly polarized photons between the valence and conduction bands in the GaAs, which had been treated to assure a surface with negative electron affinity [14,15]. The light source was a dye laser operated at 710 nm and pulsed to match the linac (1.5 μs pulses at 120 pulses per second). Linearly polarized light from the laser was converted to circularly polarized light by a Pockels cell, a crystal with birefringence proportional to the applied electric field. The plane of polarization of the light incident on the Pockels cell could be varied by rotating a calcite prism. Reversing the sign of the high voltage pulse driving the Pockels cell reversed the helicity of the photons which in turn reversed the helicity of the electrons. This reversal was done randomly on a pulse to pulse basis. The rapid reversals minimized the effects of drifts in the experiment, and the randomization avoided changing the helicity synchronously with periodic changes in experimental parameters. Pulsed beam currents of several hundred milliamperes were achieved, with intensity fluctuations of a few percent.

The longitudinally polarized electrons were accelerated with negligible depolarization as confirmed by earlier tests [16] [‡1]. Both the sign and the magnitude of the polarization of the beam at the target were measured periodically by observing the asymmetry in Møller (elastic electron–electron) scattering from a magnetized iron foil [16]. The polarization, $|P_e|$, averaged 0.37. Each measurement had a statistical error less than 0.01; we estimate an overall systematic uncertainty of 0.02. The beam intensity at the target varied between 1 and 4×10^{11} electrons per pulse.

A schematic of the apparatus is shown in fig. 1. The target was a 30 cm cell of liquid deuterium. The spectrometer consisted of a dipole magnet, followed by a single quadrupole and a second dipole. The scattering angle was 4° and the momentum setting was about 20% below the beam energy (see table 1 for the kinematic settings). The acceptance was ±7.4 mrad in scattering angle, ±16.6 mrad in azimuth and about ±30% in momentum, as determined from a Monte Carlo model of the spectrometer.

Two separate electron detectors intercepted electrons analyzed by the spectrometer. The first was a nitrogen-filled Cerenkov counter operated at atmospheric pressure. The second was a lead-glass shower counter with a thickness of nine radiation lengths (the TA counter). Approximately 1000 scattered electrons per pulse entered the counters.

The high rates were handled by integrating the outputs of each phototube rather than by counting individual particles. For each pulse, i, the integrated output of each phototube, N_i, was divided by the integrated beam intensity (charge), Q_i, to form the yield for that pulse, $Y_i = N_i/Q_i$. For the distributions of the Y_i we verified experimentally that the (charge weighted) means of the distributions, $\langle Y \rangle$, were independent of Q, within errors of about ±0.3%, and that the (charge weighted) standard deviations, ΔY, were consistent with the statistical fluctuations expected from the number of scattered electrons per pulse. For a run with n beam pulses the statistical uncertainty on $|Y|$ was given by $\Delta Y/\sqrt{n}$.

As a check on our procedures we measured the asymmetry for a series of runs using the unpolarized beam from the regular SLAC gun for which the asymmetry should be zero. For a given run the experimental asymmetry was given by:

$$A_{exp} = [\langle Y(+)\rangle - \langle Y(-)\rangle] / [\langle Y(+)\rangle + \langle Y(-)\rangle] , \qquad (2)$$

‡1 The present experiment used the same target as ref. [16], but used a different spectrometer and detectors.

Volume 77B, number 3 PHYSICS LETTERS 14 August 1978

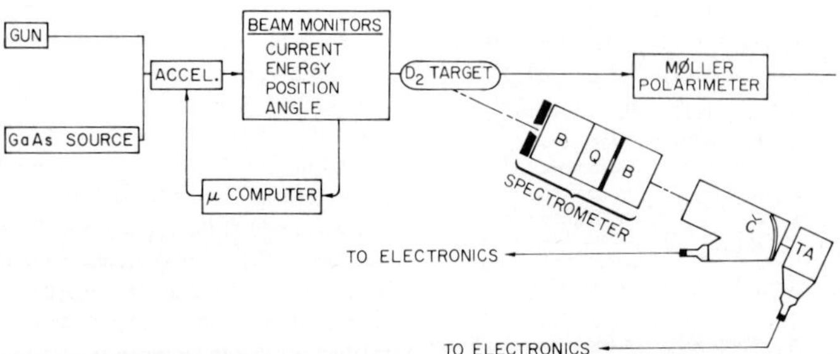

Fig. 1. Schematic layout of the experiment. Electrons from the GaAs source or the regular gun are accelerated by the linac. After momentum analysis in the beam transport system the beam passes through a liquid deuterium target. Particles scattered at 4° are analyzed in the spectrometer (bend-quad-bend) and detected in two separate counters (a gas Cerenkov counter, and a lead-glass shower counter). A beam monitoring system and a polarization analyzer are only indicated, but they provide important information in the experiment.

where + and – were assigned by the same random number generator that determined the sign of the voltage applied to the Pockels cell. For the shower counter we obtained a value of $(-2.5 \pm 2.2) \times 10^{-5}$ for A_{exp} divided by 0.37, the average value of $|P_e|$ for polarized beams from the GaAs source. The individual values were distributed about zero consistent with the calculated statistical errors. We conclude that asymmetries can be measured in this apparatus to a level of about 10^{-5}.

The same procedures were next applied to a similar series of runs using polarized beams. The helicity of the electrons coming from the source depended on the orientation of the linearly polarizing prism as well as on the sign of the voltage on the Pockels cell. Rotation of the plane of polarization by rotating the calcite prism through an angle ϕ_p caused the net electron helicity to vary as $\cos(2\phi_p)$. We chose three operating conditions:

(a) prism orientation at 0°, producing + (–) helicity electrons for + (–) Pockels cell voltage;

(b) prism orientation at 45°, producing unpolarized electrons for either sign of Pockels cell voltage; and

(c) prism orientation at 90°, producing – (+) helicity electrons for + (–) Pockels cell voltage.

Positive helicity indicates that the spin is parallel to the direction of motion. As the prism is rotated by 90°, A_{exp} should change sign since it is defined only with respect to the sign of the voltage on the Pockels cell. We may define a physics asymmetry, A, whose sign depends on the helicity of the beam at the target

$$A_{exp} = |P_e| A \cos(2\phi_p), \tag{3}$$

where ϕ_p is the angle of orientation of the calcite prism.

Fig. 2 shows the results at 19.4 GeV for $A_{exp}/|P_e|$. For the 45° point we used a value of 0.37 for $|P_e|$. These data are in satisfactory agreement with expecta-

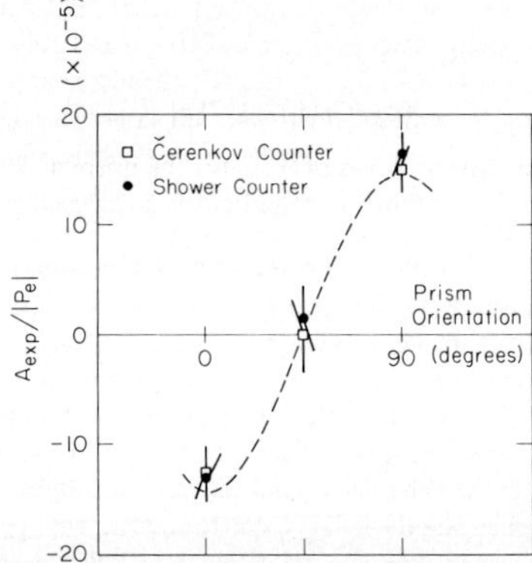

Fig. 2. The experimental asymmetry shows the expected variation (dashed line) as the beam helicity changes due to the change in orientation of the calcite prism. The data are for 19.4 GeV and deuterium. Since the same scattered particles strike both counters, they are not statistically independent. No systematic errors are shown. No corrections have been made for helicity dependent differences in beam parameters.

tions, and serve to separate effects due to the helicity of the beam from possible systematic effects associated with the reversal of the Pockels cell voltage. Only statistical errors are shown. The results at 45° are consistent with zero and indicate that other sources of error in A_{exp} must be small. Furthermore, the asymmetries measured at 0° and 90° are equal and opposite, within errors, as expected. Fig. 2 shows data from both the Cerenkov counter and the shower counter. Although these two separate counters were not statistically independent, they were analyzed with independent electronics and responded quite differently to potential backgrounds. The consistency between these counters serves as a check that such backgrounds are small.

At 19.4 GeV with the prism at 0° the helicity at the target was positive for positive Pockels cell voltage. However, this helicity depended on beam energy, owing to the $g-2$ precession of the spin in the transport magnets which deflected the beam through 24.5° before reaching the target. Because of the anomalous magnetic moment of the electron, the electron spin direction precessed relative to the momentum direction by an angle

$$\theta_{prec} = \frac{E_0}{m_e c^2}\frac{g-2}{2}\theta_{bend} = \frac{E_0\,(\mathrm{GeV})}{3.237}\,\pi\ \mathrm{rad}, \qquad (4)$$

where m_e is the mass and g the gyromagnetic ratio of the electron. Thus we expect

$$A_{exp} = |P_e|\,A\cos[(E_0\,(\mathrm{GeV})/3.237)\pi], \qquad (5)$$

where the signs of values of A_{exp} for the prism at 90° have been reversed before combining with values for the prism at 0°. Fig. 3 shows the results for the kinematic points in table 1 as a function of beam energy. At each point Q^2 is different. Since we expect A to be proportional to Q^2, we divide A_{exp} by Q^2 ‡2. Fig. 3 also shows the expected curve normalized to the point at 19.4 GeV. The data clearly follow the $g-2$ modulation of the helicity. At 17.8 GeV the spin is transverse; any effects from transverse components of the spin are expected to be negligible, in agreement with our data.

We conclude from figs. 2 and 3 that the observed asymmetries are due to electron helicity. Nevertheless,

‡2 This fact is true in all models. It arises because the electromagnetic amplitude has a $1/Q^2$ dependence, giving an asymmetry proportional to Q^2.

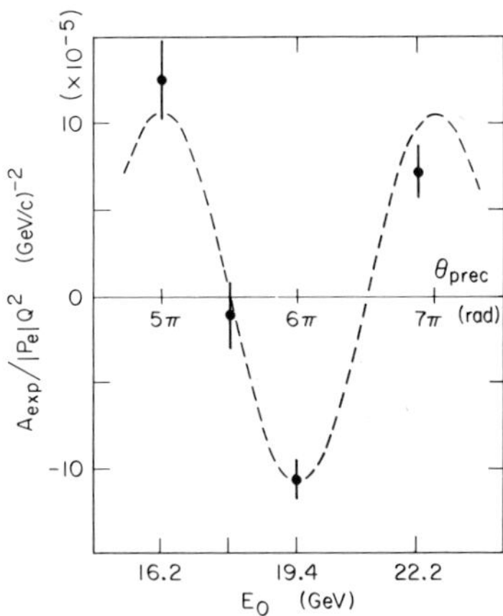

Fig. 3. The experimental asymmetry shows the expected variation (dashed line) as the beam helicity changes as a function of beam energy due to the $g-2$ precession in the beam transport system. The data are for the shower counter and the deuterium target. No systematic errors are shown. No corrections have been made for helicity dependent differences in beam parameters.

it is essential to search for and set limits on asymmetries due to effects other than helicity. Systematic effects due to slow drifts in phototube gains, magnet currents, etc., were minimized by the rapid, random reversals of polarization, and had negligible effects on A_{exp}. Effects due to random fluctuations in the beam parameters were small compared to the 3% pulse to pulse fluctuations due to counting statistics in the detectors. This was verified experimentally by measuring A_{exp} with unpolarized beams from the regular SLAC gun, and also by generating "fake" asymmetries using pulses of the same helicity from the polarized data runs themselves.

A more serious source of potential error came from small systematic differences between the beam parameters for the two helicities. Small changes in position, angle, current or energy of the beam can influence the measured yields. If these changes are correlated with reversals of the beam helicity, they may cause apparent parity violating asymmetries. Using an extensive beam monitoring system based on microwave cavities, measurements were made for each beam pulse of the average energy and position [17]. Angles were deter-

Volume 77B, number 3 PHYSICS LETTERS 14 August 1978

mined from cavities 50 m apart. The beam charge was determined using the standard toroid monitors [18]. The resolutions per pulse were about 10 μm in position, 0.3 μrad in angle, 0.01% in energy, and 0.02% in beam intensity. A microcomputer driven feedback system used position and energy signals to stabilize the average beam position, angle, and energy. Using the measured pulse to pulse beam information together with the measured sensitivities of the yield to each of the beam parameters, we made corrections to the asymmetries for helicity dependent differences in beam parameters. For these corrections, we have assigned a systematic error equal to the correction itself. The most significant imbalance was less than one part per million in E_0 which contributed -0.26×10^{-5} to A/Q^2.

We combine the values of A/Q^2 from the shower counter for the two highest energy points to obtain

$$A/Q^2 = (-9.5 \pm 1.6) \times 10^{-5}\ (\mathrm{GeV}/c)^{-2}\ \text{(deuterium)}. \quad (6)$$

We do not include the point at 16.2 GeV because it contains fairly strong elastic and resonance contributions. The sign implies a greater yield from electrons with spin antiparallel to momentum. For this combined point the average value of $y = 1 - E'/E_0$ is 0.21 and the average value of Q^2 is 1.6 $(\mathrm{GeV}/c)^2$. The quoted error, based on preliminary analysis, is derived from a statistical error of $\pm 0.86 \times 10^{-5}$ added linearly to estimated systematic uncertainties of 5% in the value of $|P_e|$, and of 3.3% from asymmetries in beam parameters. We determined experimentally that the π^- background contributed less than 0.1×10^{-5} to A/Q^2. The result in eq. (6) includes normalization corrections of 2% for the π^- background, and 3% for radiative corrections.

Any observation of non-conservation of parity in interactions involving electrons adds new information on the nature of neutral currents and gauge theories. Certain classes of gauge theory models predict no observable parity violations in experiments such as ours. Among these are those left–right symmetric models in which the difference between neutral current neutrino and anti-neutrino scattering cross sections is explained as a consequence of the handedness of the neutrino and anti-neutrino, while the underlying dynamics are parity conserving. Such models are incompatible with the results presented here.

The simplest gauge theories are based on the gauge

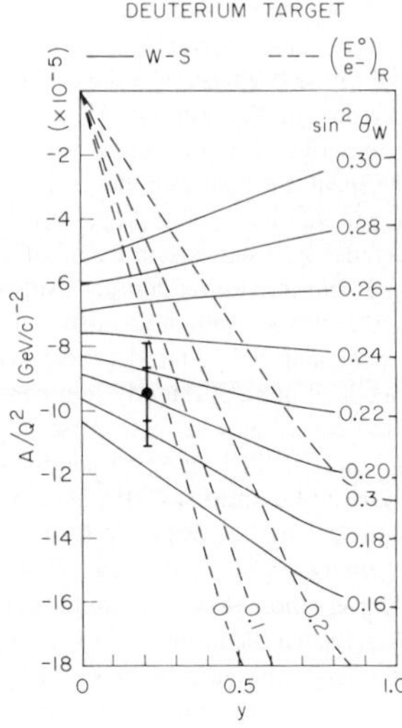

Fig. 4. Comparison of our result for deuterium with two SU(2) × U(1) predictions using the simple quark-parton model for nucleons. The outer error bars correspond to the error quoted in the text (eq. (6)). The inner error bars correspond to the statistical error. The y-dependence of A/Q^2 for various values of $\sin^2\theta_W$ is shown for two models: Weinberg–Salam (solid lines) and the hybrid model (dashed line).

group SU(2) × U(1). Within this framework the original Weinberg–Salam (W–S) model makes specific weak isospin assignments: the left-handed electron and quarks are in doublets, the right-handed electron and quarks are singlets [19]. Other assignments are possible, however. In particular, the "hybrid" or "mixed" model that assigns the right-handed electron to a doublet and the right-handed quarks to singlets has not been ruled out by neutrino experiments.

To make specific predictions for parity violation in inelastic electron scattering, it is necessary to have a model for the nucleon, and the customary one is the simple quark-parton model. The predicted asymmetries depend on the kinematic variable y as well as on the weak isospin assignments and on $\sin^2\theta_W$, where θ_W is the Weinberg angle. Fig. 4 compares our result for two SU(2) × U(1) models. The simplest model (W–S) is in good agreement with our measurement for $\sin^2\theta_W = 0.20 \pm 0.03$ which is consistent with the values obtained in neutrino experiments. The hybrid model is consistent with our data only for values of $\sin^2\theta_W \lesssim 0.1$.

We took a limited amount of data at 19.4 GeV using a liquid hydrogen target with the result

$$A/Q^2 = (-9.7 \pm 2.7) \times 10^{-5}\,(\mathrm{GeV}/c)^{-2}\ \text{(hydrogen)}, \tag{7}$$

where the error contains both statistical and systematic uncertainties. A proton target provides a different mix of quarks and is expected to give a slightly smaller asymmetry than deuterium [7]. Our results are not inconsistent with this expectation.

It is a pleasure to acknowledge the support we received from many people at SLAC. In particular we would like to thank M.J. Browne, G.J. Collet, R.L. Eisele, Z.D. Farkas, H.A. Hogg, C.A. Logg and H.L. Martin for especially significant contributions.

References

[1] A. Love et al., Nucl. Phys. B49 (1972) 513.
[2] E. Derman, Phys. Rev. D7 (1973) 2755.
[3] W.W. Wilson, Phys. Rev. D10 (1974) 218.
[4] S.M. Berman and J.R. Primack, Phys. Rev. D9 (1974) 2171; D10 (1974) 3895 (erratum).
[5] M.A.B. Beg and G. Feinberg, Phys. Rev. Lett. 33 (1974) 606.
[6] S.M. Bilenkii et al., Sov. J. Nucl. Phys. 21 (1975) 657.
[7] R.N. Cahn and F.J. Gilman, Phys. Rev. D17 (1978) 1313; further references to the theory may be found in this reference.
[8] Y.B. Bushnin et al., Sov. J. Nucl. Phys. 24 (1976) 279.
[9] M.J. Alguard et al., Phys. Rev. Lett. 37 (1976) 1258, 1261; 41 (1978) 70.
[10] W.B. Atwood et al., SLAC preprint SLAC-PUB-2123 (1978).
[11] L.L. Lewis et al., Phys. Rev. Lett. 39 (1977) 795.
[12] P.E.G. Baird et al., Phys. Rev. Lett. 39 (1977) 798.
[13] L.M. Barkov and M.S. Zolotorev, Zh. Eskp. Teor. Fiz. Pis'ma 26 (1978) 379.
[14] E.L. Garwin, D.T. Pierce and H.C. Siegmann, Swiss Physical Society Meeting (1974), Helv. Phys. Acta 47 (1974) 393 (abstract only); the full paper is available as SLAC-PUB-1576 (1975) (unpublished).
[15] D.T. Pierce et al., Phys. Lett. 51A (1975) 465; Appl. Phys. Lett. 26 (1975) 670.
[16] P.S. Cooper et al., Phys. Rev. Lett. 34 (1975) 1589.
[17] Z.D. Farkas et al., SLAC-PUB-1823 (1976).
[18] R.S. Larsen and D. Horelick, in: Proc. Symp. on Beam intensity measurement, DNPL/R1, Daresbury Nuclear Physics Laboratory (1968); their contribution is available as SLAC-PUB-398.
[19] S. Weinberg, Phys. Rev. Lett. 19 (1967) 1264; A. Salam, in: Elementary particle theory: relativistic groups and analyticity, Nobel Symp. No. 8, ed. N. Svartholm (Almqvist and Wiksell, Stockholm, 1968) p. 367.

VOLUME 42, NUMBER 21 PHYSICAL REVIEW LETTERS 21 MAY 1979

Experimental Test of Special Relativity from a High-γ Electron $g-2$ Measurement

P. S. Cooper, M. J. Alguard,[a] R. D. Ehrlich,[b] V. W. Hughes, H. Kobayakawa,[c] J. S. Ladish,[d] M. S. Lubell, N. Sasao,[e] K. P. Schüler, and P. A. Souder
J. W. Gibbs Laboratory, Yale University, New Haven, Connecticut 06520

and

D. H. Coward, R. H. Miller, C. Y. Prescott, D. J. Sherden, and C. K. Sinclair
Stanford Linear Accelerator Center, Stanford, California 94305

and

G. Baum and W. Raith
University of Bielefeld, Bielefeld, West Germany

and

K. Kondo
University of Tsukuba, Ibaraki, Japan

(Received 19 January 1979)

We report a verification of the theory of special relativity at a value of $\gamma \approx 2.5 \times 10^4$ based upon a comparison of electron $g-2$ measurements at meV and GeV kinetic energies. Specially we obtain a measure of the equivalence between the quantities $\gamma \equiv (1-\beta^2)^{-1/2}$ and $\tilde{\gamma} \equiv (p/m_0)dp/dE$.

A recent publication[1] has pointed out that an experimental test of special relativity is provided by comparing the values of the electron g-factor anomaly a $[a \equiv \frac{1}{2}(g-2)]$ for electrons with different velocities or γ values. Special relativity predicts that the value of a should be independent of the electron velocity. Newman *et al.*,[1] refer to two measurements of a, one done with electrons of 1 meV kinetic energy ($\gamma - 1 = 10^{-9}$) and the other done with electrons of 100 keV kinetic energy ($\gamma = 1.2$). These two measured values agree. We point out here that another measurement of a has been done with electrons of about 12 GeV kinetic energy ($\gamma \sim 2.5 \times 10^4$), which is relevant to this test of special relativity.[2]

The high-γ $g-2$ measurement[3] was obtained as a by-product of the measurement of the polarization of the high-energy longitudinally polarized electron beam at the Stanford Linear Accelerator Center (SLAC). After acceleration to high energy the longitudinally polarized beam was deflected through the beam switchyard by an angle $\theta_c = 24.5°$ into the experimental area, with the spin precessing relative to the momentum by an angle

$$\theta_a = \gamma a \theta_c. \tag{1}$$

The longitudinal component of the beam polarization is then given by

$$P(E) = P_0 \cos(\pi E/E_0 + \varphi_0), \tag{2}$$

in which P_0 is the magnitude of the initial vector polarization, $\vec{P}_0$, of the electron beam before the magnetic deflection, φ_0 is projected angle of $\vec{P}_0$ with respect to the electron momentum in the plane of the bent trajectory, E is the electron energy, and E_0 is defined as

$$E_0 = \left(\frac{180°}{24.5°}\right)\frac{m_0 c^2}{a} \simeq 3.2 \text{ GeV}, \tag{3}$$

where m_0 is the electron rest mass.

The longitudinal polarization of the deflected beam was measured by Møller scattering[4] from a Supermendur target foil magnetized to saturation in a 90-G longitudinal magnetic field and inclined at 20° with respect to the beam direction in order to provide a large component of longitudinal polarization. Reversal of the 90-G field reversed the polarization of the target. The Møller-scattered electrons were observed by conventional particle-detection techniques with the SLAC 8-GeV/c spectrometer.[5]

The results of the Møller measurement are shown in Fig. 1 together with the fitted curve $P(E)$ given by Eq. (2) with P_0 and a as free parameters and φ_0 fixed at zero. The data points shown are taken from the earlier publication.[3] From the fit, the value $a = (1.1622 \pm 0.0200) \times 10^{-3}$ is obtained, where the quoted 1.7% uncertainty is the linear contribution of counting statistics (0.7%) and possible systematic effects (1.0%). The systematic contributions are the estimated 0.3% uncertainty in the absolute momentum calibration of the beam switchyard magnet system,[6] and an uncertainty of 82 mrad in the value of φ_0, which

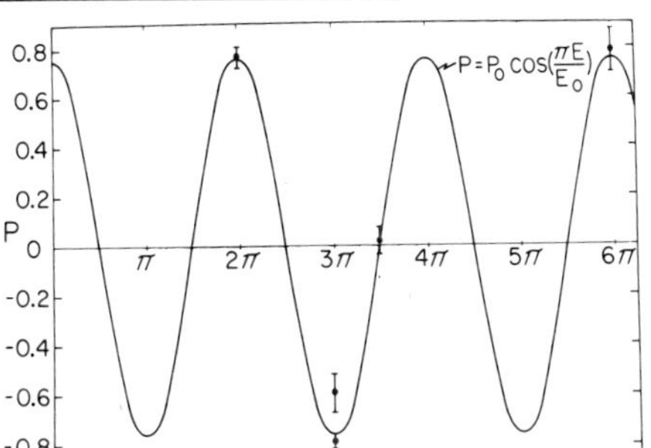

FIG. 1. The longitudinal component P of the electron beam polarization plotted as a function of E/E_0, the angle through which the spin precesses relative to the momentum during the 24.5° magnetic bend into the experimental area. The curve shown is the best fit to the data, with a and P_0 as free parameters.

results in a 0.7% uncertainty in a. The estimate of the uncertainty in φ_0 is obtained by considering the calculated upper bound to φ_0 for a single electron in the beam[7] as a 3-standard-deviation effect. Had data been taken at more than one zero crossing point, φ_0 and a could have been separately determined. However, with only one zero crossing the two parameters are highly correlated, which requires φ_0 to be estimated independently as we have done above. In this respect the measurement of a could be significantly improved by the addition of data taken at one or more of the five remaining zero-crossing points in the present SLAC energy range. Finally, the fitted value of P_0 is 0.755 ± 0.026, which agrees with both the theoretical expectations and experimental measurements of the polarization of the injected electrons.

We note that our value of a from the high-γ measurement agrees with the more precise values of a determined in the lower-γ measurements. In Table I, we summarize the lepton $g-2$ measurements for electrons and muons. For the high-γ electron $g-2$ measurement reported in this paper, we use the average value of γ over the range covered by the experiment; namely, $\gamma = 1.27\times10^4$ to $\gamma = 3.81\times10^4$. We note that this average value, $\gamma = 2.5\times10^4$, is very close to the only zero-crossing point, $\gamma = 2.22\times10^4$, at which we have obtained data. Since the measured value of $P(E)$ obtained at the crossing point most strongly influences our value of a, the choice of $\gamma = 2.5\times10^4$ to characterize our measurement of a seems well justified.

TABLE I. Lepton $g-2$ measurements.

Lepton	Reference	γ	$a\times10^3$ [a]
e^-	8	$1+10^{-9}$	1.159 652 41(20)
	9	1.2	1.159 657 70(350)
	The present work	2.5×10^4	1.1622(200)
μ^-,μ^+	10	12	1.166 16(31) [b]
	11	29.2	1.165.922(9) [b]

[a] The errors quoted are 1-standard-deviation uncertainties in the last digits.
[b] Average μ^- and μ^+.

Any discussion of the sensitivity of these measurements as a test of special relativity requires, of course, some theoretical model for, or at least a parametrization of, a breakdown of special relativity. As a theoretical problem applied to the $g-2$ measurements, some breakdown of relativistic quantum field theory may be involved. This is a very profound problem which must involve some preferred frame of reference, perhaps determined from cosmological considerations. A systematic phenomenological viewpoint might involve an analysis of the accuracy with which the coefficients of the Lorentz transformation are tested. A recent theoretical model[12,13] for a breakdown of special relativity predicts effects proportional to γ^2.

In their parametrization, Newman *et al.* introduce $\tilde{\gamma} \equiv (P/m_0)dp/dE$ which they allow to be different from $\gamma = (1-\beta^2)^{-1/2}$. Hence the cyclotron, spin, and $g-2$ precession frequencies for motions perpendicular to a magnetic field $\vec{B}$ are given, respectively, by

$$\omega_c = eB/\tilde{\gamma} m_0 c, \tag{4}$$

$$\omega_s = geB/2m_0c + (1-\gamma)\omega_c, \tag{5}$$

$$\omega_a = \omega_s - \omega_c = (\tfrac{1}{2}g - \gamma/\tilde{\gamma})eB/m_0c. \tag{6}$$

The term $(1-\gamma)\omega_c$ in Eq. (5) is the Thomas precession frequency and is regarded by Newman *et al.* as of kinematic origin and hence involves the usual γ term, whereas the term $\tilde{\gamma}$ in Eq. (4) is regarded as arising from electron dynamics and hence as possibly different. The $g-2$ experiments determine the quantity $\omega_a(eB/m_0c)^{-1}$ which in the conventional theory equals $\frac{1}{2}(g-2) = a$. Following the parametrization of Newman *et al.*, we

TABLE II. Summary of lepton $g-2$ relativity tests.

Method	References	$\gamma^{(1)}$	$\gamma^{(2)}$	C_1
μ^-,μ^+ g factor	10 and 11	12	29.2	$(1.4\pm1.8)\times10^{-8}$
e^- g factor	8 and 9	1	1.2	$(-2.6\pm1.8)\times10^{-8}$
	8 and the present work	1	$2.5\times10^{4\,a}$	$(-1.0\pm8.0)\times10^{-10}$

[a]The g factor was measured over the γ interval $(1.3-3.8)\times10^4$.

set

$$\omega_a(eB/m_0c)^{-1}=\tfrac{1}{2}g-\gamma/\tilde{\gamma}=a \tag{7}$$

and regard the various $g-2$ measurements done at different electron velocities as determining $\gamma/\tilde{\gamma}$.

In the accompanying Letter by Combley *et al.*,[14] a more general phenomenological viewpoint of a breakdown of special relativity is taken and four distinct γ factors—γ_t, γ_E, γ_M, and γ_T—are introduced for the transformation of time, electromagnetic fields and mass, and for determining the Thomas precession. With certain assumptions about relations among these four γ factors, the parametrization of Combley *et al.* reduces to that of Newman *et al.*

We use the phenomenological model of Newman *et al.*, and, in addition, as do Combley *et al.*, and, in addition, as do Combley *et al.*,[14] assume a power-series expansion for $\gamma/\tilde{\gamma}$ of the form

$$\gamma/\tilde{\gamma}=1+C_1(\gamma-1)+\ldots, \tag{8}$$

which preserves the nonrelativistic equivalence of γ and $\tilde{\gamma}$ in the limit $\gamma\to1$. In order to define a figure of merit, we retain only the leading nonconstant term in Eq. (8). Then for each lepton, $g-2$ measurements at two values of γ suffice to determine C_1 according to

$$C_1=(a^{(2)}-a^{(1)})/(\gamma^{(1)}-\gamma^{(2)}) \tag{9}$$

for measurements $a^{(1)}$ and $a^{(2)}$ at $\gamma^{(1)}$ and $\gamma^{(2)}$, respectively. In Table II, we present the values of C_1 derived from various pairs of lepton $g-2$ measurements given in Table I. Implicit in this parametrization are the assumptions that any violation of special relativity vanishes as one approaches the nonrelativistic limit, and that g is a constant independent of γ. We note that although our measurement of a is relatively imprecise, our value of γ is comparatively very large. Thus our experiment provides a sensitive determination of the coefficients in a power-series expansion such as given by Eq. (8).

As can be seen from Table II, the limit on $|C_1|$ of $<1.7\times10^{-9}$ measurement is the most sensitive upper limit obtained to date. Within the framework of the relativity-breaking model expressed by Eq. (8), we have thus demonstrated the equivalence of γ and $\tilde{\gamma}$. Of course, a linear dependence on $\gamma-1$ is but one possible choice. Indeed, Rédei,[12,13] in a discussion of the validity of special relativity at small distances and the existence of a universal length, suggests that for the lifetime of the muon a modification with a leading term quadratic in γ should be introduced. In the context of higher-order terms we wish to point out that the relative sensitivity of our measurement is enhanced by any higher-order dependence on $\gamma-1$.

In conclusion, we emphasize that we have included in our discussion only those tests of special relativity which are directly comparable to ours. For reference to other tests see Newman *et al.*[1] and Bailey *et al.*[15]

We wish to acknowledge useful discussions with D. M. Eardley, W. Lysenko, and L. Michel. This research (Yale Report No. COO-3075-228) was supported in part by the U. S. Department of Energy, the German Federal Ministry of Research and Technology, The University of Bielefeld, and the Japan Society for the Promotion of Science.

---

[a]Present address: Department of Electrical Engineering, Stanford University, Stanford, Calif. 94305.

[b]Present address: Department of Physics, Cornell University, Ithaca, N. Y. 14850.

[c]Present address: Nagoya University, Nagoya, Japan.

[d]Present address: Los Alamos Scientific Laboratory, Los Alamos, N. M. 87545.

[e]Present address: Kyoto University, Kyoto, Japan.

[1]D. Newman *et al.*, Phys. Rev. Lett. 40, 1355 (1978).

[2]P. S. Cooper *et al.*, Bull. Am. Phys. Soc. 22, 72 (1979).

[3]P. S. Cooper *et al.*, Phys. Rev. Lett. 34, 1589 (1975); P. S. Cooper, Ph.D. thesis, Yale University, 1975 (un-

published).

[4]See, for example, J. D. Bjorken and S. D. Drell, *Relativistic Quantum Mechanics* (McGraw Hill, New York, 1964), p. 140.

[5]SLAC Users Handbook, 1971 (unpublished), Sect. D 3.

[6]J. L. Harris *et al.*, in *Two Mile Accelerator*, edited by R. B. Neal (Benjamin, New York, 1968), p. 585.

[7]M. J. Alguard *et al.*, in *Proceedings of the Ninth International Conference on High Energy Accelerators, Stanford, California, 1974, CONF-740322* (National Technical Information Service, Springfield, Va., 1974), p. 313; M. J. Alguard, J. E. Clendenin, R. D. Ehrlich, V. W. Hughes, J. S. Ladish, M. S. Lubell, K. P. Schüler, G. Baum, W. Raith, and R. H. Miller, "A Source of Highly Polarized Electrons at the Stanford Linear Accelerator Center" (to be published).

[8]R. S. Van Dyck, Jr., P. B. Schwinberg, and H. G. Dehmelt, Phys. Rev. Lett. 38, 310 (1977).

[9]J. C. Wesley and A. Rich, Phys. Rev. A 4, 1341 (1971).

[10]J. Bailey *et al.*, Nuovo Cimento A 9, 369 (1972).

[11]J. Bailey *et al.*, Phys. Lett. 68B, 191 (1977).

[12]L. B. Rêdei, Phys. Rev. 145, 999 (1966).

[13]L. B. Rêdei, Phys. Rev. 162, 1299 (1967).

[14]F. Combley, F. J. M. Farley, J. H. Field, and E. Picasso, preceding Letter [Phys. Rev. Lett. 42, 1383 (1979)].

[15]J. Bailey *et al.*, Nature (London) 268, 301 (1977).

Nuclear Physics **A518** (1990) 371-388
North-Holland

HIGH-ENERGY PHYSICS WITH POLARIZED ELECTRONS AND MUONS

Vernon W. HUGHES

Yale University, Physics Department, J.W. Gibbs Laboratory, New Haven, CT 06520, USA

Received 22 May 1990
(Revised 17 August 1990)

Abstract: In this paper I will review the topic of high-energy physics with polarized electrons and muons. The two main topics will be the spin-dependent structure functions of the nucleon and parity-violating electroweak interference observed in the scattering of polarized electrons. Both topics now relate to particle physics and to nuclear physics.

1. The SLAC experiment on the proton spin dependent structure function*

It is well known that we can study the structure function of the proton by deep inelastic inclusive ep scattering in which only the outgoing electron is measured. The kinematic variables are Q^2 (4-momentum transfer squared), ν (energy loss) and the dimensionless variable $x = Q^2/2M\nu$, in which M is the proton mass. The proton tensor $W_{\mu\nu}$ involves four independent structure functions $-F_1$ and F_2, which are spin-independent structure functions, and g_1 and g_2, which are spin dependent. In the deep inelastic or scaling regime where Q^2 and ν are large compared to the proton mass, the photon can be considered to act on an individual quark and these functions depend only on the variable x, which can be interpreted as the fraction of the momentum of the proton in the infinite momentum frame that is carried by the struck quark. If both the electron and proton are polarized, we can determine the spin-dependent structure functions from the spin-dependent asymmetries in the differential scattering cross sections. In particular the asymmetry A between the antiparallel and parallel ep-spin cases determines the virtual photon–proton asymmetry A_1 which leads to the spin-dependent structure function g_1. The definitions and relationships are given in eq. (1), where D and η are kinematic factors.

$$A \equiv \left[\frac{d^2\sigma}{d\Omega\, dE'}(\uparrow\downarrow) - \frac{d^2\sigma}{d\Omega\, dE'}(\uparrow\uparrow)\right] \Big/ \left[\frac{d^2\sigma}{d\Omega\, dE'}(\uparrow\downarrow) + \frac{d^2\sigma}{d\Omega\, dE'}(\uparrow\uparrow)\right],$$

$$A = D(A_1 + \eta A_2) \simeq DA_1\,, \qquad g_1(x) = \frac{A_1(x) F_2(x)}{2x(1+R(x))},$$

$$R = \sigma_L/\sigma_T\,. \tag{1}$$

* The review article in ref. [1]) includes a quite complete list of references on this topic.

0375-9474/90/$03.50

Further aspects of polarized deep inelastic scattering can be discussed in terms of the absorption of the polarized virtual photon by the polarized proton. For a z-axis collision between a polarized photon and a polarized proton, the total absorption cross section is $\sigma_{3/2}$ for the case of total angular momentum component $+\frac{3}{2}$ and $\sigma_{1/2}$ for total angular momentum component $\sigma_{1/2}$. The asymmetry $A_1 = (\sigma_{1/2} - \sigma_{3/2})/(\sigma_{1/2} + \sigma_{3/2})$ is the virtual photon–proton asymmetry corresponding to the cases of components with total photon and proton angular momentum along the collision axis equal to $\frac{3}{2}$ and $\frac{1}{2}$. At the quark level there is incoherent absorption of the photons by the quarks, and in the absence of orbital angular momentum the virtual polarized photon can be only absorbed when the quark and photon spins are oppositely directed, in order to conserve the component of total angular momentum. A_1 is proportional to the probability that a quark of type i has its spin along the proton-spin direction multiplied by the square of its charge and summed over all types of quarks, or equivalently to the spin-dependent structure function g_1, as given in eq. (2).

$$A_1(x) = \frac{\sum_i e_i^2(q_i^{\uparrow}(x) - q_i^{\downarrow}(x))}{\sum_i e_i^2(q_i^{\uparrow}(x) + q_i^{\downarrow}(x))}, \tag{2a}$$

$$g_1(x) = \tfrac{1}{2}\sum_i e_i^2(q_i^{\uparrow}(x) - q_i^{\downarrow}(x)). \tag{2b}$$

For quark models of the proton $A_1 > 0$.

In 1971 SLAC approved experiment E80 to do the first measurement of polarized deep inelastic ep scattering. The polarized-electron source for this experiment was built at Yale and was based on our atomic physics research at Yale on the production of polarized electrons with an atomic beam (fig. 1). By magnetic deflection in an inhomogeneous magnetic field atoms with a particular component of electronic polarization were selected and photo-ionization of these atoms then led to a source of polarized electrons. The direction of $\boldsymbol{H}$ in the photo-ionization region determines the electron-spin direction, which can therefore be modulated by changing the direction of the current in the large coil. Lithium-6 atoms were chosen for technical reasons and an intense UV flash lamp was used for photo-ionization. The source was installed at SLAC in 1973.

The electron polarization was measured at the injection energy of 70 keV by Mott scattering and at the high-energy output (10–20 GeV) of the linac by Möller scattering by polarized electrons in an iron foil (fig. 2). The analyzing power is large, indeed about 0.8 for 90° scattering in the c.m. system, and the cross section is also large. Because the electron beam from the linac must be deflected about 24° into the experimental area (end station A), the polarization of the beam on the target for a given polarization of the linac output depends on the energy because of the electron g-2 precession. The data agree well with expectation. The operating characteristics of the polarized-electron source (designated PEGGY) are given table 1.

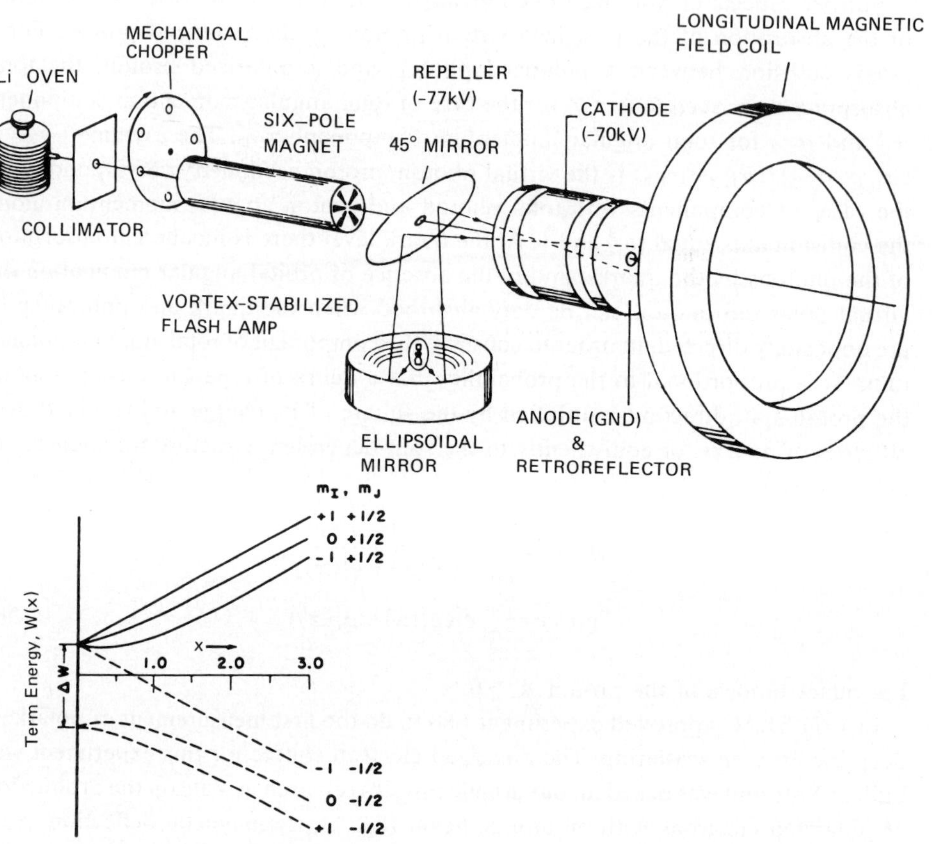

Fig. 1. Schematic diagram of PEGGY showing the principal components of the lithium atomic beam, the uv optics and the ionization region electron optics; Energy level and magnetic moments of ^{6}Li (nuclear spin $I=1$) in the ground $^2S_{1/2}$ atomic state as a function of magnetic field H.

The polarized-proton target utilized the method of dynamic nuclear polarization with the hydrocarbon butanol and the paramagnetic impurity porphyrexide (fig. 3). At the high field ($B \simeq 5$T) and low temperature ($T \simeq 1$ K) the unpaired electron spins in the porphyrexide are highly polarized according to the Boltzman factor. Due to spin–spin coupling of the electrons and protons, microwave transitions can be driven which transfer the electron polarization to the protons. The target volume was 25 cm^3. The operating characteristics of the target are given in the table 2.

The data for SLAC E80 were taken with the high-resolution 8 GeV/c SLAC spectrometer. Most of the data on polarized electron–proton scattering were obtained in a subsequent experiment, SLAC E130, for which the large acceptance spectrometer shown in fig. 4 was used. This spectrometer consisted of two dipoles, PWC chambers, a gas Cerenkov threshold counter and a Pb glass shower counter. The spectrometer

TABLE 1

Operating characteristics of polarized electron beam

Characteristic	Value
Pulse length	1.5 μs
Repetition rate	180 pps
Average intensity at GeV energies	5×10^8 e^- per pulse
Pulse-to-pulse intensity variation	<5%
Polarization	0.8 ± 0.03
Polarization reversal time	3s
Intensity difference upon reversal	<5%

acceptance $\delta p/p$ was 50%, its resolution $\Delta p/p$ was 1%, and the π^-/e^- rejection factor was 10^{-3}.

The counting rate asymmetry Δ is given by $\Delta = P_e P_p fA$, in which P_e = polarization of electron beam, P_p = polarization of the H-protons in butanol, f = fraction of

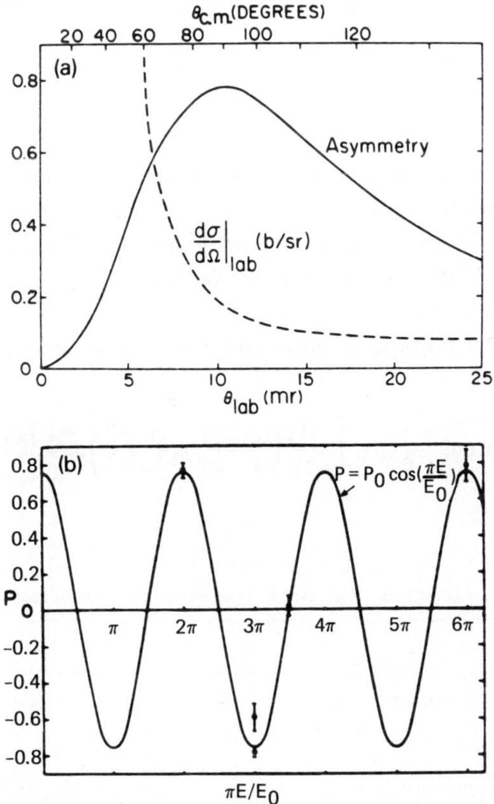

Fig. 2 Measurement of electron beam polarization. (a) The Möller asymmetry and laboratory cross section plotted versus scattering angle for the representative incident energy of 9.712 GeV. (b) The longitudinal component, P, of the beam polarization plotted versus $\pi E/E_0$ with $E_0 = 3.237$ GeV.

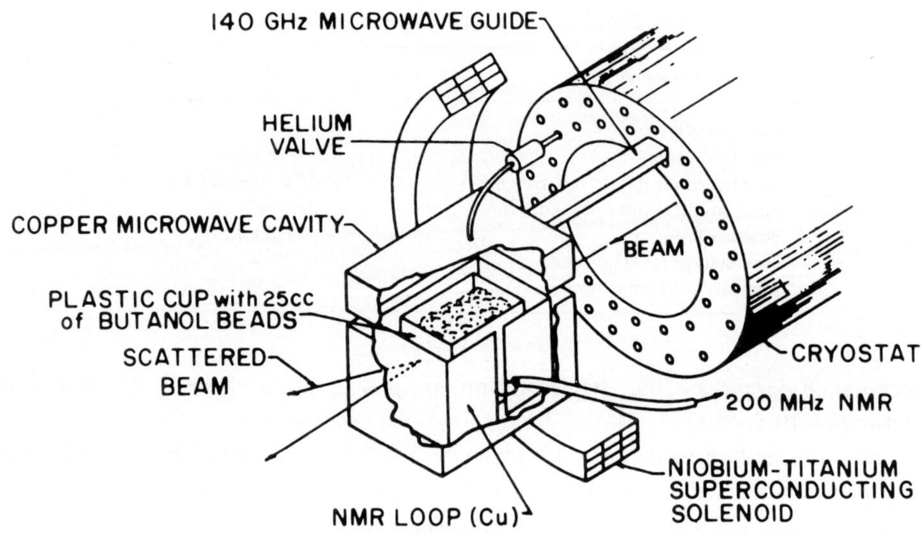

Fig. 3. Schematic diagram of the Yale–SLAC polarized proton target.

nucleons in butanol associated with H, and A = intrinsic ep asymmetry. Since $P_e \approx 0.8$; $P_p \approx 0.6$; $f \simeq 0.1$, the asymmetries Δ are in the range of 0.001 to 0.01 and statistical errors are dominant.

Fig. 5 shows the kinematic range covered which included elastic, resonance region, and deep inelastic scattering. Fig. 6 indicates that A_1^p obeys the scaling relation i.e. it depends only on x and not on Q^2.

There are several important sum rules for spin-dependent structure functions given in eq. (3).

$$S_{Bj} = \int_0^1 dx\,(g_1^p - g_1^n) = \int_0^1 \frac{dx}{2x}\left(\frac{A_1^p F_2^p}{1+R^p} - \frac{A_1^n F_2^n}{1+R^n}\right) = \frac{1}{6}\left|\frac{g_A}{g_V}\right| = 0.209\,(1)\,.$$

TABLE 2

Operating characteristics of polarized proton target

Characteristic	Value
Magnetic field (superconducting)	50 kG
Temperature	1 K
Target material	25 cm^3 of butanol-porphyrexide
Maximum polarization, P_p	0.75
Depolarizing dose $(1/e)^p$	3×10^{14} e$^-$ cm^{-2}
Polarizing time (1/e)	~4 min
Anneal or target change time	~45 min

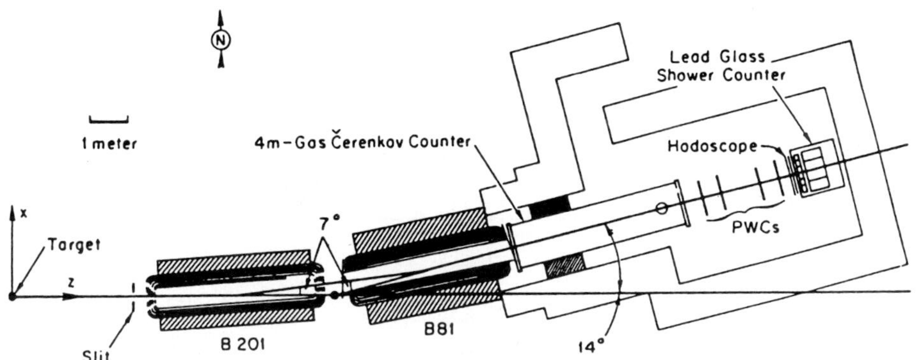

Fig. 4. Electron spectrometer used in SLAC E130 experiment.

With QCD correction

$$S_{\mathrm{Bj}}=\int_0^1 \mathrm{d}x\,(g_1^{\mathrm{p}}-g_1^{\mathrm{n}})=\frac{1}{6}\left|\frac{g_{\mathrm{A}}}{g_{\mathrm{V}}}\right|(1-\alpha_{\mathrm{s}}(Q^2)/\pi)=0.191\ (2)\,,$$

$$S_{\mathrm{EJ}}^{\mathrm{p}}=\Gamma_1^{\mathrm{p}}=\int_0^1 \mathrm{d}x\, g_1^{\mathrm{p}}=\frac{1}{12}\left|\frac{g_{\mathrm{A}}}{g_{\mathrm{V}}}\right|\left[1+\frac{5}{3}\,\frac{3\mathrm{F}/\mathrm{D}-1}{\mathrm{F}/\mathrm{D}+1}\right]+\mathrm{O}(\alpha_{\mathrm{s}})=0.189\pm0.005\,,$$

$$S_{\mathrm{EJ}}^{\mathrm{n}}=\Gamma_1^{\mathrm{n}}=\int_0^1 \mathrm{d}x\, g_1^{\mathrm{n}}=\frac{1}{12}\left|\frac{g_{\mathrm{A}}}{g_{\mathrm{V}}}\right|\left[-1+\frac{5}{3}\,\frac{3\mathrm{F}/\mathrm{D}-1}{\mathrm{F}/\mathrm{D}+1}\right]+\mathrm{O}(\alpha_{\mathrm{s}})=-0.002\pm0.005\,, \qquad (3)$$

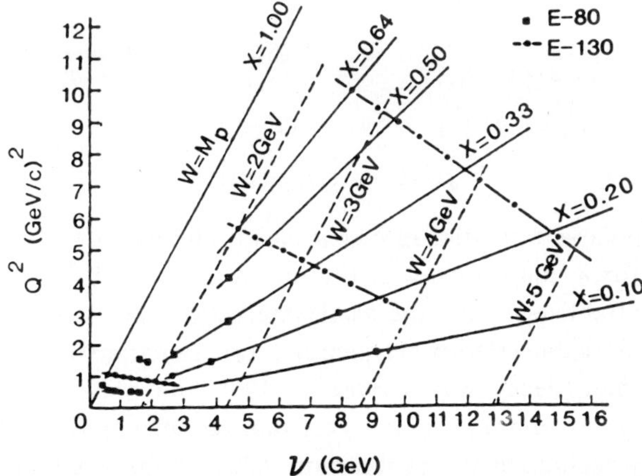

Fig. 5. Kinematic points measured.

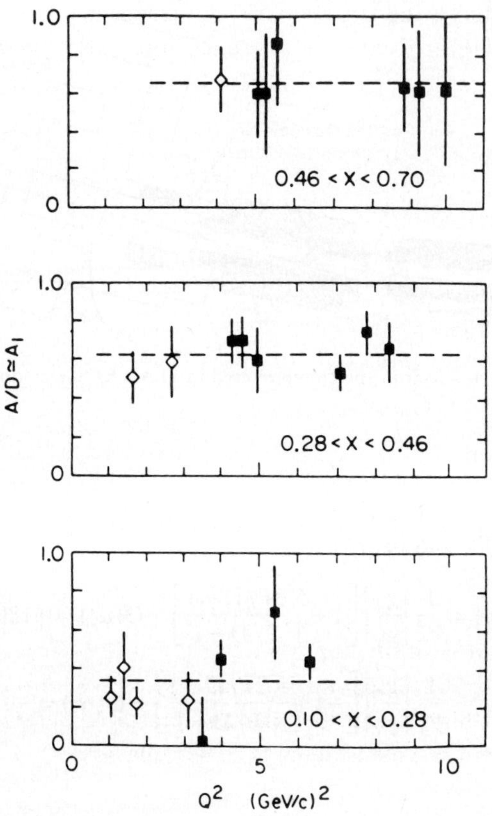

Fig. 6. Radiatively corrected values of $A/D \simeq A_1$ obtained in SLAC E80 (open diamonds) and in SLAC E130 (closed squares) plotted vs Q^2. The fits to horizontal lines demonstrate scaling of A_1.

where

$$g_1 = \frac{A_1 F_2}{2x(1+R)}, \qquad g_A = 1.254\,(6)\,, \qquad R = \frac{\sigma_L}{\sigma_T},$$

$$\alpha_s = 0.27\,(2) \qquad \text{at} \qquad Q^2 = 11\,(\text{GeV}/c)^2\,,$$

$$F/D = 0.631 \pm 0.018\,.$$

The Bjorken polarization sum rule is a basic relation, originally derived from current algebra for a quark model of the nucleon and with incorporation of the view, now well established, that the weak interactions of quarks and leptons are the same. This remarkable relation between structure functions and constants characterizing nuclear beta decay was derived in 1966 and played a seminal role in stimulating our experimental program. This sum rule can be derived in the quark-parton model by evaluating the expectation value of the axial vector beta decay operator for $n \to p$ decay. It is now recognized as a rigorous consequence of QCD in the scaling limit and the $O(\alpha_s)$ correction has been computed by perturbative

QCD. First moments of the spin-dependent structure functions for proton and neutron separately were given by Ellis and Jaffe. These sum rules are model dependent, with the principal assumption being that the strange quark sea is unpolarized.

The measured values of A_1^p are shown in fig. 7. At low x, where the scattering is expected to be predominantly by unpolarized sea quarks, A_1^p is small. As x increases, scattering by the valence quarks which should carry the spin of the proton becomes more important and A_1^p increases. Indeed as $x \to 1$, perturbative QCD predicts that $A_1^p \to 1$ since the struck quark should carry the entire spin of the proton. A test of the Ellis–Jaffe sum rule can be made with these data. The value obtained for the first moment of the proton spin dependent structure function is

$$\Gamma_1^p = \int_0^1 g_1^p(x)\,dx = 0.17 \pm 0.05 . \tag{4}$$

The error in the experimental value is relatively large because the data only extend down to $x = 0.1$ and hence a large extrapolation is required to determine the first moment of g_1. Agreement between experiment and theory is satisfactory within the experimental error.

Comparison of the data with various theoretical models is shown in fig. 8. Best agreement is obtained with the quark model of Carlitz and Kaur. It is an unsymmetrical model which satisfied the Ellis–Jaffe sum rule, Regge theory at low x, perturbative QCD at high x and includes an adjustable parameter to account for the transfer of the spin of valence quarks to sea quarks.

2. The SLAC-Yale parity-nonconservation experiment

In 1972 when we were developing the polarized electron source at Yale for our approved SLAC experiment E80 to study the spin-dependent structure function of

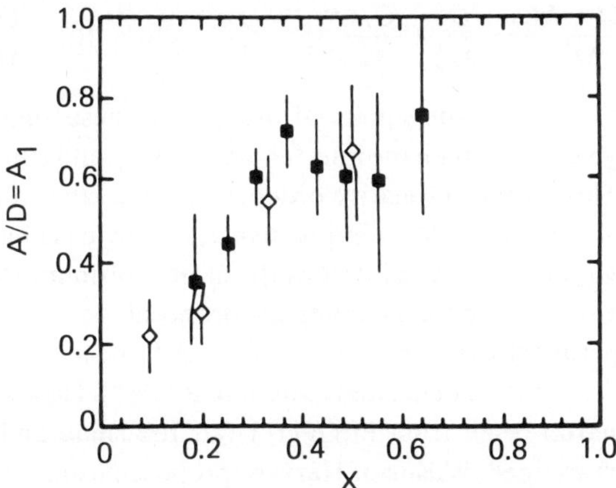

Fig. 7. Measured values of $A/D \simeq A_1$ versus x.

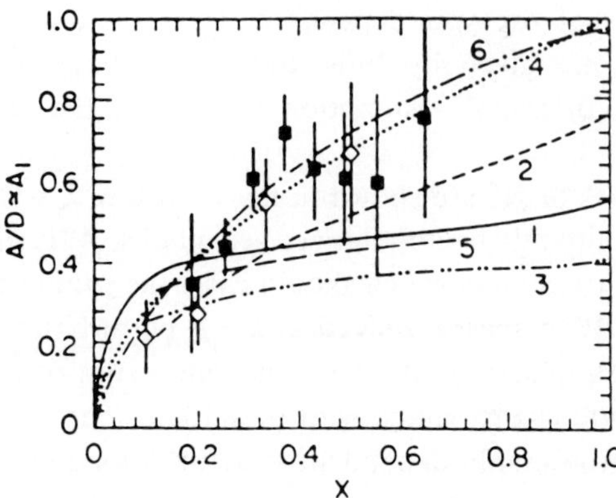

Fig. 8. Experimental values A_1 compared with theories. 1 - Symmetrical valence-quark model. 2 - Current quarks. 3 - Orbital angular momentum. 4 - Unsymmetrical model. 5 - MIT bag model. 6 - Source theory.

the proton, Prescott of SLAC proposed that we test parity conservation in deep inelastic electron scattering with our polarized electron source by looking for a helicity dependence of the differential scattering cross section [2,3]). The motivation for this proposal was to search for an electromagnetic interaction involving a new neutral axial vector electromagnetic current. No weak interaction or electroweak interference was mentioned. Actually, as is now well known, the real motivation turned out to be very strong when in 1973 neutral currents were discovered, and then establishing electroweak interference became urgent. The Feynman diagrams of fig. 9 illustrate that scattering can occur either through γ or Z exchange. Parity nonconservation would be indicated by a pseudoscalar term with $\boldsymbol{\sigma}_e \cdot \boldsymbol{p}_e$ in which $\boldsymbol{p}_e$ = incident electron momentum and $\boldsymbol{\sigma}_e$ = electron spin. Eq. (5) gives the relevant relations

$$\mathrm{d}\sigma^{+(-)} \propto |M_{\mathrm{e.m.}} + M_{\mathrm{W}}^{+(-)}|^2, \qquad A = \frac{\mathrm{d}\sigma^+ - \mathrm{d}\sigma^-}{\mathrm{d}\sigma^+ + \mathrm{d}\sigma^-},$$

$$|A| \simeq \frac{M_{\mathrm{e.m.}} M_{\mathrm{W}}}{M_{\mathrm{e.m.}}^2} \simeq \frac{Q^2}{M_Z^2} \simeq \frac{G_F Q^2}{4\pi\alpha}, \qquad |A| \simeq (10^{-5}\text{–}10^{-4}) \frac{Q^2}{M_{\mathrm{p}}^2}. \tag{5}$$

The scattering cross section is the square of the sum of these amplitudes. If weak Z-exchange is parity violating, then the interference term should contribute a helicity dependence to $\mathrm{d}\sigma$. This term is of relative order $A_{\mathrm{weak}}/A_{\mathrm{e.m.}}$ and would be expected to have the magnitude indicated. Most impressively, Zeldovich [4]) had suggested this experiment and viewpoint in 1959, as well as the atomic bismuth PNC experiment. None of us at SLAC or Yale were aware of his proposal.

The first PNC experiment was done with the PEGGY I beam, first as a byproduct of E80 [5]) and then in a separate dedicated experiment E95 [6]). However, the intensity of the polarized electron beam from PEGGY I was too small and the asymmetry level of only 10^{-3} was attained. Wilson of Harvard proposed using a large-solid-angle spectrometer for backward scattering using the PEGGY beam. But the experiment

Fig. 9. Feynman diagrams indicating a cross section asymmetry A between + and − helicity polarized electrons scattered from unpolarized protons. The asymmetry arises from interference between the electromagnetic amplitude $M_{e.m.}$ associated with γ exchange and the weak amplitude M_W associated with Z exchange. M is the proton mass in GeV and Q^2 is the 4-momentum transfer squared in $(GeV/c)^2$.

actually done at SLAC, E122, involved the development of a new high intensity source of pulsed polarized electrons (PEGGY II). This was based on photoemission of valence electrons from the semiconductor GaAs by polarized laser light (fig. 10). The operating characteristics are given in the accompanying table 3.

The experimental setup is shown in fig. 11. A liquid deuterium target was used because its nucleon density is greater than that of liquid hydrogen and D is an isoscalar nucleus. The spectrometer consisted of two bending magnets and a quadrupole magnet. Both an N_2 gas threshold counter and a Pb glass shower counter were used to detect scattered electrons and discriminate against pions. Because of the high counting rates both detectors were used in an integrating mode. To avoid helicity dependence of the characteristics of the polarized electron beam - its position, angle and energy - extensive measurements of the linac output beam were made with beam position monitors. This information was used in a computer controlled feedback system to control the beam (fig. 12).

Fig. 13 shows the ed scattering cross section and the spectrometer acceptance. The scattering observed was predominantly in the deep inelastic regime.

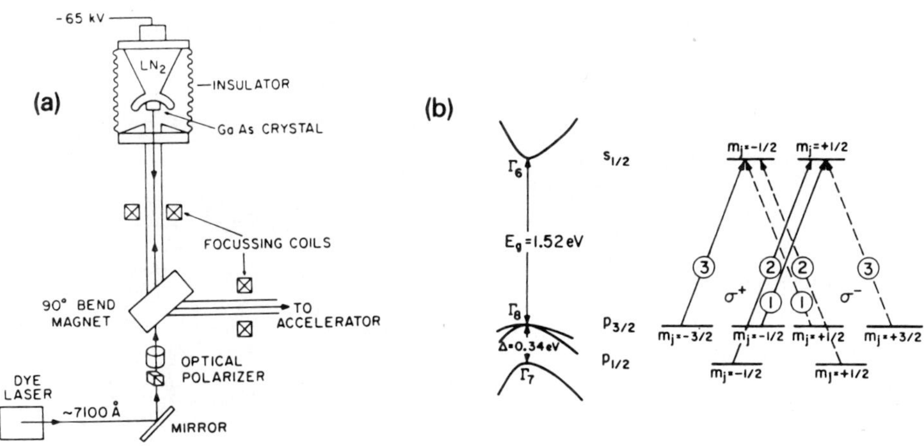

Fig. 10. (a) Schematic diagram of PEGGY II, showing the GaAs crystal in the electron gun and the laser optics. (b) Energy bands of GaAs at the Γ point (left) and transitions between the $S_{1/2}$ levels and the $P_{1/2}$ and $P_{3/2}$ levels. Solid (broken) lines indicate transitions for $\sigma^+(\sigma^-)$ light, and the circled numbers indicate the relative transition strengths. Operating characteristics shown in the table 3.

TABLE 3

PEGGY II operating characteristics

Characteristic	Value
Pulse length	1.5 μs
Repetition rate	120 pps
Electron intensity (at high energy)	(1 to 4) $\times 10^{11}$ e^-/pulse
Pulse to pulse intensity variation	~3%
Electron polarization	0.37, average
Polarization reversal time	pulse to pulse

Fig. 14 shows the experimental results. The measured asymmetry versus the beam polarization is shown, where as indicated earlier (fig. 2a) the beam polarization on the target in end station A depends on energy due to the electron g-2 spin precession. The measured asymmetry values are given below for ed and also for ep scattering.

$$e^- + D \to e^- + X \text{ (DIES)},$$

$$\frac{A}{Q^2} = (-9.5 \pm 1.6) \times 10^{-5}\,(\text{GeV}/c)^{-2},$$

$$[\text{Statistical error} = 0.86 \times 10^{-5}] \qquad [\text{Systematic error} \simeq 0.7 \times 10^{-5}],$$

$$\langle Q^2 \rangle = 1.6\,(\text{GeV}/c)^2 \qquad \langle y \rangle = \nu / E_0 = 0.21\,, \tag{6a}$$

$$e^- + P \to e^- + X \text{ (DIES)},$$

$$\frac{A}{Q^2} = (-9.7 \pm 2.7) \times 10^{-5}\,(\text{GeV}/c)^{-2},$$

$$[\text{Statistical error} = 1.6 \times 10^{-5}], \qquad [\text{Systematic error} = 1.1 \times 10^{-5}],$$

$$\langle Q^2 \rangle \simeq 1.6\,(\text{GeV}/c)^2, \qquad \langle y \rangle = 0.21\,. \tag{6b}$$

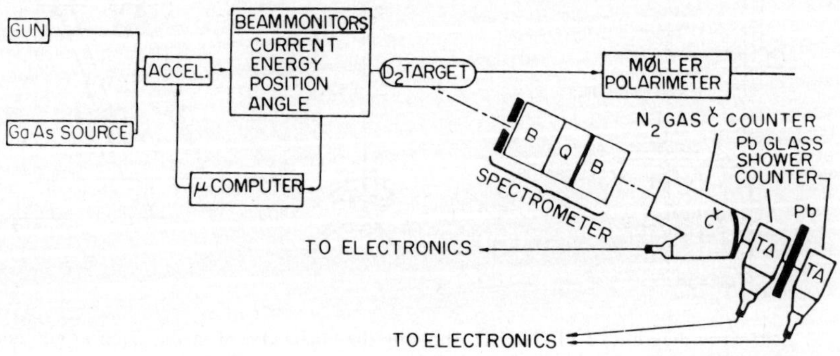

Fig. 11. Block diagram of the apparatus used to detect parity nonconservation in deep inelastic scattering of polarized electrons.

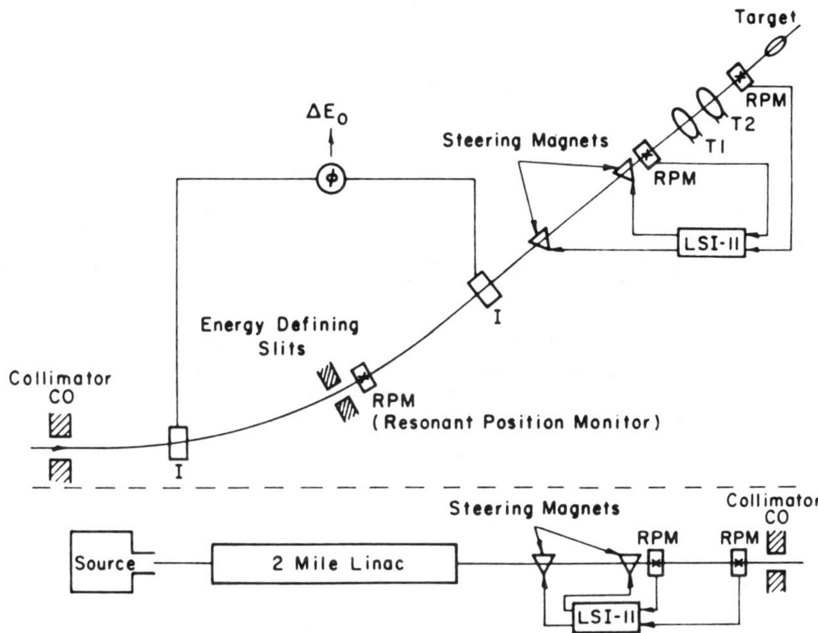

Fig. 12. Detail of the beam monitoring system, including resonant position monitors, ΔE monitor, and toroids T_1 and T_2.

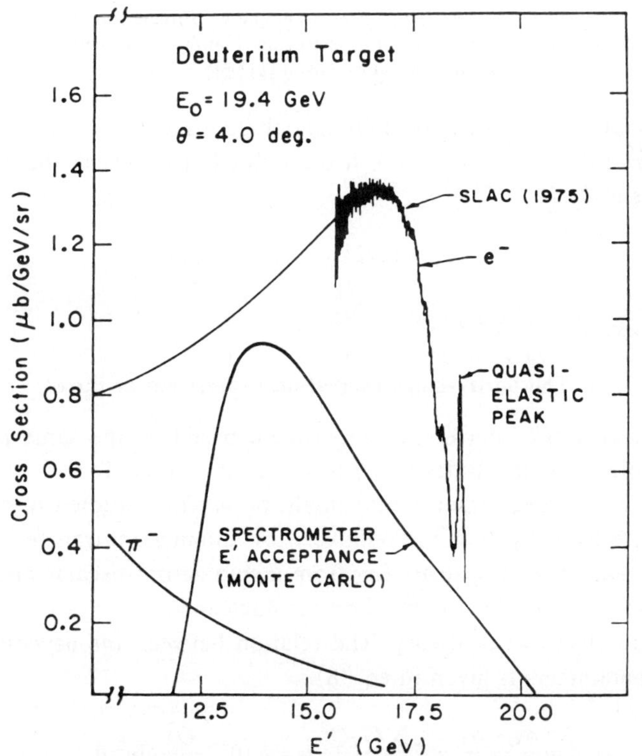

Fig. 13. Acceptance of the high rate spectrometer used for the parity nonconservation experiment.

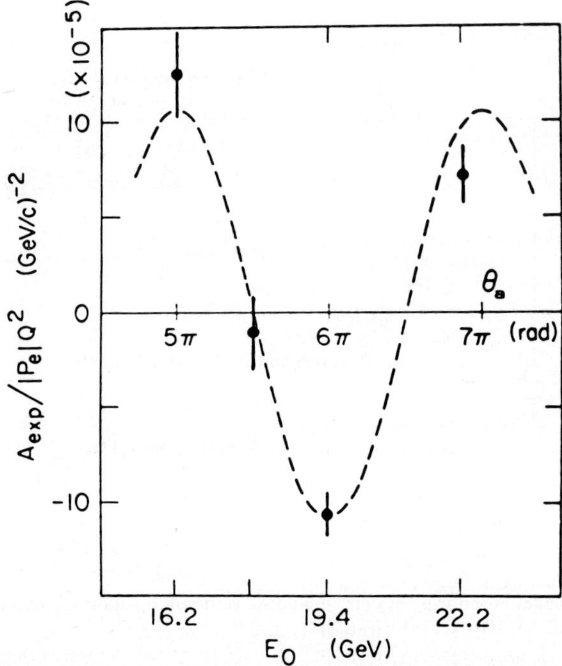

Fig. 14. Experimental asymmetry as a function of beam energy. Broken line was calculated assuming that the theoretical asymmetry is independent of energy but that the beam helicity changes in the 24.5° bending magnet due to the g-2 precession.

This experimental result was the first unambiguous observation of electroweak interference and did much to establish the unified electroweak theory. The value obtained for $\sin^2\theta_W$ is

$$\sin^2\theta_W = 0.221 \pm 0.015 \pm \underset{\text{(theor.)}}{[0.013]}\,. \tag{7}$$

3. The parity-nonconservation experiment at Bates

Another electroweak interference experiment based on the same principle was recently completed at the Bates electron linac laboratory. In 1975 Feinberg suggested [7]) that electroweak interference might be usefully studied in the scattering of polarized electrons by nuclei at relatively low momentum transfer. In particular he suggested that for elastic scattering from a spin zero, isoscalar nucleus such as ^{12}C, the asymmetry would not depend on the nuclear form factor and could cleanly test the unified electroweak theory. The relation between the asymmetry and the electroweak parameters is given in eq. (8).

$$A = \frac{\sigma_R - \sigma_L}{\sigma_R + \sigma_L} = \tilde{\gamma}\frac{3}{2}\frac{G_F Q^2}{\sqrt{2}\pi\alpha} = 4\times 10^{-4}\frac{Q^2}{M_p^2}\sin^2\theta_W\,. \tag{8}$$

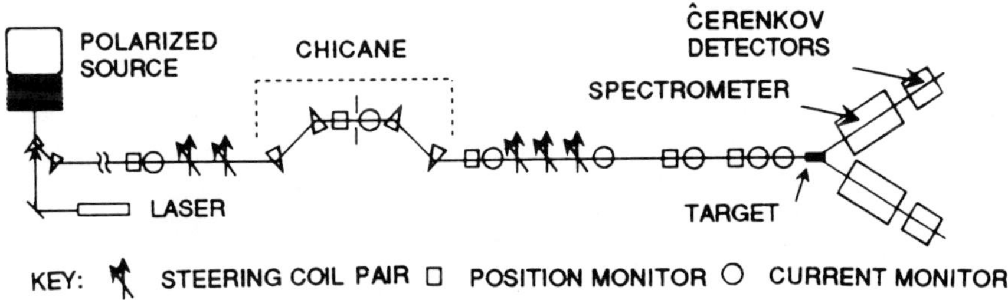

Fig. 15. Schematic diagram of the Bates parity nonconservation experiment which measured helicity-dependent elastic scattering of polarized electrons by ^{12}C.

The experimental arrangement is shown in fig. 15. The polarized electron source, built at Yale [8]), was based on photo-emission from GaAs by polarized laser light. The duty factor was about 1% and the average current was 30–60 μA. The energy of the electron beam was 250 MeV and Q^2 was 0.02 $(GeV/c)^2$. A pair of quadrupole spectrometers and integrating Cerenkov detectors were used.

The measured asymmetry was small, $A = 0.56 \pm 0.14$ ppm, and the error is dominantly due to limited statistics. A value for the isoscalar vector hadronic coupling constant $\tilde{\gamma}$ was determined [9]). The result also determines $\sin^2 \theta = 0.202 \pm 0.049$. The values are consistent with the standard theory [10]).

4. The CERN polarized muon–proton scattering experiment*

Finally we review briefly the recent CERN experiment by the EMC group on the measurement of the proton-spin-dependent structure function using a high-energy polarized muon beam. We will also mention future plans at CERN for further measurements of the spin-dependent structure functions of both the proton and the neutron.

The experimental setup is shown in fig. 16 with the polarized-proton target, the EMC forward spectrometer and the incident polarized muon beam. The muon beam energy was between 100 and 200 GeV, and principally inclusive scattering was studied. The EMC polarized target is shown in fig. 17. This enormous target used irradiated NH_3 as the target material. Shown in table 4 are the projected operating conditions for a modified target with a hydrocarbon as the target material to be used in a new experiment by the spin muon collaboration (SMC).

The virtual photon–proton asymmetry A_1^p determined from the measured asymmetries are shown in fig. 18 together with the SLAC data points. In the region of

* Ref. [11]) includes a quite complete list of references on this topic.

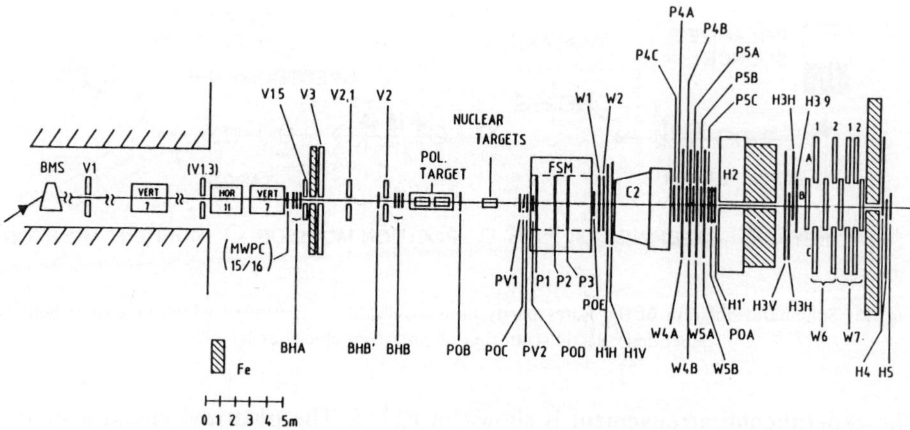

Fig. 16. The EMC forward spectrometer for the CERN EMC polarized-target experiment.

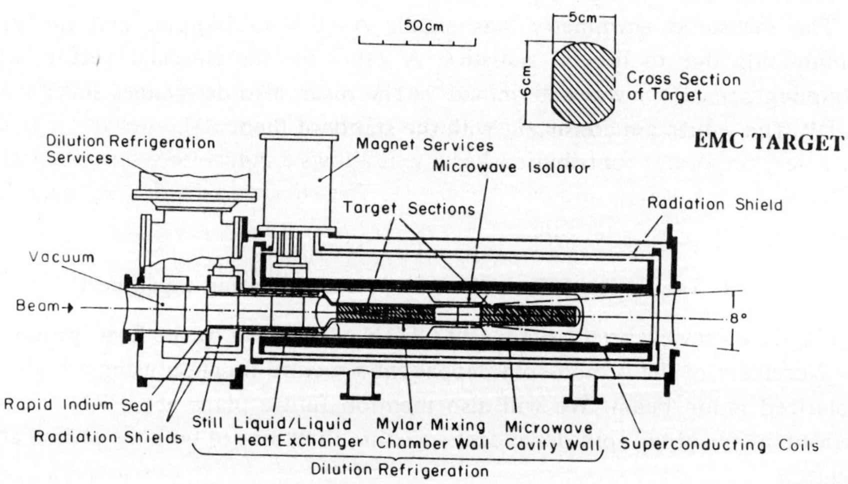

Fig. 17. The EMC polarized target.

TABLE 4

Projected operating conditions for SMC polarized target

Target material	Butanol and deuterated butanol (Doped with EHBA-Cr (V))
Magnet field	2.5 T
Temperature	0.5 K for DNP mode
	0.05 K for frozen spin mode
Polarization	0.8 for proton; 3% rel. error
	0.4 for deuteron; 3% rel. error

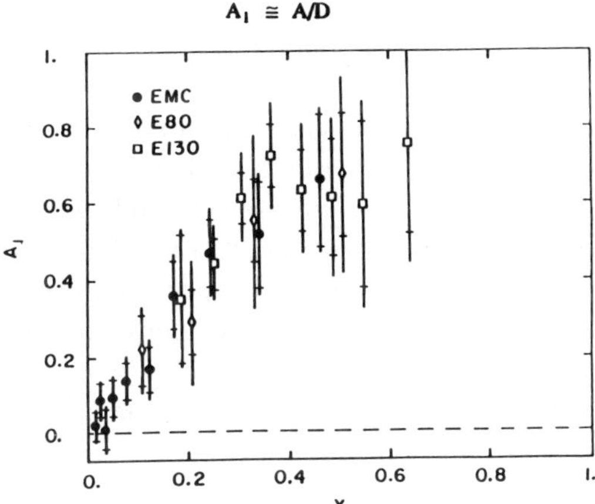

Fig. 18. Compilation of all the data of A_1^p as a function of x. The EMC points are shown as full circles while the SLAC points are shown as open diamonds (experiment E80) and open squares (E130). Inner error bars are the statistical errors and the outer error bars are the total errors (statistical plus systematic added in quadrature). The systematic errors include uncertainties in the values of R and A_2.

overlap the EMC and the SLAC data agree well. The principal contribution of the CERN data is to determine A_1 to lower values of x, indeed down to $x = 0.01$. This is a very important contribution because it allows a much more sensitive test of the Ellis–Jaffe sum rule as indicated in fig. 19. The conclusion is that the data disagree with the Ellis–Jaffe sum rule.

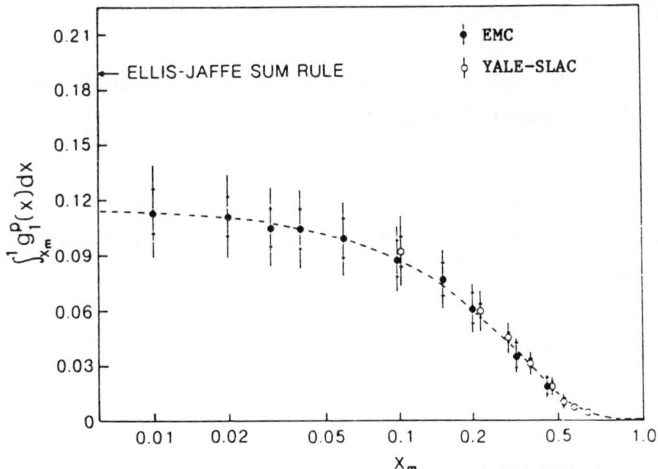

Fig. 19. The value of the integral $\int_{x_m}^1 g_1^p\,dx$ as a function of x_m, the value of x at the low edge of each bin.

The implication based on the naive quark–parton model is that very little of the proton spin is contributed by the quark spins. This surprising conclusion has attracted great interest and has led to many theoretical papers. Possible reasons for the discrepancy and mechanisms by which gluons or orbital angular momentum could contribute to the proton spin have been discussed extensively.

The present high interest in the nucleon-spin-dependent structure functions has led to plans for new experiments. In particular, a new collaboration at CERN (SMC) will do CERN experiment NA47 [12]) to measure the neutron-spin-dependent structure function and also to improve knowledge of A_1^p. This should allow a test of the fundamental Bjorken polarization sum rule. In addition, an experiment is planned at HERA to use the electron ring with an internal polarized H or D gas target to measure A_1^p and A_1^n. Also an experiment has been approved at SLAC to use a polarized ^{3}He target to measure A_1^n.

A summary of the present situation and a listing of some important experimental and theoretical problems is given below.

EMC(CERN) data on polarized μ-p DIS extended SLAC polarized ep data to $x \cong 0.01$; CERN and SLAC data agree from $x = 0.1$ to 0.7:

(i) Violation of Ellis–Jaffe sum rule for Γ_1^p.

(ii) If the Bjorken sum rule is valid, the implication is that A_1^n and Γ_1^n are much larger than previously predicted.

(iii) According to interpretation with naive quark–parton model, quark spin seems to carry very little of the proton spin.

Important experimental problems:

(i) Measure A_1^n, Γ_1^n for the first time and improve data on A_1^p and Γ_1^p;

(ii) If the Bjorken sum rule is valid, the implication is that A_1^n and Γ_1^n are much larger than previously predicted.

(iii) According to interpretation with naive quark–parton model, quark spin seems to carry very little of the proton spin.

Important experimental problems:

(i) Measure A_1^n, Γ_1^n for the first time and improve data on A_1^p and Γ_1^p; Particularly to test the Bjorken polarization sum rule;

(ii) Measure A_2^p and A_2^n;

(iii) Measure spin-dependent effects in exclusive channels, e.g. (J/Ψ).

Theoretical problems:

(i) Ellis–Jaffe sum rule;

(ii) Carriers of proton spin;

(iii) Nucleon spin dependent structure functions.

Implications:

(i) Polarized hadron–hadron scattering;
(ii) Parity violation in atoms;
(iii) Dark matter in universe.

It is quite clear that the subject of the nucleon spin dependent structure functions is a rich and active one.

This paper was adapted from the lecture given on the occasion of the award of the 1990 Bonner Prize in Nuclear Physics by the American Physical Society. The discussion emphasizes those aspects of the topics in which Yale physicists were involved.

I am most pleased to be able to contribute this paper in honor of and on the happy occasion of Torleif Ericson's 60th birthday. For many years I have been fortunate to be a friend of Torleif's. Often our scientific interests have overlapped and I have learned a great deal from Torleif. For the past few years together with Darragh Nagle we have worked on a book entitled "*The meson factories*" which was just submitted for publication to the University of Califnoria Press in their Los Alamos Series. The topic of this paper "High Energy Physics with Polarized Electrons and Muons" relates to both particle physics and nuclear physics and in this respect illustrates well the focus of interest of much of Torleif's work.

Research supported in part by the Department of Energy under Contract No. DE-AC02-76-ER03075.

References

1) V.W. Hughes and J. Kuti, Ann. Rev. Nucl. Part. Sci. **33**, (1983) 611
2) C.Y. Prescott *et al.*, Phys. Lett. **B77** (1978) 347
3) C.Y. Prescott *et al.*, Phys. Lett. **B84** (1979) 524
4) Y.B. Zel'dovich, Sov. Phys. JETP **36** (1959) 682
5) M.J. Alguard *et al.*, Phys. Rev. Lett. **37** (1976) 1261
6) W.B. Atwood *et al.*, Phys. Rev. **D18** (1978) 2223
7) G. Feinberg, Phys. Rev. **D12** (1975) 3575
8) G.D. Cates *et al.*, Nucl. Instr. Meth. **A278** (1989) 293
9) P.A. Souder *et al.*, Phys. Rev. Lett. **65** (1990) 694
10) P.Q. Hund and J.J. Sakurai, Ann. Rev. of Nucl. and Part. Sci. **31** (1981) 375
11) J. Ashman *et al.*, Nucl. Phys. **B328** (1989) 1
12) V.W. Hughes (SMC spokesman) Measurement of the spin-dependent structure functions of the neutron and proton, 12/22/88, CERN Experiment NA47

VOLUME 44, NUMBER 8 PHYSICAL REVIEW LETTERS 25 FEBRUARY 1980

Neutrino Experiment to Test the Nature of Muon-Number Conservation

S. E. Willis[a] and V. W. Hughes
Yale University, New Haven, Connecticut 06520

and

P. Némethy
Yale University, New Haven, Connecticut 06520, and Lawrence Berkeley Laboratory, Berkeley, California 94720

and

R. L. Burman, D. R. F. Cochran, J. S. Frank, and R. P. Redwine[b]
Los Alamos Scientific Laboratory, Los Alamos, New Mexico 87545

and

J. Duclos
Centre d'Etudes Nucléaires de Saclay, F-91190 Gif-sur-Yvette, France

and

H. Kaspar
Swiss Institute for Nuclear Research, CH-5234 Villigen, Switzerland

and

C. K. Hargrove
National Research Council of Canada, Ottawa, Ontario K1A 0R6, Canada

and

U. Moser
University of Berne, CH-3012 Berne, Switzerland, and Yale University, New Haven, Connecticut 06520

(Received 26 December 1979)

This paper reports on a search for $\bar{\nu}_e$ from $\mu^+ \to e^+ \bar{\nu}_e \nu_\mu$, allowed by multiplicative but not additive muon conservation, and for ν_e from $\mu^+ \to e^+ \nu_e \bar{\nu}_\mu$, allowed by both. Neutrinos from the Clinton P. Anderson Meson Physics Facility have been used, together with a six-ton Cherenkov counter filled with H_2O (D_2O) to look for $\bar{\nu}_e p \to ne^+$ ($\nu_e d \to ppe^-$). The branching ratio $(\mu^+ \to e^+ \bar{\nu}_e \nu_\mu)/(\mu^+ \to \text{all}) = -0.001 \pm 0.040$ is in excellent agreement with the additive law. The cross section $\langle\sigma(\nu_e d \to ppe^-)\rangle = (0.52 \pm 0.18) \times 10^{-40}$ cm^2 agrees with theory.

Muon conservation, distinct from total lepton conservation, was introduced to account for the absence of $\mu \to e\gamma$, $\mu \to 3e$, $\mu Z \to eZ$, and $\nu_\mu Z \to eZ'$. Muon and electron numbers are defined by $L_\mu = +1$ (-1) for μ^-, ν_μ $(\mu^+, \bar{\nu}_\mu)$ and $L_e = +1$ (-1) for e^-, ν_e $(e^+, \bar{\nu}_e)$. In place of the usual, additively conserved quantum numbers, $\sum L_\mu = \text{const}$ with $\sum(L_\mu + L_e) = \text{const}$, one could, as Feinberg and Weinberg[1] pointed out, introduce a multiplicatively conserved muon number $\prod (-1)^{L_\mu} = \text{const}$ with $\sum(L_\mu + L_e) = \text{const}$. Most recently Derman[2] has considered multiplicative muon conservation in the context of gauge theories.

Both formulations prohibit the reactions above, but they are not equivalent. In particular, the additive law forbids muon decay with inverted neutrinos,

$$\mu^+ \to e^+ + \bar{\nu}_e + \nu_\mu, \qquad (1)$$

allowed by the multiplicative law. Both laws allow the decay

$$\mu^+ \to e^+ + \nu_e + \bar{\nu}_\mu. \qquad (2)$$

In order to test whether muon conservation is a multiplicative law, we have built an apparatus which is sensitive to either $\bar{\nu}_e$ from (1) or ν_e from (2) and used it to look at neutrinos from μ^+ decay at the neutrino area of the Clinton P. Anderson Meson Physics Facility (LAMPF) at Los Alamos. We utilized the neutrino reactions

$$\bar{\nu}_e p \to ne^+ \qquad (3)$$

and

$$\nu_e d \to ppe^- \qquad (4)$$

on protons and deuterons in a six-ton Cherenkov counter filled alternately with H_2O and D_2O. By comparing the rates of neutrino events in the water and heavy water, we measured the branching ratio for the exotic μ^+ decay mode (1),

$$R \equiv (\mu^+ \to e^+ \bar{\nu}_e \nu_\mu)/(\mu \to \text{all}), \quad (5)$$

in a largely bias-free fashion. Previous information on R comes from Eichten *et al.*,[3] with $R < 0.25$, and recently from Blietschau *et al.*,[4] with $R = 0.13 \pm 0.15$.

Our source of μ^+ decays was the beam stop at LAMPF with an incident proton beam of 780 MeV producing π^+ and π^- mesons. The sequential decays of stopped π^+ and μ^+ yield neutrinos, while the π^- are mostly absorbed upon stopping, leaving a contamination of $\mu^-/\mu^+ < 0.2\%$. A measurement of $\pi^+ \to \mu^+ \to e^+$ in a simulated beam stop by Chen *et al.*[5] gave (μ^+ decays)/proton $= 0.057 \pm 0.004$ at 720-MeV proton energy. Extrapolated to the LAMPF energy, this gives a rate of (μ^+ decays)/proton $= 0.069 \pm 0.007$ and a neutrino flux of about 2×10^7 cm^{-2} sec^{-1} into our detector.

Our experimental apparatus is shown schematically in Fig. 1. The LAMPF neutrino area was a steel and concrete blockhouse with a 1.2-m steel roof, separated from the beam stop by 6.3 m of steel shielding. The neutrino detector, described in detail elsewhere,[6] was a (180 cm)^3 nondirectional water Cherenkov counter used as an electron total-energy calorimeter. It contained 6000 liters of water with a dissolved wavelength shifter, had diffuse reflector walls, and was viewed by 96 12.5-cm phototubes. Its resolution was $\sigma = 12\%$ at the typical e^+ or e^- energy of 40 MeV. The expected event rates, for 300 μA of protons on the beam stop, were $70 \times R$/day on H_2O and 20/day on D_2O. To first order the rate on D_2O is independent of R, since deuterium provides a target for either ν_e or $\bar{\nu}_e$.

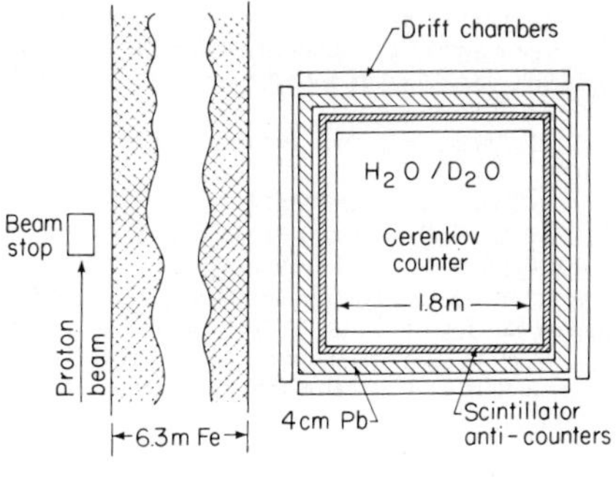

FIG. 1. Schematic plan view of apparatus.

There are no significant neutrino reactions competing with (3) and (4). Muon neutrinos from pion and muon decay at rest are below threshold for charged-current reactions. Inverse β decay by ν_e (or $\bar{\nu}_e$) on the oxygen in the water is expected to be a very small background since the cross sections are greatly reduced by Pauli exclusion effects and the negative Q values involved.[7]

We reduced the cosmic background by 10^4 with the cosmic-ray shield shown in Fig. 1. Plastic scintillators completely surrounded the Cherenkov detector to veto charged cosmic rays. Neutral backgrounds from muon and electron bremsstrahlung were attenuated by covering layers of lead and drift chambers. In addition, we accepted neutrino events only during the beam spill (6% duty factor) for a final cosmic background of 120/day (30–60 MeV). To subtract this background, which dominated our observed neutral events, we also accumulated data between beam spills, renormalizing to the live time during the beam. This beam-in, beam-out subtraction, monitored continuously, was bias-free to 0.2%.

Beam-associated neutron backgrounds were studied with partial shielding, 4 and 5 m of steel, between the detector and the beam stop. Exponential extrapolation of the observed rate of neutron-induced high-energy events gave 1.3 ± 0.2 d^{-1} for our full shielding. We tolerated a large flux of few-megaelectronvolt γ rays from low-energy neutron capture, present even in the final shielding configuration. The pileup and resolution tail of these events did not extend above our chosen energy threshold of 25 MeV; 22% of $\bar{\nu}_e$ and 31% of ν_e events fall below this cut.

Data were accumulated at accelerator currents between 225 and 500 μA for a total of 1270 and 400 C of protons on the beam stop for H_2O and D_2O, respectively. Figures 2 and 3 show the background-subtracted energy spectra for D_2O and H_2O, respectively; the dashed lines are the expected neutrino event spectra for D_2O and for H_2O ($R = 1$). Fits to the expected spectra yield R_D = (observed rate)/(expected rate) $= 1.09 \pm 0.37$ for D_2O and R_H = (observed rate)/[expected rate ($R = 1$)] $= -0.001 \pm 0.044$ for H_2O. The expected rates use cross-section calculations by O'Connell.[8] Both R_H and R_D have had small corrections applied for beam-associated backgrounds. These corrections are $\Delta R_D = -0.06$ and $\Delta R_H = -0.025$ for neutrons, and $\Delta R_D = -0.02$ and $\Delta R_H = -0.004$ for neutrino events on oxygen and on the counter walls. The errors are dominated by the statistical error of the cosmic subtraction but include

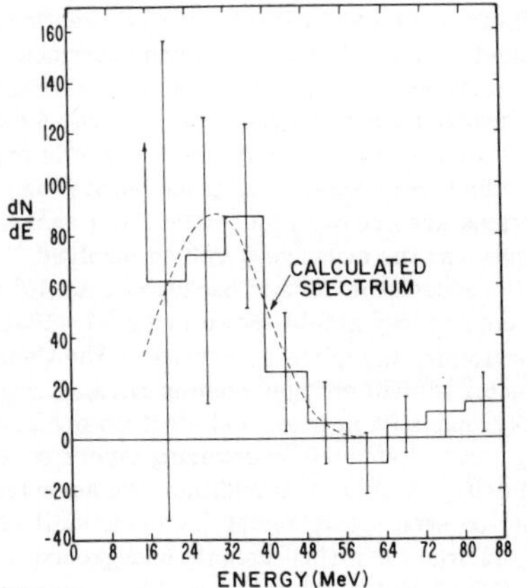

FIG. 2. Background-subtracted energy spectrum for D_2O data.

systematic errors as well.

After the error on R_D is increased for the 10% uncertainty in neutrino flux, the D_2O result translates to a spectrum-averaged cross section

$$\langle\sigma(\nu_e d \to ppe)\rangle = (0.52 \pm 0.18)\times 10^{-40}\ \text{cm}^2,$$

in good agreement with O'Connell's predicted value[8] of $\langle\sigma\rangle = 0.48\times 10^{-40}$ cm^2. This is the first measurement of inverse β decay by low-energy ν_e rather than $\bar{\nu}_e$. The reaction is the inverse of the reactions $pp \to de^+\nu_e$ and $ppe^- \to d\nu_e$, the primary energy sources in the sun.

For calculating the branching ratio (5) we eliminate the uncertainties of detector acceptance and neutrino flux by taking the ratio of the H_2O and D_2O results, $R = R_H/R_D$, with the errors added in quadrature, to obtain

$$R = -0.001 \pm 0.040.$$

We have excellent agreement with the additive law and see no evidence for a multiplicative one. The error includes systematic and statistical contributions. The result translates to an upper limit $R < 0.065$ (90% confidence level).

An upper limit on a $\bar{\nu}_e$ signal is also a limit on possible neutrino oscillations of the type $\bar{\nu}_\mu \to \bar{\nu}_e$. For a maximal mixing parameter, we get a limit of < 0.64 eV^2 (90% confidence level) on the square of the mass difference between the neutrino eigenstates. This limit is consistent with other recent experiments.[9]

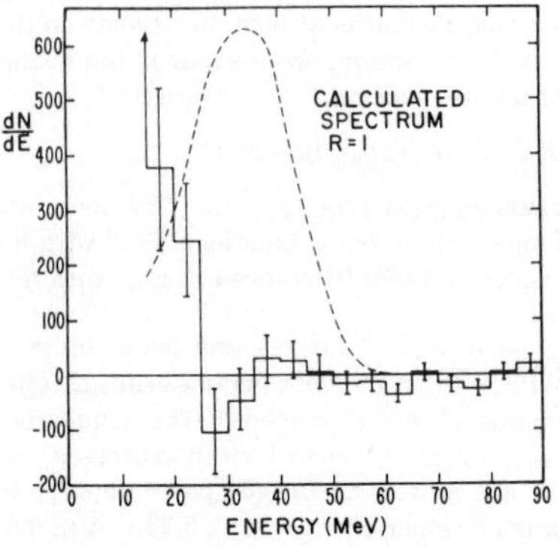

FIG. 3. Background-subtracted energy spectrum for H_2O data.

We acknowledge the support of LAMPF during the experiment and we thank J. Kukulka who wrote our data acquisition program. One of us (P.N.) thanks Leon Lederman and Albert Messiah for helpful discussions. This research was supported in part by the U. S. Department of Energy under Contracts No. EY-76-C-3075, No. W-7405-ENG-36, and No. W-7405-ENG-48.

---

[a]Present address: Physics Department, Fermilab, Batavia, Ill. 60510.

[b]Present address: Department of Physics and Laboratory for Nuclear Science, MIT, Cambridge, Mass. 02139.

[1]G. Feinberg and S. Weinberg, Phys. Rev. Lett. 6, 381 (1961). For a review, see also B. Pontecorvo, Zh. Eksp. Teor. Fiz. 53, 1717 (1967) [Sov. Phys. JETP 26, 984 (1968)].

[2]E. Derman, Phys. Rev. D 19, 317 (1979).

[3]T. Eichten *et al.*, Phys. Lett. 46B, 281 (1973).

[4]J. Blietschau *et al.*, Nucl. Phys. B133, 205 (1978). The relation between our R and their r is $R = 1 - r$. The paper gives four values, $r = 0.9 \pm 0.3$, 1.0 ± 0.6, 0.8 ± 0.2, and 1.3 ± 0.6; we have quoted the weighted average of these results.

[5]H. H. Chen *et al.*, Nucl. Instrum. Methods 160, 393 (1979).

[6]P. Némethy *et al.*, to be published.

[7]T. W. Donnelly, Phys. Lett. 43B, 93 (1973); J. B. Langworthy *et al.*, Nucl. Phys. A280, 351 (1977).

[8]J. S. O'Connell, Los Alamos Scientific Laboratory Report No. LA-5175-MS, 1973 (unpublished). See also N. T. Nguyen, Nucl. Phys. A254, 485 (1975).

[9]E. Bellotti *et al.*, Nuovo Cimento Lett. 17, 533 (1976).

VOLUME 44, NUMBER 13 PHYSICAL REVIEW LETTERS 31 MARCH 1980

ERRATA

NEUTRINO EXPERIMENT TO TEST THE NATURE OF MUON-NUMBER CONSERVATION. S. E. Willis, V. W. Hughes, P. Némethy, R. L. Burman, D. R. F. Cochran, J. S. Frank, R. P. Redwine, J. Duclos, H. Kaspar, C. K. Hargrove, and U. Moser [Phys. Rev. Lett. 44, 522 (1980)].

In line 4 of the abstract change $\nu_e p \rightarrow ne^+$ to $\overline{\nu}_e p \rightarrow ne^+$. In line 5 of the abstract, change $(\mu^+ \rightarrow e^+ \nu_e \nu_\mu)$ to $(\mu^+ \rightarrow e^+ \overline{\nu}_e \nu_\mu)$. On page 523, left-hand column, line 7, read "Blietschau *et al.*"

Physica Scripta. Vol. T22, 111–118, 1988.

The Muon Anomalous Magnetic Moment*

Vernon W. Hughes

Gibbs Laboratory, Yale University, Physics Department, New Haven, CT 06520, U.S.A.

Received November 19, 1987; accepted January 4, 1988

Abstract

A brief review is given of the experimental and theoretical status of the muon g-2 factor. The theoretical motivation for an improved determination of g-2 is presented , and a new AGS experiment under development at Brookhaven National Laboratory is discussed.

1. Introduction

The muon anomalous magnetic moment or its g-2 factor has played a central role in establishing the validity of quantum electrodynamics (QED) and the behaviour of the muon as a heavy electron [1, 2]. The electron anomalous magnetic moment or its g-2 factor was one of the early discoveries which led to modern QED including its renormalization prescriptions, and at present its g-2 value has been measured with extreme precision [3] and provides one of the most sensitive tests of QED [4, 5] and a precise value [6] for the fine structure constant α. Indeed the electron and muon g-2 factors play somewhat complementary roles. The electron g-2 factor now provides a critical test of pure QED involving only the electron-positron and photon fields. On the other hand, because of its higher mass the muon g-2 factor is much more sensitive to otheer particles or fields (characteristically by a factor of $(m_\mu/m_e)^2 = 4 \times 10^4$) and hence provides the possibility of testing sensitively other interactions including the electroweak interaction and speculative theories beyond the standard model.

The experimental value for the muon g-2 factor has been determined in three progressively more precise measurements at CERN, [7–9], the latest one [9] achieving a precision of 7.3 ppm (Tables I and II). The theoretical value for g-2 (Table I) has improved steadily as higher order QED radiative contributions have been evaluated, and as knowledge of the virtual hadronic contributions to g-2 has been improved both by further measurements of the relevant quantity $R(s) = \sigma(e^+e^- \to \text{hadrons})/\sigma(e^+e^- \to \mu^+\mu^-)$ and by calculations [10]. The theoretical value of g-2 is now known [5, 11] to 1.3 ppm, a factor of 6 better than the experimental value. The present agreement of theory and experiment establishes that QED applies for the muon up to $Q^2 \simeq 1000\ (\text{GeV}/c)^2$ and determines the hadronic contribution to the vacuum polarization to about 12%. Furthermore, one of the most sensitive limits on muon substructure ($\Lambda > 800\,\text{GeV}$) is provided, as well as limits to various speculative modern theories.

The next stage in research on muon g-2, a measurement to 0.35 ppm at the AGS, has received first stage approval at Brookhaven National Laboratory and is now under development [12]. It will determine the contribution to g-2 of the newly discovered vector bosons W and Z, which mediate the weak interactions. This virtual radiative contribution is based on the renormalizability of the theory and is equivalent in the electroweak theory to the Lamb shift in QED.

Historically, the Lamb shift and the electron and muon g-2 values played a key role in the discovery and establishment of modern quantum electrodynamics. Precision measurements of hyperfine structure intervals and the Zeeman effect in hydrogen, muonium and positronium have provided important tests of QED bound state theory as well as values for the magnetic moments of fundamental particles. High energy experiments can explore the physics of W, Z, and other particles by direct production and are certainly the definitive way to identify new particles and new physics which are energetically accessible. On the other hand, it is difficult in these experiments to study the properties of more massive particles that cannot be produced by the accelerators. In principle, the existence of heavier particles can be detected through their effects on the behavior of lighter observed particles. Precise, sensitive experiments, such as the muon g-2 measurement, will give us a useful insight into the domain of physics which is not now accessible to high-energy experiments. In this sense low-energy, high-precision experiments play a role complementary to that of high energy experiments.

A high precision measurement of a fundamental quantity such as muon g-2 for which a precise value can be calculated from basic theory provides an important calibration point for modern particle theory. Not only is this valuable for the insight it provides about the very high energy regime beyond present accelerators, but also it may reveal new and deeper aspects of physics within the accessible energy regime, as was so dramatically illustrated for quantum electrodynamics by the Lamb shift in hydrogen and anomalous g-value of the electron.

2. Theory of muon g-2

The g-2 value for the electron has been measured to an accuracy of 4 parts per billion (ppb) in an experiment in which a single electron is stored in a Penning trap at low temperature [3]. The experimental value is

$$a_e(\text{expt}) = 1\ 159\ 652\ 193\ (4) \times 10^{-12} (3.4\,\text{ppb}) \qquad (1)$$

where $a_e = (g_e - 2)/2$.

The theoretical value has been computed [4] through terms

* Research supported in part by DOE under contract DE-AC02-ER03075 and DE-FG02-84ER40243.

of order α^4:

$$a_e(\text{theor}) = 0.5\left(\frac{\alpha}{\pi}\right) - 0.328\,478\,966\left(\frac{\alpha}{\pi}\right)^2 + 1.176\,5(13)\left(\frac{\alpha}{\pi}\right)^3 - 0.8(2.5)\left(\frac{\alpha}{\pi}\right)^4. \quad (2)$$

Using the condensed matter value of α based on the ac-Josephson effect and the quantized Hall effect [13],

$$\alpha^{-1} = 137.035\,981\,5\,(123)\,(0.090\,\text{ppm}), \quad (3)$$

we find

$$a_e(\text{theor}) = 1\,159\,652\,263\,(113) \times 10^{-12}(0.092\,\text{ppm}) \quad (4)$$

in good agreement with a_e(expt). Alternatively, comparison of a_e(expt) and a_e(theor) yields the most precise value [5] for the fine structure constant α:

$$\alpha^{-1} = 137.035\,989(8)\,(27)\,(0.020\,\text{ppm}). \quad (5)$$

Only the electron and photon contribute significantly to a_e(theor); the contributions of virtual μ and τ leptons as well as virtual hadrons and $W^{\pm}$, Z particles are less than the present 4 ppb experimental error.

The theoretical value for the anomalous g-value of the muon, $a_\mu = (g_\mu - 2/2$, in the standard theory [5, 11] and the present experimental value [9] are given in Table I. Here a_μ(QED) arises from the virtual radiative contributions of QED involving photons and charged leptons, a_μ(had) arises from virtual hadron contributions, and a_μ(weak) arises from the virtual radiative contributions involving the W, Z and Higgs particles. In contrast to the electron anomaly, which is dominated by the QED effect, the muon g-2 is much more sensitive to physics at smaller distances because of the larger mass scale.

The weak contribution [14] arises from single loop Feynman diagrams involving W, Z and ϕ as shown in Fig. 1 and is analogous to Lamb shift or g_e-2 radiative corrections in QED. The uncertainty arises from the quoted error in $\sin^2\theta_W$. The expected size of uncalculated higher order weak contributions is also about 10^{-11}. An essential feature of this calculation is the renormalizability of the electroweak theory. We note that recent experimental evidence only establishes that $m_\phi > 3.6$ GeV (private communication from S. Dawson and W. Marciano) and hence a slightly revised value for a_μ(weak) may soon become necessary.

SINGLE LOOP DIAGRAMS CONTRIBUTING TO THE MUON g-FACTOR

$$\Delta a_\mu(W) = \frac{G_F m_\mu^2}{8\pi^2\sqrt{2}} \times \frac{10}{3} = +3.89 \times 10^{-9}$$

$$\Delta a_\mu(Z) = \frac{G_F m_\mu^2}{8\pi^2\sqrt{2}} \times \frac{1}{3}\left[(3-4\cos^2\theta_W)^2 - 5\right] = -1.94 \times 10^{-9}$$

$$\Delta a_\mu(\phi) = \frac{G_F m_\mu^2}{2\pi^2\sqrt{2}} \int_0^1 \frac{y^2(2-y)\,dy}{y^2 + (1-y)(m_\phi/m_\mu)^2}$$

$$= \frac{G_F m_\mu^2}{2\pi^2\sqrt{2}} \left(\frac{m_\mu}{m_\phi}\right)^2 \ln\left[\left(\frac{m_\phi}{m_\mu}\right)^2\right] \leq 0.01 \times 10^{-9}, \quad \text{IF } m_\phi \gg m_\mu$$

$$= \frac{3G_F m_\mu^2}{4\pi^2\sqrt{2}} \qquad m_\phi \ll m_\mu$$

$$\sin^2\theta_W = 0.226 \pm 0.004; \quad m_\phi > 7\ \text{GeV}$$

$$a_\mu(\text{WEAK}) = (1.95 \pm 0.01) \times 10^{-9}$$

Fig. 1. Weak interaction contribution to a_μ.

Other tests of electroweak radiative corrections in the Glashow, Salam, Weinberg unified electroweak theory have been made. The first was the radiative correction to the semi-leptonic case of neutron beta decay [15, 16], and two others [17] at two times the standard deviation level concern deep inelastic neutrino scattering and mass measurements of W and Z. From the new BNL muon g-2 experiment the electroweak radiative correction to the g-value for the muon due to virtual W and Z particles should be determined with an accuracy of about 20%.

The hadronic vacuum polarization contribution to a_μ is substantial and its uncertainty is the dominant error in a_μ(theor). The strong interactions contribute to a_μ principally through the vacuum polarization diagram which can be expressed as shown in Fig. 2. Since it is not yet possible to compute $R(s)$ from QCD theory, measured values of $R(s)$ from e^+e^- colliders are used to evaluate a_μ(had). A careful tabulation of measured values of $R(s)$ and of their contributions to a_μ(had) has been made [10], and a graph of the relative contributions is shown in Fig. 2 in the range of s up to $s = 1$, including the ρ and ω resonances, from which come the principal contribution to and the error in a_μ(had). Higher order hadronic contributions have been evaluated and found to be small [10]. Other evaluations of a_μ(had) have also been done recently [18, 19]. The quantity a_μ(had) is now known a factor of about 6 better than when the CERN experiment was reported [9]. However, it will be necessary to determine a_μ(had) another factor of 3.5 better, that is to 0.6%, in order to utilize fully the projected accuracy of 0.35 ppm in a new measurement of a_μ for deriving new physical information. It now seems possible to obtain the required data for $\sqrt{s} \lesssim$ 1 GeV with the VEPP 2 e^+e^- collider at Novosibirsk [18, 20]. (Fig. 3). Fixed target experiments such as the CERN experiment [21] in which energetic e^+ on a H target was studied to

Table 1. *Theoretical and experimental values for a_μ*

a_μ(theor)	=	a_μ(QED) + a_μ(had) + a_μ(weak)
a_μ(QED)	=	$C_1(\alpha/\pi) + C_2(\alpha/\pi)^2 + C_3(\alpha/\pi)^3 + C_4(\alpha/\pi)^4 + a_\mu(\tau)$
C_1 = 0.5;		C_2 = 0.765 857 577;
C_3 = 24.0725(123);		C_4 = 137.96 (250)
$a_\mu(\tau)$	=	420×10^{-12}
α^{-1}	=	137.035 989 5 (61) (0.045 ppm) (1986 value) [6]
a_μ(QED)	=	1 165 848 124 (152) (52) × 10^{-12} (0.14 ppm) ↑ ↑ Calculation α
a_μ(had)	=	6940 (142) × 10^{-11}, or (59.5 ± 1.2) ppm in a_μ(theor)
a_μ(weak)	=	195 (1) × 10^{-11}, or (1.7 ± 0.01) ppm in a_μ(theor)
a_μ(theor)	=	116 591 947 (143) × 10^{-11} (1.3 ppm)
a_μ(expt)	=	11 659 230 (84) × 10^{-10} (7.2 ppm)
a_μ(expt) − a_μ(theor)	=	35 (85) × 10^{-10}
Fundamental constant		
μ_μ/μ_p	=	3.183 345 47(47)(0.15 ppm) (1986 value [6]

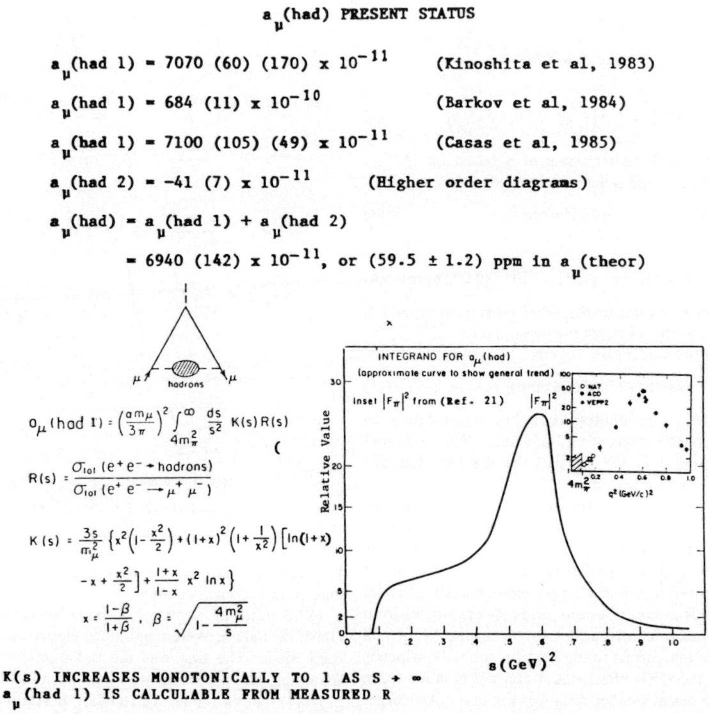

a_μ(had) PRESENT STATUS

a_μ(had 1) = 7070 (60) (170) x 10^{-11} (Kinoshita et al, 1983)

a_μ(had 1) = 684 (11) x 10^{-10} (Barkov et al, 1984)

a_μ(had 1) = 7100 (105) (49) x 10^{-11} (Casas et al, 1985)

a_μ(had 2) = -41 (7) x 10^{-11} (Higher order diagrams)

a_μ(had) = a_μ(had 1) + a_μ(had 2)

= 6940 (142) x 10^{-11}, or (59.5 ± 1.2) ppm in a_μ(theor)

$$a_\mu(\text{had 1}) = \left(\frac{\alpha m_\mu}{3\pi}\right)^2 \int_{4m_\pi^2}^{\infty} \frac{ds}{s^2} K(s) R(s)$$

$$R(s) = \frac{\sigma_{tot}(e^+e^- \to \text{hadrons})}{\sigma_{tot}(e^+e^- \to \mu^+\mu^-)}$$

$$K(s) = \frac{3s}{m_\mu^2}\left\{x^2\left(1-\frac{x^2}{2}\right) + (1+x)^2\left(1+\frac{1}{x^2}\right)\left[\ln(1+x) - x + \frac{x^2}{2}\right] + \frac{1+x}{1-x}x^2 \ln x\right\}$$

$$x = \frac{1-\beta}{1+\beta}, \quad \beta = \sqrt{1-\frac{4m_\mu^2}{s}}$$

K(s) INCREASES MONOTONICALLY TO 1 AS S → ∞
a_μ(had 1) IS CALCULABLE FROM MEASURED R

Fig. 2. Principal hadronic contribution to a_μ.

measure $e^+ + e^- \to \pi^+ + \pi^-$ and $e^+ + e^- \to \mu^+ + \mu^-$ also provide a determination of $R(s)$. An improved experiment of this type may be done at CERN [22] to provide accurate data on $R(s)$ at invariant mass $\sqrt{s} < 500$ MeV. At Fermilab such an experiment could reach $\sqrt{s} \simeq 800$ MeV.

The primary purpose of the AGS experiment is to see whether the measurement of g-2 confirms the present theoretical picture of the standard model including the weak contribution. It is not at all a foregone conclusion, however, that the measurement simply confirms the electroweak effect (Fig. 1), as described by the GSW theory (or any grand unification theories which keep the electroweak sector intact). These may be nothing but the "low-energy limit" of a more fundamental theory which gives a completely different prediction for electroweak radiative contributions. A further discussion of speculative topics including lepton substructure and supersymmetry which will be tested is given in Ref. [23].

Since the discovery of the muon it has been a mystery that the muon seems to be identical with the electron except for its mass. After many years of investigation, this mystery has not yet been resolved. Instead, in the standard theory, it is simply accepted as a fact and used to rationalize the existence of the second and third generations of elementary particles. In this sense the mystery has merely deepened, and one of the goals of composite models is the attempt to solve this mystery. New experimental information treating this subject will be of fundamental importance to our understanding of the structure of matter, and in this context the muon g-2 has a crucial role to play.

3. The experimental determination of the muon g-2

The goal of the proposed new AGS experiment is a measurement of a_μ to 0.35 ppm, a factor of 20 improvement over the CERN experiment which achieved 7.2 ppm. A number of approaches to the measurement of muon g-2, are listed in Table III. After considering various approaches the group now proposing the new AGS experiment [12] chose a method similar to that used at CERN [9] but allowing either for pion or muon injection into the storage ring (i.e., the 4th choice listed, 4B or 4C). It will involve the storage of polarized muons with momentum of about 3 GeV/c in a 14 m diameter ring with a homogeneous magnetic field of about 15 kG and with a quadrupole electric field to provide vertical focusing. Initially pions will be injected into the ring and polarized muons from pion decays will be captured. The g-2 precession is observed through the parity-violating correlation between the muon spin direction and the direction of emission of the decay electron.

The famous CERN experiment [9] measured the g-2 precession frequency ω_a which is the difference between the spin precession frequency ω_s and the cyclotron frequency ω_c

Table II. *CERN experiments*

$$\omega_c = \frac{eB}{mc\gamma}$$

$$\omega_s = \frac{eB}{mc\gamma} + \frac{e}{mc}aB$$

$$\omega_a = \omega_s - \omega_c = \frac{e}{mc}aB$$

The first experiment

Method: G-2 precession in a bending magnet ($B = 1.6$ T), with trajectories and focussing determined by B inhomogeneties.

Synchrocyclotron; $p_\mu = 90$ MeV/c.

$a_\mu = (1162 \pm 5) \times 10^{-6}$ (4300 ppm)

G. Charpak, F. J. M. Farley, R. L. Garwin, T. Muller, J. C. Sens, and A. Zichichi, Nuovo Cimento **37**, 1241 (1965).

The second experiment

Method: G-2 precession in a muon storage ring ($B = 1.7$ T) with weak focussing due to inhomogeneous B.

Proton synchrotron; $p_\mu = 1.3$ GeV/c, $\gamma = 12$

$a_\mu = (116\,616 \pm 31) \times 10^{-8}$ (270 ppm)

J. Bailey, W. Bartl, G. von Bochmann, R. C. A. Brown, F. J. M. Farley, M. Giesch, H. Jöstlein, S. van der Meer, E. Picasso, and R. W. Williams, Il. Nuovo Cimento **A9** 369 (1972.

The third experiment

Method: G-2 precession in a muon storage ring ($B = 1.5$ T) with electric quadrupole focussing at the magic γ, where g-2 precession and electric focussing are independent.

Proton synchrotron; $p_\mu = 3.1$ GeV/c, $\gamma = 29$.

$a_\mu = 1\,165\,924(8.5) \times 10^{-9}$ (7.3 ppm)

J. Bailey, K. Borer, F. Combley, H. Drumm, C. Eck, F. J. M. Farley, J. H. Field, W. Flegal, P. M. Hattersley, F. Krienen, F. Lange, G. Lebée, E. McMillan, G. Petrucci, E. Picasso, O. Runolfsson, W. von Rüden, R. W. Williams and S. Wojcicki, Nucl. Phys. B150, 1 (1979).

in a known magnetic field B:

$$\omega_s = \frac{eB}{mc\gamma} + \frac{e}{mc}aB \tag{6}$$

$$\omega_c = \frac{eB}{mc\gamma} \tag{7}$$

$$\omega_a = \omega_s - \omega_c = \frac{e}{mc}aB, \tag{8}$$

in which $a = (g\text{-}2)/2$ is the anomalous g-value of the muon and m is the muon rest mass (Table II).

The design of the CERN experiment and also of the proposed AGS experiment is based upon an important observation about the spin motion in a magnetic and electric field. The equation for the g-2 precession frequency is given by [24]

$$\omega_a = \frac{d\theta_R}{dt} = \frac{e}{mc}\left[aB - \left(a - \frac{1}{\gamma^2 - 1}\right)\left|\boldsymbol{\beta} \times \boldsymbol{E}\right|\right] \tag{9}$$

in which $\theta_R = (\boldsymbol{s}, \boldsymbol{\beta})$ is the angle between the muon spin ($\boldsymbol{s}$) direction in its rest frame and the muon velocity ($\boldsymbol{\beta}$) direction in the laboratory frame, and other quantities refer to the laboratory frame. If the bracketed factor $a - [1/(\gamma^2 - 1)]$ is zero, which is true when γ has the "magic" value 29.3 and $p_\mu = 3.094$ GeV/c, then the g-2 precession frequency is determined by $\boldsymbol{B}$ alone and is independent of $\boldsymbol{E}$. This observation allows for a separated function muon storage ring in which a homogeneous magnetic field B determines the g-2 precession frequency, and an electric quadrupole field provides vertical weak-focusing for the muons.

The overall arrangement of the planned AGS experiment with its muon storage ring is shown in Fig. 4. Polarized muons from pion decay in flight will be trapped in a large ring magnet with a uniform magnetic field B of 1.5 T, vertical focusing being provided by electric quadrupoles distributed more or less uniformly around the ring. The precession frequency ω_a of the spin relative to the momentum vector will be measured by observing the decay electrons emerging on the inside of the ring with electron calorimeters. As the angular

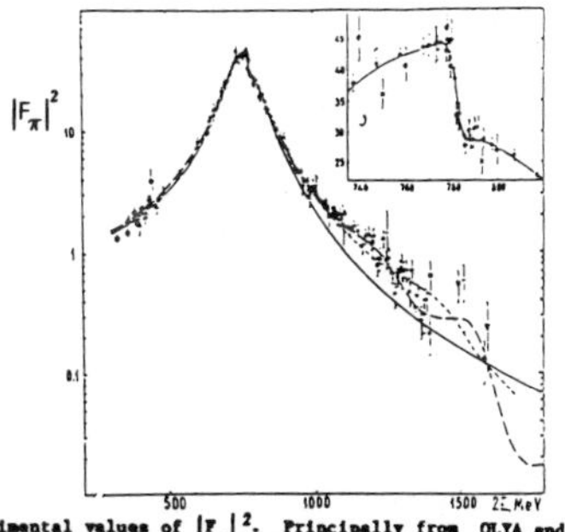

(a) Experimental values of $|F_\pi|^2$. Principally from OLYA and CMD detectors of Novosibirsk. Solid curve corresponds to the Gounaris–Sakurai Formula taking into account $\rho - \omega$ interference, dotted line – parametrization with ω, ρ, $\rho'(1250)$, $\rho''(1600)$ mesons.

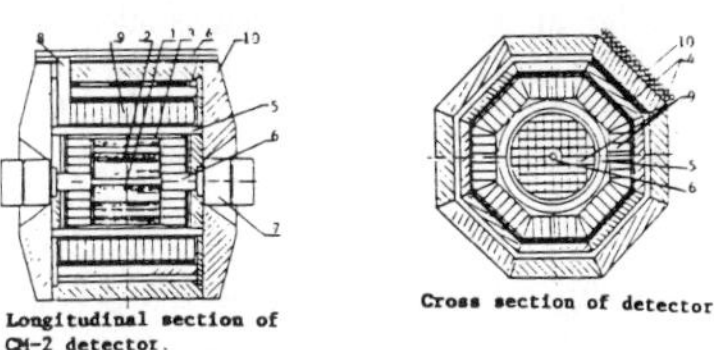

1. Interaction point of the beams.
2. Drift Chamber.
3. Z Chamber.
4. Range System.
5. Main Magnet.
6. Compensation Magnet.
7. Magnetic lenses of the Storage Ring.
8. Current leads and liquid helium for magnets
9. CsI crystals.
10. Iron yoke.

(b) New Detector at Novosibirsk

Fig. 3. Storage ring experiments in Novosibirsk to measure $R(s)$. (a) Results of previous experiments; (b) A new detector for more precise measurements.

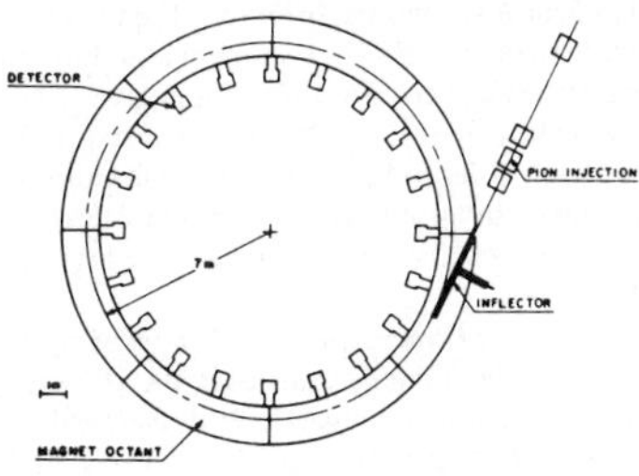

Fig. 4. General arrangement of AGS muon g-2 experiment.

Table III. *Other approaches to muon g-2*

(1) Low momentum (surface) μ^+; 30 MeV/c; LAMPF
Magnetic bottle storage (Hughes, Lysenko, Nemethy).

(2) High momentum $\mu^\pm$; ~30 GeV/c; ISR, AGS
AG type machine, momentum compaction factor $\alpha = 1$;
Comparison of μ and deuteron (Farley).

(3) Directly drive g-2 transition at ω_a by RF magnetic field with μ in storage ring.[a]

(4) Magic* γ; $p_\mu \simeq 3$ GeV/c
(A) $B = 5$ T; π injection
(B) $B = 1.5$ T; π injection
(C) $B = 1.5$ T; μ injection

[a] Kh. A. Simonyan and Yu. F. Orlov, JETP **18**, 123 (1964).
* See Section 3 for a discussion of this value.

distribution in the decay rotates with the muon spin, the counting rate for the high energy electrons (emitted forwards) will be modulated at the precession frequency. The g-2 precession frequency ω_a will be deduced from the decay electron counting data fitted to the formula

$$N = N_0 \exp(-t/\tau)\{1 - A\cos(\omega_a t + \phi)\}, \tag{10}$$

in which τ = muon laboratory lifetime and ϕ is a phase angle. The asymmetry parameter A is the product of the stored muon polarization and the assymmetry in μ–e decay which is a function of electron energy and the angular and energy spread of the electrons that are accepted by the detection system. The statistical error in ω_a is given by

$$\frac{\Delta\omega_a}{\omega_a} = \frac{\sqrt{2}}{\omega_a A\tau N_e^{1/2}} \tag{11}$$

in which N_e = total number of detected electrons.

In addition to values of ω_a and B, determination of a from eq. (9) requires a value for the constant (e/mc). Since B is measured by a proton resonance frequency ω_p(corrected for molecular and bulk magnetic shielding), the constant needed is [6, 25]:

$$\lambda = \mu_\mu/\mu_p, \tag{12}$$

and a is then given by

$$a = R/(\lambda - R), \tag{13}$$

where $R = \omega_a/\omega_p$.

Improvement in the g-2 measurement at BNL as compared to the CERN result will be due principally to improved statistics obtainable with the much higher proton beam intensity at the AGS (~100 times). In addition we plan to know the mean field $\bar{B}$ around the orbit to 0.1 ppm, some factor of 10 better than in the CERN experiment and thus reduce their principal systematic error to the 0.1 ppm level. The precision in measurement of the time interval t in eq. (10) must be about 30 ps in order to keep this contribution to the error in determining ω_a small compared to the statistical error in ω_a. Other small systematic errors must also be kept at or below the 0.1 ppm level.

The superferric storage ring (Fig. 5) now being designed at BNL [26] is to be homogeneous over the storage region to 1 ppm and the effective magnetic field averaged around the ring known to 0.1 ppm. In order to achieve a highly stable magnetic field we have chosen an iron magnet with superconducting coils. The field is of course determined principally by the iron configuration. Its value is chosen to be about 1.5 T in order to stay just below saturation. The use of superconducting coils should be advantageous for stability because the coil temperature is well stabilized and coil and iron temperatures can be well isolated. Also because of the long time constant of the coil, ripple effects from the power supply should not be troublesome. Shimming of the magnet to achieve the 1 ppm homogeneity will involve iron shims about the yoke and in the spaces between the pole pieces and the yoke, as well as mechanical grinding of the pole faces and the use of pole face current windings. For field measurement and feedback control, extensive NMR measurements with fixed probes and also with an NMR trolley (Fig. 6) capable of moving about the ring inside the vacuum chamber will be used.

Decay electrons will be detected with electron calorimeters (Fig. 7). The special requirements of the detectors will be the ability to handle high instantaneous rates including a large initial background and to measure time intervals with ~30 ps accuracy.

Table IV gives general features of the new AGS experiment compared to the CERN experiment, and Table V gives the specific parameters of the AGS experiment. The projected counting rates and errors are given in Table VI. The CERN experiment was dominated by the statistical error of 7.0 ppm, and the largest systematic error was an uncertainty of 1.5 ppm in $\bar{B}$. For the AGS experiment the primary proton beam average intensity will be about 100 times that available at CERN, so the statistical error in determining ω_a is projected

Table IV. *AGS g-2 experiment*

General approach
Similar to CERN experiment.
A muon storage ring operating at the magic $\gamma = 29.3$ with $B \simeq 1.5$ T and electrostatic quadrupole focusing.
Pion injection.

Improvements

1. Primary proton beam intensity [injected π beam] (with AGS booster) × 100 [× 200]
2. Storage ring magnet: Field homogeneity and control:
   Homogeneity (≃1 ppm) [CERN ≃ 10–15 ppm]
   Control (≃0.1 ppm) [CERN ≃ 0.5 ppm]
   Achieved by:
   (a) Superconducting coils
   (b) Larger magnet gap
   (c) Aximuthal symmetry in iron contruction
   (d) Small air gaps (~1 cm) between pole pieces and yoke
   (e) Extensive shimming features
   (f) NMR feedback and control
3. Magnetic field measurement (NMR) Accuracy (0.1 ppm)
   Achieved by:
   (a) NMR trolley (movable within vacuum chamber)
   (b) Many (~200) fixed NMR probes outside vacuum chamber
   (c) Insertable NMR probe for absolute calibration
4. Detector system
   Increased acceptance, data rate capacity, and time measurement accuracy.
   Achieved by:
   (a) Larger solid angle and thinner-walled vacuum vessel
   (b) Detector segmentation
   (c) Improved electronics
   (d) Improved digitrons

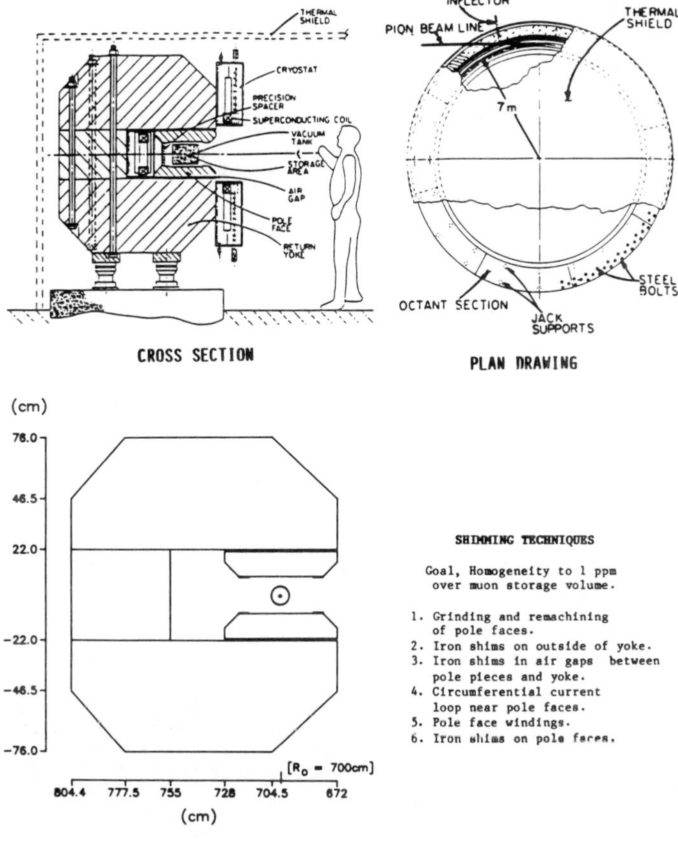

Fig. 5. Superferric storage ring magnet.

to be 0.3 ppm. An improved storage ring magnet and NMR system should reduce the systematic error in $\bar{B}$ to about 0.1 ppm. Other systematic errors should be small enough to allow a determination of a to 0.35 ppm.

For a number of reasons π injection into the storage ring has been our chosen method. This was the method used at CERN where the ratio of captured decay muons to injected pions was about 10^{-4}. However, we are studying seriously the alternative of μ injection into the ring with capture achieved by a fast kicker magnetic field, which is provided by current pulses in conducting plates. Muon injection has the important advantage of avoiding the large initial π induced background present with π injection.

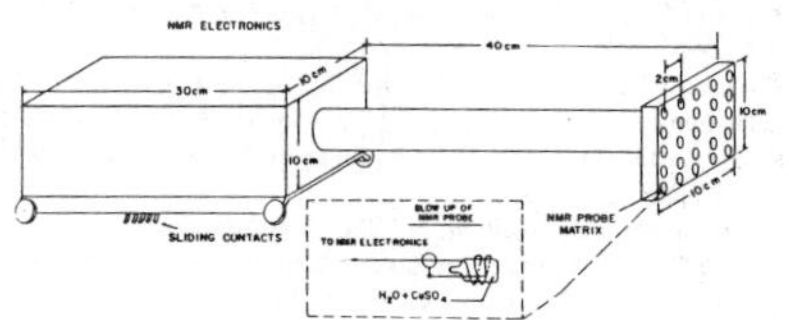

Fig. 6. Trolley and NMR probe matrix.

This AGS experiment is now being actively developed, with data-taking projected in less than 4 years.

In addition to the primary experiment to measure a_μ very precisely, the system we are developing can be used to compare a_μ^+ and a_μ^- as a test of CPT, to measure the lifetime of energetic μ in the storage ring, and to search for an electric dipole moment of the muon, as was done at CERN [9] but with much improved precision.

It is interesting to remark that the CERN *g*-2 storage ring magnet has had a varied and successful but unanticipated history after its use to measure muon *g*-2. First at CERN it was converted into a strong focussing ring and used to do the initial cooling experiment (ICE) both on stochastic cooling [27] and electron cooling [28], which was an essential step in the development of the CERN $\bar{p}p$ collider [29] and thence in

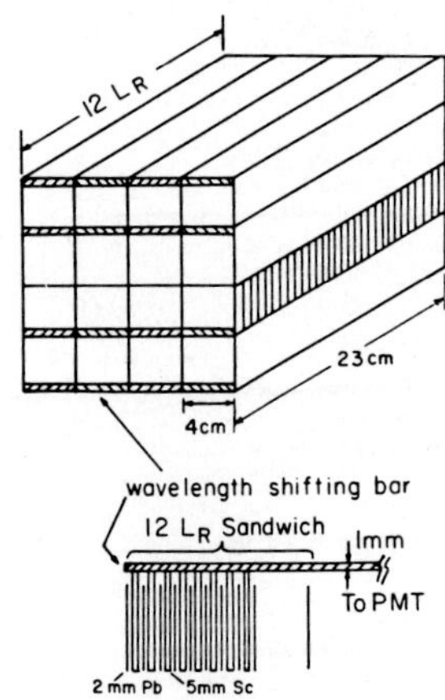

Fig. 7. Electron calorimeter.

the discovery of the W and Z particles [30]. The ICE storage ring with the help of stochastic cooling was also used to measure the antiproton lifetime and set a lower limit of 32 h [31].

At present this famous CERN ring magnet is at the Gustaf Werner Institute of Uppsala University in Sweden and is an integral part of the CELSIUS nuclear physics facility [32].

Table V. *Parameters of AGS experiment*

Magnet	
Orbit radius	7.0 m
Central magnetic field	1.47 T
Magnet gap	18 cm
Pole width	56 cm
Storage systems	
Storage aperture diameter	9 cm
Vertical focusing by electric quadrupole field (pulsed 1 ms)	40 kV
Particle injection	
Pulsed magnetic inflector	
π-μ decay	
Muon motion in storage ring	
Gamma, γ	29.3
Momentum, p_μ	3.094 GeV/c
Lifetime, τ	64.4 µs
Orbital frequency, f_c	6.81 MHz
Orbital period, τ_c	147 ns
g-2 precession frequency, f_a	0.2327 MHz
g-2 period, τ_a	4.3 µs
Number electron shower detectors	80
Intensity data	
Proton per rf bunch with booster	4.2×10^{12}
Bunches injected per ring fill	2
Fills per AGS cycle	3
Pions injected per fill	1.07×10^8
Muons stored per fill	1.4×10^4
Electron counts per fill	2.8×10^3
Running time	1500 hrs
Statistical error in a_μ	0.3 ppm
Systematic errors	0.2 ppm
Overall error	0.35 ppm

Table VI. *AGS experiment errors*

Counting Rates and Statistical Errors	
Storage aperture diameter	90 mm
Protons per rf bunch with booster	4.2×10^{12}
Bunches ejected per ring fill	2
Protons per fill	8.4×10^{12}
Pion $\Delta p/p$	$\pm 0.6\%$
Pions at inflector exit per fill (1.28×10^7 per 10^{12} protons)	1.07×10^8
Muons stored per fill (at 134 ppm capture efficiency)	14×10^3
Electrons counted above 1.6 GeV per fill (20% of the decays)	2.8×10^3
Fills per AGS cycle (1.4 s)	3
Fraction of AGS protons used	50%
Fills per hour	7714
Electron counts per hour	22×10^6
Running time for 0.3 ppm (1 std. dev.)	1288 hours

Source	Systematic Errors Comments	Error (ppm)
Magnetic	Includes absolute calibration of NMR probes and averaging over space, time, and muon distribution.	0.1
Electric field correction	0.7 ppm correction	0.03
Pitch correction	0.4 ppm correction	0.02
Particle losses		0.05
Timing error		0.01
	Total (in quadrature)	0.12

One fascinating experiment being considered there with this ring is a measurement of the electron neutrino mass by observing the decays of tritons (^{3}H) stored in the ring into ^{3}He and e^- and thus measuring the endpoint of the triton spectrum with high precision [33]. This approach would avoid all the complexities otherwise associated with the final atomic or molecular electronic states associated with the ^{3}He.

In view of this impressive history of the CERN g-2 storage ring, it is convincing to believe that other exciting experiments and uses will be found for the Brookhaven precise muon g-2 storage ring after completion of the Brookhaven g-2 experiment.

References

1. Hughes, V. W. and Kinoshita, T., in Muon Physics I (Edited by V. W. Hughes and C. S. Wu), p.11, Academic Press, New York (1977).
2. Hughes, V. W. and Kinoshita, T., Comments Nucl. Part. Phys. **14**, 341 (1985).
3. Schwinberg, P. B., Van Dyck, Jr. R. S. and Dehmelt, H. G., Phys. Rev. Lett. **47**, 1679 (1981); Van Dyck, Jr., R. S., Schwinberg, P. B. and Dehmelt, H. G., in Atomic Physics 9 (Edited by R. S. Van Dyck, Jr. and E. N. Fortson), p. 53, World Scientific Publ. Co., Singapore (1984); Dehmelt, H. G., This volume.
4. Kinoshita, T. and Lindquist, W. B., Phys. Rev. Lett. **47**, 1573 (1981); Phys. Rev. **D27**, 853, 867, 877, 886 (1983).
5. Kinoshita, T., IEEE Trans. Instrum. Meas. **IM-36**, No. 2, 201 (1987); Kinoshita, T. and Sapirstein, J., in Atomic Physics 9 (Edited by R. S. Van Dyck, Jr. and E. N. Fortson), p. 38, World Scientific Publ. Co., Singapore (1984).
6. Cohen, E. R. and Taylor, B. N., The 1986 Adjustment of the Fundamental Physical Constants, CODATA Bulletin **63** (A Report of the CODATA Task Group on Fundamental Constants, Pergamon Press, November 1986).
7. Charpak, G., *et al.*, Nuovo Cim. **37**, 1241 (1965).

8. Bailey, J. *et al.*, Il Nuovo Cimento **9A**, 369 (1972).
9. Bailey, J. *et al.*, Nucl. Phys. **B150**, 1 (1979).
10. Kinoshita, T., Nizic, B. and Okamoto, Y., Phys. Rev. **D31**, 2108 (1985).
11. Kinoshita, T., Nizic, B. and Okamoto, Y., Phys. Rev. Lett. **52**, 717 (1984).
12. A new precision measurement of the muon *g*-2 value at the level of 0.35 ppm., AGS Proposal 821, September, 1985; revised September, 1986, V. W. Hughes, spokesman. E. Hazen, C. Heisey, B. Kerosky, F. Krienen, E. K. McIntyre, D. Magaud, J. P. Miller, B. L. Roberts, D. Stassinopoulos, L. R. Sulak, W. Worstell – Boston University; H. N. Brown, E. D. Courant, G T. Danby, C. R. Gardner, J. W. Jackson, M. May, A. Prodell, R. Shutt, P. A. Thompson – Brookhaven National Laboratory; J. A. Johnson, M. S. Lubell – City College of New York; A. M. Sachs – Columbia University; T. Kinoshita – Cornell University; D. Winn – Fairfield University; M. Janousch, H.-J. Mundinger, G. zu Putlitz, J. Rosenkranz, W. Schwarz – University of Heidelberg; W. P. Lysenko – Los Alamos National Laboratory; A. Rich – University of Michigan; J. J. Reidy – University of Mississippi; F. Combley – Sheffield University; K. Nagamine, K. Nishiyama – University of Tokyo; K. Endo, H. Hirabayashi, S. Kurokawa, T. Sato – KEK; K. Ishida – Riken; L. M. Barkov, B. I. Khazin, E. A. Kuraev, Ya. M. Shatunov – Institute of Nuclear Physics, Novosibirsk, USSR; J. M. Bailey, S. K. Dhawan, A. A. Disco, F. J. M. Farley, V. W. Hughes, Y. Kuang, H. Venkataramania – Yale University.
13. Taylor, B. N., J. Res. Natl. Bur. Stand. **90**, 91 (1985).
14. Jackiw, R. and Weinberg, S., Phys. Rev. **D5**, 2396 (1972); Altarelli, G., Cabibbo, N. and Maiani, L., Phys. Lett. **40B**, 415 (1972); Bars, I. and Yoshimura, M., Phys. Rev. **D6**, 374 (1972); Fujikawa, K., Lee, B. W. and Sanda, A. I., Phys. Rev. **D6**, 2923 (1972); Bardeen, W. A., Gastmans, R. and Lautrup, B. E., Nucl. Phys. **B46**, 319 (1972).
15. Sirlin, A., 50 Years of Weak Interactions (Edited by D. Cline and G. Riedasch), p. 93, HEP Group, Univ. of Wisconson, Madison Wisconson (1984).
16. It is a pleasure to thank A. Sirlin for helpful and informative communications and discussions of this topic.
17. Marciano, W., XXIII Int. Conf. on High Energy Physics (Edited by S. C. Loken), p. 999, World Scientific Pub. Co. (1987).
18. Barkov, L. M. *et al.*, Nucl. Phys. **B256**, 365 (1985).
19. Casas, J. A. *et al.*, Phys. Rev. **D32**, 736 (1985).
20. Aulchenko, V. M. *et al.*, Nucl. Inst. Meth. **A252** 299 (1986); Private communication from L. Barkov.
21. Amendolia, S. R. *et al.*, Phys. Lett. **138B**, 454 (1984); **146B**, 116 (1984).
22. Private communication from I. Mannelli and L. Foa.
23. Hughes, V. W., in Fundamental Symmetries (ed. by P. Bloch, P. Povlopoulos and R. Klapisch), p. 271, Plenum, N.Y., 1987.
24. Bargmann, V. *et al.*, Phys. Rev. Lett. **2**, 435 (1959); Farley, F. J. M., Cargese Lectures in Physics, (1968) Vol. 2, p. 55, Gordon and Breach; Bailey, J. and Picasso, E., Progr. in Nuclear Physics **12**, 43 (1970); Jackson, J. D., Classical Electrodynamics, p. 556, John Wiley & Sons, New York (1975).
25. Mariam, F. G. *et al.*, Phys. Rev. Lett. **49**, 993 (1982).
26. Brown, D. *et al.*, Ultraprecise superferric storage ring magnet for the muon *g*-2 experiment, 10th Int. Conf. on Magnet Technology, Boston MA., September (1987) (to be published).
27. Carron, G. *et al.*, Phys. Lett. **77B**, 353 (1978); Carron, G. *et al.*, IEEE, NS **26**, 3456 (1979).
28. Bell, M. *et al.*, Phys. Lett. **87B**, 275 (1979).
29. Gareyte, J., 11th International Conference on High-Energy Accelerators (Edited by W. S. Newman), p. 79, (Birkhäuser Verlag, Basel (1980).
30. Arnison, G. *et al.*, Phys. Lett. **126B**, 398 (1983); Bagnaia, P. *et al.*, Phys. Lett. **129**, 130 (1983).
31. Bregman, M. *et al.*, Phys. Lett. **78B**, 174 (1978).
32. The Svedberg Laboratory, Uppsala University, Brochure; Uppsala Accelerator News (Edited by G. Tibell), Vol. 2, No. 1 (1986); No. 4 (1987).
33. Kullander, S., Neutrino Mass Measurements, Workshop on the Physics Program at CELSIUS, November (1983), GWI-PH 5/83.

Publication List
Vernon W. Hughes

1. *Ultrasonic Delay Lines. I.* H. B. Huntington, A. G. Emslie, and V. W. Hughes J. of the Franklin Inst. **245**, 1-23 (1948).

2. *Generation of Triangular Waveforms.* V. W. Hughes and R. M. Walker, pp. 254-288; *Pulse-Recurrence-Frequency Division.* A. H. Frederick, V. W. Hughes and E. F. MacNichol, Jr., pp. 567-601; *Electrical Delay Lines.* V. W. Hughes, pp. 730-750; *Supersonic Delay Device.* V. W. Hughes and H. B. Huntington, pp. 751-765, ed. by B. Chance, V. W. Hughes, E. F. MacNichol, D. Sayre and F. C. Williams, **Massachusetts Institute of Technology Radiation Laboratory Series, 19**, (McGraw-Hill Book Co., New York, 1949).

3. *The Radiofreqeuncy Spectrum of $Rb^{85}F$ and $Rb^{87}F$ by the Electric Resonance Method.* V. W. Hughes and L. Grabner, Phys. Rev. **79**, 314-322 (1950).

4. *Energy Levels, Selection Rules, and Line Intensities for Molecular Beam Electric Resonance Experiments with Diatomic Molecules.* V. W. Hughes and L. Grabner, Phys. Rev. **79**, 829-836 (1950).

5. *The Radiofrequency Spectrum of $K^{39}F$ by the Electric Resonance Method.* L. Grabner and V. W. Hughes, Phys. Rev. **79**, 819-828 (1950).

6. *Further Evidence for a Two Quantum Transition in Molecular Spectroscopy.* L. Grabner and V. W. Hughes, Phys. Rev. **82**, 561 (1951).

7. *Effect of Nuclear Structure on the Hyperfine Structure of He^3.* V. W. Hughes and G. Weinreich, Phys. Rev. **91**, 196-197 (1953).

8. *The Magnetic Moment of the Helium Atom in the Metastable Triplet State.* V. W. Hughes, G. Tucker, E. Rhoderick and G. Weinreich, Phys. Rev. **91**, 828-841 (1953).

9. *Relativistic Contributions to the Magnetic Moment of $^3S_1$ Helium.* W. Perl and V. W. Hughes, Phys. Rev. **91** 842-852 (1953); Phys. Rev. **89**, 886-887 (1953).

10. *Hyperfine Structure of Helium-3 in the Metastable Triplet State.* G. Weinreich and V. W. Hughes, Phys. Rev. **95**, 1451-1460 (1954).

11. *Hyperfine Structure of Helium-3 in the Metastable Triplet State.* W. B. Teutsch and V. W. Hughes, Phys. Rev. **95**, 1461-1463 (1954).

12. *Static Magnetic Field Quenching of the Orthopositronium Decay: Angular Distribution Effect.* V. W. Hughes, S. Marder and C. S. Wu, Phys. Rev. **98**, 1840-1848 (1955).

13. *Two-Quantum Transition in the Microwave Zeeman Spectrum of Atomic Oxygen.* V. W. Hughes and J. S. Geiger, Phys. Rev. **99**, 1842-1845 (1955).

14. *Effect of an Electric Field on Positronium Formation in Gases: Experimental.* S. Marder, V. W. Hughes, C. S. Wu and W. Bennett, Phys. Rev. **103**, 1258-1265 (1956).

15. *Effect of an Electric Field on Positronium Formation in Gases: Theoretical.* W. B. Teutsch and V. W. Hughes, Phys. Rev. **103**, 1266-1281 (1956).

16. *Experimental Limit for the Electron-Proton Charge Difference.* V. W. Hughes, Phys. Rev. **105**, 170-172 (1957).

17. *Electron g Value in the Ground State of Deuterium.* J. S. Geiger and V. W. Hughes, Phys. Rev. **105**, 183-188 (1957).

18. *Positronium Formation in Gases.* V. W. Hughes, J. Appl. Phys. **28**, 16-22 (1957).

19. *Hyperfine Structure of Positronium in Its Ground State.* V. W. Hughes, S. Marder and C. S. Wu, Phys. Rev. **106**, 934-947 (1957).

20. *Information Obtainable on Polarization of u^+ and Asymmetry of e^+ in Muonium Experiments.* G. Breit and V. W. Hughes, Phys. Rev. **106**, 1293-1295 (1957).

21. *Considerations of Depolarization of Positive Muons in Gases; Effect of Molecular Ions.* V. W. Hughes, Phys. Rev. **108**, 1106-1107 (1957).

22. *Magnetic Moment of Helium in Its $^3S_1$ Metastable State.* C. W. Drake, V. W. Hughes, A. Lurio and J. A. White, Phys. Rev. **112**, 1627-1637 (1958).

23. *Electron Magnetic Moment and Atomic Magnetism.* V. W. Hughes, **Recent Research in Molecular Beams**, ed. by I. Estermann, (Academic Press, Inc., NY, 1959) pp. 65-92.

24. *Microwave Zeeman Spectrum of Atomic Oxygen.* H. E. Radford and V. W. Hughes, Phys. Rev. **114**, 1274-1279 (1959).

25. *Atomic and Molecular Beams Spectroscopy.* P. Kusch and V. W. Hughes, **Handbuch der Physik, Vol. 37/1**, ed. by S. Flügge, (Springer-Verlag, Berlin, 1959).

26. *Considerations on the Design of a Molecular Frequency Standard Based on the Molecular Beam Electric Resonance Method.* V. W Hughes, Rev. Sci. Instr. **30**, 689-693 (1959).

27. *Hyperfine Structure of the Metastable Triplet State of Helium Three.* J. A. White, L. Y. Chow, C. Drake and V. W. Hughes, Phys. Rev. Lett. **3**, 428-429 (1959).

28. *Molecular Beam Electric Resonance Method with Separated Oscillating Fields.* J. C. Zorn, G. E. Chamberlain and V. W. Hughes, **Quantum Electronics**, ed. by C. H. Townes, (Columbia Univ. Press, NY, 1960) pp. 156-159.

29. *Narrow Linewidths for Decaying States by the Method of Separated Oscillating Fields.* V. W. Hughes, **Quantum Electronics**, ed. by C. H. Townes, (Columbia Univ. Press, NY, 1960) pp. 582-587.

30. *Upper Limit for the Anisotropy of Inertial Mass from Nuclear Resonance Experiments.* V. W. Hughes, H. G. Robinson and V. Beltran-Lopez, Phys. Rev. Lett. **4**, 342-344 (1960).

31. *Production and Detection of a Polarized Deuteron Beam Using the Atomic Beam Magnetic Resonance Method.* V. W. Hughes, C. W. Drake, Jr., D. C. Bonar, J. S. Greenberg, G. F. Pieper, **Proceedings of the International Symposium on Polarization Phenomena of Nucleons**, ed. by P. Huber and K. P. Meyer, (Birkhauser Verlag, Basel, 1960) pp. 89-107, 435.

32. *Formation of Muonium and Observation of Its Larmor Precession.* V. W. Hughes, D. W. McColm, K. Ziock and R. Prepost, Phys. Rev. Lett. **5**, 63-65 (1960).

33. *Atomic g_J Values for Neon and Argon in Their Metastable $^3P_2$ States; Evidence for Zero Spin of $_{10}Ne^{20}$.* A. Lurio, G. Weinreich, C. W. Drake, V. W. Hughes and J. A. White, Phys. Rev. **120**, 153-157 (1960).

34. *Mott-Scattering Analysis of Longitudinal Polarization of Electrons from Co^{60}.* J. S. Greenberg, D. P. Malone, R. L. Gluckstern and V. W. Hughes, Phys. Rev. **120**, 1393-1405 (1960).

35. *Experimental Limits for the Electron-Proton Charge Difference and for the Neutron Charge.* J. C. Zorn, G. E. Chamberlain and V. W. Hughes, **Proceedings of the 1960 Annual International Conference on High Energy Physics at Rochester**, ed. by E. C. G. Sudarshan, J. H. Tinlot, A. C. Melissionos, (University of Rochester, NY, 1960) pp. 790-792.

36. *Positronium* V. W. Hughes **Encyclopedia of Sciences and Technology** Vol. X, ed. by Daniel N. Lapides, (McGraw Hill, NY, 1960) pp. 524-525.

37. *Observation of the Hyperfine Structure Splitting of Muonium by Use of a Static Magnetic Field.* R. Prepost, V. W. Hughes and K. Ziock, Phys. Rev. Lett. **6**, 19-21 (1961).

38. *Production and Detection of an Accelerated Beam of Completely Polarized Deuterons.* C. W. Drake, D. C. Bonar, R. D. Headrich and V. W. Hughes, Rev. Sci. Instr. **32**, 995-996 (1961).

39. *Microwave Zeeman Spectrum of Atomic Fluorine.* H. E. Radford, V. W. Hughes and V. Beltran-Lopez, Phys. Rev. **123**, 153-160 (1961).

40. *Atomic Processes Involving Muonium and Anti-Muonium.* V. W. Hughes, D. McColm, K. Ziock and R. Prepost, **2nd International Conference on the Physics of Electronic and Atomic Collisions**, (W. A. Benjamin, Inc., NY, 1961) pp. 166-169.

41. *Theoretical Values for Magnetic Moments of Mu-Mesonic Atoms.* K. W. Ford, V. W. Hughes and J. G. Wills, Phys. Rev. Lett. **7**, 134-135 (1961).

42. *A Polarized Ion Source Using the Atomic-Beam Magnetic Resonance-Method.* C. W. Drake, D. C. Bonar, R. D. Headrick and V. W. Hughes, **International Conference on High Energy Accelerators**, ed. by M. H. Blewett, (Division of Technical Information, U.S. Atomic Energy Commission, Brookhaven National Laboratory, NY, 1961) pp. 379-384.

43. *Hyperfine Structure of Muonium.* K. Ziock, V. W. Hughes, R. Prepost, J. Bailey and W. Cleland, Phys. Rev. Lett. **8**, 103-105 (1962).

44. *Hyperfine Structure of Muonium.* J. Bailey, W. Cleland, V. W. Hughes, R. Prepost and K. Ziock, **International Conference on High Energy Physics at CERN**, ed. by J. Prentki, (CERN, Geneva, 1962) pp. 473-476.

45. *Muon Resonance.* V. W. Hughes, **International Conference of Paramagnetic Resonances**, ed. by W. Low, (Academic Press, NY, 1963) pp. 382-396.

46. *Theoretical Values for Magnetic Moments of Mu-Mesonic Atoms.* K. W. Ford, V. W. Hughes and J. G. Wills, Phys. Rev. **129**, 194-201 (1963).

47. *Experimental Limits for the Electron-Proton Charge Difference and for the Charge of the Neutron.* J. C. Zorn, G. E. Chamberlain and V. W. Hughes, Phys. Rev. **129**, 2566-2576 (1963).

48. *A Very High Intensity Proton Linear Accelerator as a Meson Factory.* E. R. Beringer, W. A. Blanpied, R. L. Gluckstern, V. W. Hughes, H. B. Knowles, S. Ohnuma and G. W. Wheeler, **International Conference on Sector-Focused Cyclotrons and Meson Factories**, ed. by F. T¿ Howard and N. Vogt-Nilsen, (CERN, Geneva, 1963) pp. 365-371.

49. *Photoproduction of Negative and Positive Pions from Carbon at Forward Angles.* W. A. Blanpied, J. S. Greenberg, V. W. Hughes, D. C. Lu and R. C. Minehart, Phys. Rev. Lett. **11**, 477-479 (1963).

50. *Sources of Polarized Electrons.* V. W. Hughes, **Proceedings of the Conference on Photon Interactions in the BeV-Energy Range**, ed. by B. T. Feld, (M. I. T. Cambridge, Mass., 1963) pp. VI. 13-15.

51. *Electromagnetic Pair Production.* V. W. Hughes, **Proceedings of the Conference on Photon Interactions in the BeV-Energy Range**, ed. by B. T. Feld, (M. I. T. Press, Cambridge, Mass., 1963) pp. VIII. 1-7.

52. *Atomic Beam Source of Polarized Electrons for High Energy Accelerators.* V. W. Hughes, R. L. Long and W. Raith, **Proceedings of the International Conference on High Energy Accelerators**, ed. by A. A. Kolomensky, A. B. Kusnetsov, and A. N. Lebedev, Atomizdat, Moscow 1964) pp. 988-992.

53. *New Value for the Fine-Structure Constant α from Muonium Hyperfine Structure Interval.* W. E. Cleland, J. M. Bailey, M. Eckhause, V. W. Hughes, R. M. Mobley, R. Prepost and J. E. Rothberg, Phys. Rev. Lett. **13**, 202-205 (1964).

54. *Muonium and Positronium Physics.* J. M. Bailey and V. W. Hughes, **Proceedings of the Third International Conference on the Physics of Electronic and Atomic Collisions**, ed. by M. R. C. McDowell (North-Holland Publishing Co., Amsterdam, 1964) pp. 839-846.

55. *The Lyttleton-Bondi Universe and Charge Equality.* V. W. Hughes, **Gravitation and Relativity**, ed. by H. -Y. Chiu and W. F. Hoffmann, (W. A. Benjamin, Inc., N.Y., 1964) pp. 259-278.

56. *Mach's Principle and Experiments on Mass Anisotropy.* V. W. Hughes, **Gravitation and Relativity**, ed. by H. -Y. Chiu and W. F. Hoffmann, (W. A. Benjamin, Inc., N.Y., 1964) pp. 106-120.

57. *Status of Knowledge of the Fine-Structure Constant, Particularly as It Relates to Proton Structure.* V. W. Hughes, **Proceedings of the International Conference on Nuclear Structure**, ed. by R. Hofstadter and L. Schiff, (Stanford University Press, California, 1964) pp. 235-244.

58. *Evidence for the Photoproduction of the $Y = 0$ States with Masses Greater Than 1900 MeV.* W. A. Blanpied, J. S. Greenberg, V. W. Hughes, P. Kitching, D. C. Lu, and R. C. Minehart, Phys. Rev. Lett. **14**, 741-744 (1965).

59. *Polarized Electrons from a Polarized Atomic Beam.* W. Raith, R. L. Long, Jr., V. W. Hughes and M. Posner, **Proceedings of the IVth International Conference on the Physics of Electronic and Atomic Collisions**, (Science Bookcrafters, Inc., N.Y. 1965) pp. 256-260.

60. *Polarized Electrons from a Polarized Atomic Beam.* R. L. Long, Jr., W. Raith and V. W. Hughes, Phys. Rev. Lett. **15**, 1-4 (1965).

61. *Atomic Interactions of Muonium.* R. M. Mobley, J. M. Bailey, W. E. Cleland, V. W. Hughes and J. E. Rothberg, **Proceedings of the IVth International Conference on the Physics of Electronic and Atomic Collisions**, (Science Bookcrafters, Inc., N.Y. 1965) pp. 194-197.

62. *Polarization of Photoelectrons from Magnetized Nickel.* R. L. Long, Jr., V. W. Hughes, J. S. Greenberg, I. Ames and R. L. Christensen, Phys. Rev. **138**, A1630-A1635 (1965).

63. *Production of Stopped Pions and Muons from a Multi-BeV Proton Synchrotron.* V. W. Hughes and R. D. Edge, IEEE Trans. Nuclear Sci. **NS-12**, 943-948 (1965).

64. *Removal of the RF Microstructure from a Proton Linear Accelerator Beam.* R. D. Edge, V. W. Hughes and J. Sandweiss, IEEE Trans. Nuclear Sci. **NS-12**, 949-953 (1965).

65. *A Polarized Electron Source for High Energy Accelerators.* V. W. Hughes, R. L. Long, Jr., M. Posner and W. Raith, **Proceedings of the International Symposium on Electron and Photon Interactions at High Energies**, Vol. II, ed. by G. Höhler, G. Kramer and U. Meyer-Berkhout, (Springer- Verlag, Berlin, 1966) pp. 440-444.

66. *The Ratio of the Cross Sections for Photoproduction of Asymmetric Muon and Electron Pairs in Hydrogen and Carbon.* V. W. Hughes, W. A. Blanpied, J. S. Greenberg, P. Kitching, D. C. Lu and R. C. Minehart, **Proceedings of the International Symposium on Electron and Photon Interactions at High Energies**, Vol. II, ed. by G. Höhler, G. Kramer and U. Meyer-Berkhout, (Spring-Verlag, Berlin, 1966) pp. 361-368.

67. *The Photoproduction of Charged Pions in Hydrogen and Carbon for Photon Energies up to 6 BeV.* W. A. Blanpied, J. S. Greenberg, V. W. Hughes, P. Kitching, D. C. Lu and R. C. Minehart, **Proceedings of the International Symposium on Electron and Photon Interactions at High Energies**, Vol. II, ed. by G. Höhler, G. Kramer and U. Meyer-Berkhout, (Springer- Verlag, Berlin, 1966) pp. 185-192.

68. *The Photoproduction of Charged K Mesons and Evidence for New High Mass Hyperons.* J. S. Greenberg, W. A. Blanpied, V. W. Hughes, P. Kitching, D. C. Lu and R. C. Minehart, **Proceedings of the International Symposium on Electron and Photon Interactions at High Energies**, Vol. II, ed. by G. Höhler, G. Kramer and U. Meyer-Berkhout, (Springer-Verlag, Berlin, 1966) pp. 192-200.

69. *Parity Conservation in Strong Interactions.* C. W. Drake, D. C. Bonar, R. D. Headrick and V. W. Hughes, **Proceedings of the 2nd International Symposium on Polarization Phenomena of Nucleons**, ed. by P. Huber and H. Schopper, (Birkhäuser Verlag, Basel and Stuttgart, 1966) pp. 362-364.

70. *Muonium Chemistry.* R. M. Mobley, J. M. Bailey, W. E. Cleland, V. W. Hughes and J. E. Rothberg, J. Chem. Phys. **44**, 4354-4355 (1966).

71. *Recent Experiments on Muonium.* V. W. Hughes, J. Amato, R. Mobley, J. Rothberg and P. Thompson, **Proceedings of the Williamsburg Conference on Intermediate Energy Physics**, ed. by H. O. Funsten, (College of William and Mary, Virginia, 1966) pp. 377-409.

72. *The Muonium Atom.* V. W. Hughes, Scientific American **214**, 93-100 (1966).

73. *Muonium.* V. W. Hughes, Ann. Rev. Nucl. Sci. **16**, 445-470 (1966).

74. *Physics with Polarized Particles.* V. W. Hughes, **V International Conference on High Energy Accelerators**, ed. by M. Grilli, (Comitato Nazionale Per L'Energia Nucleare, Rome, 1966) pp. 531-545.

75. *Atomic Beam Study of the 2^3P State of Helium.* F. M. J. Pichanick, C. E. Johnson, R. D. Swift and V. W. Hughes, **Abstracts of the Conference on The Physics of Free Atoms**, ed. by V. W. Cohen, (University of California, Berkeley, 1966) pp. 68-80.

76. *Precision Redetermination of the Hyperfine Structure Interval of Positronium.* E. D. Theriot, Jr., R. H. Beers and V. W. Hughes, Phys. Rev. Lett. **18**, 767-769 (1967).

77. *Search for $S = +1$ Baryon States in Photoproduction.* J. Tyson, J. S. Greenberg, V. W. Hughes, D. C. Lu, R. C. Minehart, S. Mori and J. E. Rothberg, Phys. Rev. Lett. **19**, 255-259 (1967).

78. *Hyperfine Structure of the v=0, J=1 State in $Rb^{85}F$, $Rb^{87}F$, $K^{39}F$, and $K^{41}F$ by the Molecular-Beam Electric-Resonance Method.* P. A. Bonczyk and V. W. Hughes, Phys. Rev. **161**, 15-22 (1967).

79. *Detection of Positrons and of Positronium.* V. W. Hughes, **Methods of Experimental Physics**, Vol. 4A and 4B, ed. by V. W. Hughes and H. L. Schultz (Academic Press, N.Y., 1967) pp. 389, Vol. 4A.

80. *Muonium Chemistry II.* R. M. Mobley, J. J. Amato, V. W. Hughes, J. E. Rothberg and P. A. Thompson, J. Chem. Phys. **47**, 3074-3075 (1967).

81. *Muonium.* V. W. Hughes, Phys. Today **20**, 29-40 (1967).

82. *A Pulsed Source of Highly Polarized Electrons.* V. W. Hughes, M. S. Lubell, M. Posner and W. Raith, **Proceedings of the Sixth International Conference on High Energy Accelerators**, ed. by R. A. Mack, (Cambridge Electron Accelerator, Cambridge, 1967) pp. A-144-A-147.

83. *Production of Polarized Electrons by Pulsed Photoionization of a Polarized Atomic Beam of Lithium-6.* V. W. Hughes, M. S. Lubell, M. Posner and W. Raith, **Fifth International Conference on the Physics of Electronic and Atomic Collisions**, ed. by I. P. Flaks (NAUKA, Leningrad, 1967) pp. 544-545.

84. *Ratio of Λ and Σ^0 Photoproduction Cross Sections; High-Mass Hyperon Resonances.* J. S. Greenberg, V. W. Hughes, D. C. Lu, R. C. Minehart, S. Mori, J. E. Rothberg and J. Tyson, Phys. Rev. Lett. **20**, 221-223 (1968).

85. *Parity Conservation in the Reaction T(d,n) He^4.* D. C. Bonar, C. W. Drake, R. D. Headrick and V. W. Hughes, Phys. Rev. **174**, 1200-1207 (1968).

86. *Experiments on the 2^3P State of Helium I. A Measurement of the $2^3P_1-2^3P_2$ Fine Structure.* F. M. J. Pichanick, R. D. Swift, C. E. Johnson and V. W. Hughes, Phys. Rev. **169**, 55-78 (1968).

87. *Further Search for S=+1 Baryon States in Photoproduction.* S. Mori, J. S. Greenberg, V. W. Hughes, D. C. Lu, J. E. Rothberg and P. A. Thompson, Phys. Lett. **28B**, 152-154 (1968).

88. *Search for Muonium-Antimuonium Conversion.* J. J. Amoto, P. Crane, V. W. Hughes, J. E. Rothberg and P. A. Thompson, Phys. Rev. Lett. **21**, 1709-1712 (1968).

89. *Muonium.* V. W. Hughes, **Vistas in Science**, ed. by David L. Arm (Univ. of New Mexico Press, Albuquerque, 1968) pp. 237-256.

90. *A Fine Structure Constant α.* V. W. Hughes, **A Tribute to I. I. Rabi**, (Columbia University Symposium 1967).

91. *Determination of Muonium Hyperfine Structure Interval Through Measurements at Low Magnetic Fields.* P. A. Thompson, J. J. Amato, P. Crane, V. W. Hughes, R. M. Mobley, G. zu Putlitz and J. E. Rothberg, Phys. Rev. Lett. **22**, 163-167 (1969).

92. *Atoms.* V. W. Hughes, Phys. Today **22**, 33-37 (1969).

93. *The Fine-Structure Constant.* V. W. Hughes, Comments on Atomic and Molecular Physics **1**, 5-11 (1969).

94. *Tests of Quantum Electrodynamics from Radiofrequency Studies of Atoms.* V. W. Hughes, **Magnetic Resonance and Radiofrequency Spectroscopy: Proceedings of the XVth. Colloque. A. M. P. E. R. E.**, ed. by P. Averbuch (North-Holland Publ. Co., Amsterdam, 1969) pp. 1-22.

95. *Quantum Electrodynamics: Experiment.* V. W. Hughes, **Atomic Physics**, ed. by V. W. Hughes, B. Bederson, V. W. Cohen and F. M. J. Pichanick, (Plenum Press, NY, 1969) pp. 15-51.

96. *Muon g-2 Experiment at LAMPF.* V. W. Hughes, **Some Physics Uses at LAMPF 1968 Summer Study Group at Los Alamos**, LA-4080 Rev., (1969) pp. 2-4.

97. *Experimental Test for Mass Anisotropy Based on Nuclear Magnetic Resonance.* V. W. Hughes and W. L. Williams, **Gravity Research Foundation** (Essay Award First Prize, 1969).

98. *Elastic Scattering of Positive Kaons by Polarized Protons at 1.54 and 1.71 GeV/c.* G. A. Rebka, Jr., J. Rothberg, A. Etkin, P. Glodis, J. Greenberg V. W. Hughes, K. Kondo, D. C. Lu, S. Mori and P. A. Thompson, **Proceedings of the Boulder Conference on High Energy Physics**, ed. by K. T. Mahanthappa, W. D. Walker and W. E. Brittin, (Colorado Associated University Press, Boulder, 1969) pp. 531-532.

99. *Magnetic Moment and hfs Anomaly for* He^3. W. L. Williams and V. W. Hughes, Phys. Rev. **185**, 1251-1255 (1969).

100. *Search for S=+1 Baryon States in Photoproduction.* S. Mori, J. S. Greenberg, V. W. Hughes, D. C. Lu, R. C. Minehart, J. E. Rothberg, P. A. Thompson and J. Tyson, Phys. Rev. **185**, 1687-1701 (1969).

101. *Muonium and Positronium.* V. W. Hughes, **Physics of the One-and Two-Electron Atoms**, ed. by F. Bopp and H. Kleinpoppen, (North-Holland Publishing Co., Amsterdam, 1969) pp. 407-428.

102. *Atom-Antiatom Collision.* D. L. Morgan, Jr. and V. W. Hughes, **Sixth International Conference on the Physics of Electronic and Atomic Collisions**, ed. by I. Amdue, (Massachusetts Institute of Technology Press, Cambridge, 1969) pp. 830-834.

103. *Recent Muonium Hyperfine Structure Measurements.* P. Crane, J. J. Amato, V. W. Hughes, D. M. Lazarus, G. zu Putlitz and P. A. Thompson, **High-Energy Physics and Nuclear Structure**, ed. by Samuel Devons, (Plenum Press, NY, 1970) pp. 677-679.

104. *Asymmetry and Differential Cross Section for Elastic Scattering of* K^+ *Mesons by Polarized Protons at 1.54 and 1.71 GeV/c.* G. A. Rebka, Jr., J. Rothberg, A. Etkin, P. Glodis, J. Greenberg, V. W. Hughes, K. Kondo, D. C. Lu, S. Mori and P. A. Thompson, Phys. Rev. Lett. **24**, 160-164 (1970).

105. *Muonium I: Muonium Formation and Larmor Precession.* V. W. Hughes, D. W. McColm, K. Ziock and R. Prepost, Phys. Rev. **A1**, 595-553 (1970); **A2**, 551-553 (1970).

106. *Precision Redetermination of the Fine-Structure Interval of the Ground State of Positronium and a Direct Measurement of the Decay Rate of Parapositronium.* E. D. Theriot, Jr., R. H. Beers, V. W. Hughes and K. O. H. Ziock, Phys. Rev. **A2**, 707-721 (1970).

107. *Photoproducton of* Y^* *Resonances above 1800 MeV.* D. C. Lu, J. S. Greenberg, V. W. Hughes, R. C. Minehart, S. Mori, J. E. Rothberg and J. Tyson, Phys. Rev. **D2**, 1846-1851 (1970).

108. *Experiments on the 2^3P State of Helium II. Measurements of the Zeeman Effect.* S. A. Lewis, F. M. J. Pichanick and V. W. Hughes, Phys. Rev. **A2**, 86-101 (1970).

109. *Atomic Processes Involved in Matter-Antimatter Annihilation.* D. L. Morgan, Jr. and V. W. Hughes, Phys. Rev. **D2**, 1389-1399 (1970).

110. *Search for Strangeness S=+1 Baryon States.* V. W. Hughes, R. D. Ehrlich, A. Etkin, P. Glodis, K. Kondo, D. C. Lu, S. Mori, R. Patton, G. A. Rebka, Jr., J. E. Rothberg, P. A. Thompson and M. E. Zeller, **Hyperon Resonances-70**, ed. by E. C. Fowler (Moore Publishing Co., Durham, North Carolina, 1970) pp. 349-366.

111. *Determination of the Fine Structure Constant α from Helium Fine Structure.* A. Kponou, V. W. Hughes, C. E. Johnson, S. A. Lewis and F. M. J. Pichanick, **Proceedings of the International Conference on Precision Measurements and Fundamental Constants**, ed. by D. N. Langenberg and B. N. Taylor (NBS Special Publication 343, Washington, D. C. , 1971) pp. 389-391.

112. *Precision Measurement of the Fine Structure Interval of the Ground State of Positronium.* E. R. Carlson, V. W. Hughes and E. D. Theriot, Jr., **Proceedings of the International Conference on Precision Measurements and Fundamental Constants**, ed. by D. N. Langenberg and B. N. Taylor (NBS Special Publication 343, Washington, D. C. , 1971) pp. 313-316.

113. *Hyperfine Structure Interval of the Ground State of Muonium.* P. A. Thompson, D. Casperson, P. Crane, T. Crane, P. Egan, V. W. Hughes, G. zu Putlitz and R. Stambaugh, **Proceedings of the International Conference on Precision Measurements and Fundamental Constants,** ed. by D. N. Langenberg and B. N. Taylor (NBS Special Publication 343, Washington, D. C , 1971) pp. 339-343.

114. *The Breit Interaction.* V. W. Hughes, **Facets of Physics** ed. by D. A. Bromley and V. W. Hughes, (Academic Press, NY, 1970) pp. 125-140.

115. *Muonium II. Observation of Muonium Hyperfine-Structure Interval.* J. M. Bailey, W. E. Cleland, V. W. Hughes, R. Prepost and K. Ziock, Phys. Rev. **A3**, 871-884 (1971).

116. *Asymmetry Measurements for Elastic Scattering of K^+ Mesons by Polarized Protons.* R. D. Ehrlich, A. Etkin, P. Glodis, V. W. Hughes, K. Kondo, D. C. Lu, S. Mori, R. Patton, G. A. Rebka, Jr., J. E. Rothberg, P. Thompson and M. E. Zeller, Phys. Rev. Lett. **26**, 925-928 (1971).

117. *Precise Measurement of the $2^3P_0-2^3P_1$ Fine-Structure Interval of Helium.* A. Kponou, V. W. Hughes, C. E. Johnson, S. A. Lewis and F. M. J. Pichanick, Phys. Rev. Lett. **26**, 1613-1616 (1971).

118. *Stopped Muon Channel for LAMPF.* V. W. Hughes, S Ohnuma, K. Tanabe, P. Thompson and H. F. Vogel, Los Alamos Scientific Laboratory, LA-4474-MS (February, 1971), pp. 1-38.

119. *Observation of a Quadratic Term in the hfs Pressure Shift for Muonium and a New Precise Value for Muonium $\Delta\nu$.* T. Crane, D. Casperson, P. Crane, P. Egan, V. W. Hughes, R. Stambaugh, P. A. Thompson and G. zu Putlitz, Phys. Rev. Lett. **27**, 474-476 (1971).

120. *Polarized Electrons and Some of Their Uses.* V. W. Hughes, **IInd International Conference on Polarized Targets**, ed. by G. Shapiro, (University of California, Lawrence Berkeley Laboratory, 1971), pp. 191-204.

121. *Muonium Chemistry in Gases.* V. W. Hughes, **The Meeting on Muons in Solid State Physics**, (Schweizerisches Institute für Nuckearforschung (SIN) Zürich, 1971) pp. 129-134.

122. *Polarized Electrons from Photoionization of Polarized Alkali Atoms.* V. W. Hughes, R. L. Long, Jr., M. S. Lubell, M. Posner and W. Raith, Phys. Rev. **A5**, 195-222 (1972).

123. *Measurements of the Asymmetries in the Differential Cross Sections for $\bar{p}p \to \bar{p}p$ and $\bar{p}p \to \pi^-\pi^+$ Using Polarized Protons.* R. D. Ehrlich, A. Etkin, P. Glodis, V. W. Hughes, K. Kondo, D. C. Lu, S. Mori, R. Patton, G. A. Rebka, Jr., P. A. Thompson and M. E. Zeller, Phys. Rev. Lett. **28**, 1147-1150 (1972).

124. *Muonium III. Precision Measurement of the Muonium Hyperfine-Structure Interval at Strong Magnetic Field.* W. E. Cleland, J. M. Bailey, M. Eckhause, V. W. Hughes, R. Prepost, J. E. Rothberg and R. M. Mobley, Phys. Rev. **A5**, 2338-2356 (1972).

125. *Higher-Precision Determination of the Fine-Structure Interval in the Ground State of Positronium, and the Fine-Structure Density Shift in Nitrogen.* E. R. Carlson, V. W. Hughes, M. L. Lewis and I. Lindgren, Phys. Rev. Lett. **29**, 1059-1061 (1972).

126. *Status of QED Experiments.* V. W. Hughes, **Third International Conference on Atomic Physics**, ed. by S. J. Smith and G. K. Walters, (Plenum Press, NY, 1973) pp. 1-32.

127. *Observation of a Positronium Zeeman Transition in* $\gamma-Al_2O_3$. D. J. Judd, Y. K. Lee, L. Madansky, E. R. Carlson, V. W. Hughes and B. Zundell, Phys. Rev. Lett. **30**, 202-204 (1973).

128. *Atom-Antiatom Interactions.* D. L. Morgan, Jr. and V. W. Hughes, Phys. Rev. **A7**, 1811-1825 (1973).

129. *The Third International Conference on Atomic Physics.* V. W. Hughes, Comments on Atomic and Molecular Physics, **Vol. IV, Number 2**, 35-41 (1973).

130. *Higher-Order Relativistic Contributions to the Combined Zeeman and Motional Stark Effects in Positronium.* M. L. Lewis and V. W. Hughes, Phys. Rev. **A8**, 625-639 (1973).

131. *Muonium IV. Precision Measurement of the Muonium Hyperfine-Structure Interval at Weak and Very Weak Magnetic Fields.* P. A. Thompson, P. Crane, T. Crane, J. J. Amato, V. W. Hughes, G. zu Putlitz and J. E. Rothberg, Phys. Rev. **A8**, 86-112 (1973).

132. *Higher-Order Relativistic Contributions to the Zeeman Effect in Helium.* M. L. Lewis and V. W. Hughes, Phys. Rev. **A8**, 2845-2856 (1973).

133. *Positronium and Muonium.* V. W. Hughes, **Physik 1973, German Physical Society Conf.**, (Physik Verlag Hmblt, Germany, 1973), pp. 123-155.

134. *New Experimental Limit on T Invariance in Polarized-Neutron β Decay.* R. I. Steinberg, P. Liaud, B. Vignon and V. W. Hughes, Phys. Rev. Lett. **33**, 41-44 (1974).

135. *Remarks Relevant to Fine Structure Measurements of Muonic Hydrogen and Muonic Helium.* K. N. Kuang, V. W. Hughes, M. L. Lewis, R. O. Mueller, H. Rosenthal, C. S. Wu and M. Camani, **High-Energy Physics and Nuclear Structure**, ed. by B. Tibell, (North-Holland Publishing Co., 1974) pp. 312-314.

136. *Atomic Regime in Which the Magnetic Interaction Dominates the Coulomb Interaction for Highly Excited States of Hydrogen.* R. O. Mueller and V. W. Hughes, Proceedings of the National Academy of Science **71**, 3287-3289 (1974).

137. *Muonium Formation in Noble Gases and Noble-Gas Mixtures.* R. D. Stambaugh, D. E. Casperson, T. W. Crane, V. W. Hughes, H. F. Kaspar, P. A. Souder, P. A. Thompson, H. Orth, G. zu Putlitz and A. B. Denison, Phys. Rev. Lett. **33**, 568-571 (1974).

138. *Behavior of Positive Muons in Liquid Helium.* T. W. Crane, D. E. Casperson, H. Chang, V. W. Hughes, H. F. Kaspar, B. Lovett, A. Schiz, P. A. Souder, R. D. Stambaugh, G. zu Putlitz and J. P. Kane, Phys. Rev. Lett. **33**, 572-574 (1974).

139. *Evidence for Formation of the First Excited State of Positronium.* S. L. Varghese, E. S. Ensberg, V. W. Hughes and I. Lindgren, Phys. Lett. **49A**, 415-417 (1974).

140. *Polarized Electron Source for the Stanford Linear Accelerator.* M. J. Alguard, R. D. Ehrlich, V. W. Hughes, J. Ladish, M. S. Lubell, W. Lysenko, K. P. Schüler, G. Baum and W. Raith, **Proceedings of the IXth International Conference on High Energy Accelerators**, (Stanford Linear Accelerator Center, CONF-740522, 1974) pp. 309-313.

141. *Higher-Order Relativistic Contributions to the Zeeman Effect in Helium and Helium-Like Ions.* M. L. Lewis and V. W. Hughes, Phys. Rev. **A11**, 383-384 (1975).

142. *Collision Quenching of the Metastable 2S State of Muonic Hydrogen and the Muonic Helium Ion.* R. O. Mueller, V. W. Hughes, H. Rosenthal and C. S. Wu, Phys. Rev. **A11**, 1175-1186 (1975).

143. *Formation of the Muonic Helium Atom, $\alpha\mu^-e^-$, and Observation of Its Larmor Precession.* P. A. Souder, D. E. Casperson, T. W. Crane, V. W. Hughes, D. C. Lu, H. Orth, H. W. Reist, M. H. Yam and G. zu Putlitz, Phys. Rev. Lett. **34**, 1417-1420 (1975).

144. *Polarized Electron-Electron Scattering at GeV Energies.* P. S. Cooper, M. J. Alguard, R. D. Ehrlich, V. W. Hughes, H. Kobayakawa, J. S. Ladish, M. S. Lubell, N. Sasao, K. P. Schüler, P. A. Souder, G. Baum, W. Raith, K. Kondo, D. H. Coward, R. H. Miller, C. Y. Prescott, D. J. Sherden and C. K. Sinclair, Phys. Rev. Lett. **34**, 1589-1592 (1975).

145. *Beam Calculations for LAMPF Muon Channel.* W. P. Lysenko, V. W. Hughes, S. Ohnuma, P. A. Thompson and H. F. Vogel, IEEE Transactions on Nuclear Science, **Vol. NS-22**, 1593-1597 (1975).

146. *Muons: Muonic Atoms and Muonium.* V. W. Hughes, **High-Energy Physics and Nuclear Structure - 1975**, ed. by D. E. Nagle, R. L. Burman, B. G. Storms, A. S. Goldhaber and C. K. Hargrove, (AIP Conf. Proceedings No. 26, Santa Fe and Los Alamos, 1975), pp. 515-539.

147. *Positronium: Precision Determination of the Ground-State Fine-Structure Interval $\Delta\nu$ and Measurement of Density Shifts in the Noble Gases.* P. O.

Egan, W. E. Frieze, V. W. Hughes and M. H. Yam, Phys. Lett. **54A**, 412-414 (1975).

148. *A New High Precision Measurement of the Muonium Hyperfine Structure Interval* $\Delta\nu$. D. E. Casperson, T. W. Crane, V. W. Hughes, P. A. Souder, R. D. Stambaugh, P. A. Thompson, H. Orth, G. zu Putlitz, H. F. Kaspar, H. W. Reist and A. B. Denison, Phys. Lett. **59B**, 397-400 (1975).

149. **Muon Physics** Vol. II and III, ed. by V. W. Hughes and C. S. Wu, (Academic Press, NY, 1975).

150. K^- *Mass from Kaonic Atoms.* S. C. Cheng, Y. Asano, M. Y. Chen, G. Dugan, E. Hu, L. Lidofsky, W. Patton, C. S. Wu, V. W. Hughes and D. C. Lu, Nucl. Phys. **A254**, 381-395 (1975).

151. *Mass and Magnetic Moment of* Σ^- *by the Exotic Atom Method.* G. Dugan, Y. Asano M. Y. Chen, S. C. Cheng, E. Hu, L. Lidofsky, W. Patton, C. S. Wu, V. W. Hughes and D. C. Lu, Nucl. Phys. **A254**, 396-402 (1975).

152. *Mass and Magnetic Moment of the Antiproton by the Exotic Atom Method.* E. Hu, Y. Asano, M. Y. Chen, S. C. Cheng, G. Dugan, L. Lidofsky, W. Patton, C. W. Wu, V. W. Hughes and D. C. Lu, Nucl. Phys. **A254**, 403-412 (1975).

153. *E2 Dynamic Mixing in* $\bar{p}$ *and* K^- *Atoms of* ^{238}U. M. Y. Chen, Y. Asano, S. C. Cheng, G.Dugan, E. Hu, L. Lidofsky, W. Patton, C. S. Wu, V. W. Hughes and D. C. Lu, Nucl. Phys. **A254**, 413-421 (1975).

154. *Search for a Nonzero Triple-Correlation Coefficient and New Experimental Limit on T Invariance in Polarized-Neutron Beta Decay.* R. I. Steinberg, P. Liaud, B. Vignon and V. W. Hughes, Phys. Rev. **D13**, 2469-2477 (1976).

155. *The Fine Structure Constant from Helium Fine Structure.* M. L. Lewis, P. H. Serafino and V. W. Hughes, Phys. Lett. **A58**, 125-126 (1976).

156. *Elastic Scattering of Polarized Electrons by Polarized Protons.* M. J. Alguard, W. W. Ash, G. Baum, J. E. Clendenin, P. S. Cooper, D. H. Coward, R. D. Ehrlich, A. Etkin, V. W. Hughes, H. Kobayakawa, K. Kondo, M. S. Lubell, R. H. Miller, D. A. Palmer, W. Raith, N. Sasao, K. P. Schüler, D. J. Sherden, C. K. Sinclair and P. A. Souder, Phys. Rev. Lett. **37**, 1258-1261 (1976).

157. *Deep Inelastic Scattering of Polarized Electrons by Polarized Protons.* M. J. Alguard, W. W. Ash, G. Baum, J. E. Clendenin, P. S. Cooper, D. H. Coward, R. D. Ehrlich, A. Etkin, V. W. Hughes, H. Kobayakawa, K. Kondo, M. S. Lubell, R. H. Miller, D. A. Palmer, W. Raith, N. Sasao, K. P. Schüler, D. J. Sherden, C. K. Sinclair and P. A. Souder, Phys. Rev. Lett. **37**, 1261-1265 (1976).

158. *Precise Measurement of Electronic g_J Value of Helium, $g_J(^4He, 2^3S_1)$.* B. E. Zundell and V. W. Hughes, Phys. Lett. **59A**, 381-382 (1976).

159. *Precision Determination of the Fine-Structure Interval in the Ground State of Positronium III.* E. R. Carlson, V. W. Hughes and I. Lindgren, Phys. Rev. **A15**, 241-250 (1977).

160. *Precision Determination of the Fine-Structure Interval in the Ground State of Positronium. IV. Measurement of Positronium Fine-Structure Density Shifts in Noble Gases.* P. O. Egan, V. W. Hughes and M. H. Yam, Phys. Rev. **A15**, 251-260 (1977).

161. *New Precise Value for the Muon Magnetic Moment and Sensitive Test of the Theory of the hfs Interval in Muonium.* D. E. Casperson, T. W. Crane, A. B. Denison, P. O. Egan, V. W. Hughes, F. G. Mariam, H. Orth, H. W. Reist, P. A. Souder, R. D. Stambaugh, P. A. Thompson and G. zu Putlitz, Phys. Rev. Lett. **38**, 956-959 (1977); **38**, 1504 (1977).

162. *Parity Nonconservation in Hydrogen Involving Magnetic/Electric Resonance.* E. A. Hinds and V. W. Hughes, Phys. Lett. **67B**, 487-488 (1977).

163. *Operating Experience with the Polarized Electron Gun at SLAC.* M. J. Alguard, G. Baum, J. E. Clendenin, V. W. Hughes, M. S. Lubell, R. H. Miller, W. Raith, K. P. Schüler and J. Sodja, IEEE Transactions on Nuclear Sciences, **NS-24**, 1603-1604 (1977).

164. *Depolarization Effects in Pulsed Photoionization of State-Selected Lithium.* M. J. Alguard, J. E. Clendenin, P. S. Cooper, R. D. Ehrlich, V. W. Hughes and M. S. Lubell, Phys. Rev. **A16**, 209-212 (1977).

165. *Measurement of Spin-Exchange Effects in Electron-Hydrogen Collisions: Impact Ionization.* M. J. Alguard, V. W. Hughes, M. S. Lubell and P. F. Wainwright, Phys. Rev. Lett. **39**, 334-338 (1977).

166. *The Fine Structure Constant α.* V. W. Hughes, **A Festschrift for I. I. Rabi**, ed. by Lloyd Motz, (The New York Academy of Sciences Series II, Volume 38, 1977) pp. 62-76.

167. *Introduction and History.* C. S. Wu and V. W. Hughes, **Muon Physics** Vol. I, ed. by V. W. Hughes and C. S. Wu, (Academic Press, NY, 1977) pp. 2-8; *Electromagnetic Properties and Interactions of Muons.* V. W. Hughes and T. Kinoshita, **Muon Physics** Vol. I, ed. by V. W. Hughes and C. S. Wu, (Academic Press, NY, 1977) pp. 11-199.

168. *Deep Inelastic e-p Asymmetry Measurements and Comparison with the Bjorken Sum Rule and Models of Proton Spin Structure.* M. J. Alguard, W. W. Ash, G. Baum, M. R. Bergström, J. E. Clendenin, P. S. Cooper, D. H. Coward, R. D. Ehrlich, V. W. Hughes, K. Kondo, M. S. Lubell, R. H. Miller, S. Miyashita, D. A. Palmer, W. Raith, N. Sasao, K. P. Schüler, D. J. Sherden, P. A. Souder and M. E. Zeller, Phys. Rev. Lett. **41**, 70-73 (1978).

169. *Parity Non-Conservation in Inelastic Electron Scattering.* C. Y. Prescott, W. B. Atwood, R. L. A. Cottell, H. DeStaebler, E. L. Garwin, A. Gonidec, R. H. Miller, L. S. Rochestere, T. Sato, D. J. Sherden, C. K. Sinclair, S. Stein, R. E. Taylor, J. E. Clendenin, V. W. Hughes, N. Sasao, K. P. Schüler, M. G. Borghini, K. Lubelsmeyer and W. Jentschke, Phys. Lett. **77B**, 347-352 (1978).

170. *Development of a Low-Momentum "Surface" Muon Beam for LAMPF.* H. -W. Reist, D. E. Casperson, A. B. Denison, P. O. Egan, V. W. Hughes, F. G. Mariam, G. zu Putlitz, P. A. Souder, P. A. Thompson and J. Vetter, Nucl. Inst. and Meth. **153**, 61-64 (1978).

171. *Positronium Fine-Structure Interval $\Delta\nu$ in Oxide Powers.* M. H. Yam, P. O. Egan, W. E. Frieze and V. W. Hughes, Phys. Rev. **A18**, 350-353 (1978).

172. *Search for Parity Violation in Deep-Inelastic Scattering of Polarized Electrons by Unpolarized Deuterons.* W. B. Atwood, R. L. A. Cottell, H. DeStaebler, R. Miller, H. Pessard, C. Y. Prescott, L. S. Rochester, R. E. Taylor, M. J. Alguard, J. Clendenin, P. S. Cooper, R. D. Ehrlich, V. W. Hughes, M. S. Lubell, G. Baum, K. P. Schüler and K. Lubelsmeyer, Phys. Rev. **D18**, 2223-2226 (1978).

173. *A Source of Highly Polarized Electrons at the Stanford Linear Accelerator Center.* M. J. Alguard, J. E. Clendenin, R. D. Ehrlich, V. W. Hughes, J. S. Ladish, M. S. Lubell, K. P. Schüler, G. Baum, W. Raith, R. H. Miller and W. Lysenko, Nucl. Inst. and Meth. **163**, 29-59 (1979).

174. *Further Measurements of Parity Non-Conservation in Inelastic Electron Scattering.* C. Y. Prescott, W. B. Atwood, R. L. A. Cottell, H. DeStaebler, E. L. Garwin, A. Gonidec, R. H. Miller L. S. Rochester, T. Sato,

D. J. Sherden, C. K. Sinclair, S. Stein, R. E. Taylor, C. Young, J. E. Clendenin, V. W. Hughes, N. Sasao, K. P. Schüler, M. G. Borghini, K. Lubelsmeyer and W. Jentschke, Phys. Lett. **84B**, 524-528 (1979).

175. *Experimental Test of Special Relativity from a High-γ Electron g-2 Measurement.* P. S. Cooper, M. J. Alguard, R. D. Ehrlich, V. W. Hughes, H. Kobayakawa, J. S. Ladish, M. S. Lubell, N. Sasao, K. P. Schüler, P. A. Souder, D. H. Coward, R. H. Miller, C. Y. Prescott, D. J. Sherden, C. K. Sinclair, G. Baum, W. Raith and K. Kondo, Phys. Rev. Lett. **42**, 1386-1389 (1979).

176. *Polarized Lepton-Hadron Scattering.* V. W. Hughes **High Energy Physics with Polarized Beams and Polarized Targets**, ed. by G. H. Thomas, (AIP Conf. Proc. No. 51, Argonne, 1978) pp. 171-202.

177. *The Stopped Muon Channel at LAMPF.* P. A. Thompson, V. W. Hughes, W. P. Lysenko and H. F. Vogel, Nucl. Inst. and Meth. **161**, 391-411 (1979).

178. *Polarized Electroproduction.* V. W. Hughes, **Proceedings of the 19th International Conference on High Energy Physics**, ed. by S. Homma, M. Kawaguchi and H. Miyazawa, (Physical Society of Japan, 1979) pp. 286-290.

179. *Spin Effects in Electromagnetic Interactions.* P. A. Souder and V. W. Hughes, **High-Energy Physics in the Einstein Centennial Year (1979)**, ed. by B. Kursunoglu et al., (Plenum Publishing Corp., NY, 1979) pp. 395-439.

180. *Theoretical Hyperfine Structure of Muonic Helium.* K. -N. Huang and V. W. Hughes, Phys. Rev. **A20**, 706-711 (1979); **A21**, 1071 (1980).

181. *Muonium.* V. W. Hughes, **Exotic Atoms '79 Fundamental Interactions and Structure of Matter**, ed. by K. Crowe, J. Duclos, G. Fiorentini and G. Torelli (Plenum Publishing Corp., NY, 1980) pp. 3-18.

182. *Positronium.* V. W. Hughes, **Exotic Atoms '79 Fundamental Interactions and Structure of Matter**, ed. by K. Crowe, J. Duclos, G. Fiorentini and G. Torelli (Plenum Publishing Corp., NY, 1980) pp. 19-22.

183. *Neutrino Experiment to Test the Nature of Muon-Number Conservation.* S. E. Willis, V. W. Hughes, P. Nemethy, R. L. Burman, D. R. F. Cochran, J. S. Frank, R. P. Redwine, J. Duclos, H. Kaspar, C. K. Hargrove and U. Moser, Phys. Rev. Lett. **44**, 522-524 (1980); **44**, 903 (1980).

184. *Formation of the Muonic Helium Atom.* P. A. Souder, T. W. Crane, V. W. Hughes, D. C. Lu, H. Orth, H. -W. Reist, M. H. Yam and G. zu Putlitz, Phys. Rev. **A22**, 33-50 (1980).

185. *Parity Nonconservation and Neutral Current Interactions Involving Muons.* V. W. Hughes, **International Workshop on Neutral Current Interactions in Atoms**, ed. by W. L. Williams (Càrgese, 1979) pp. 327-356.

186. *Precise Measurement of the $2^3P_0-2^3P_2$ Fine Structure Interval in Helium.* W. E. Frieze, E. A. Hinds, V. W. Hughes and F. M. J. Pichanick, Phys. Lett. **A78**, 322-324 (1980).

187. *Measurement of Asymmetry in Spin-Dependent e-p Resonance-Region Scattering.* G. Baum, M. R. Bergström, J. E. Clendenin, R. D. Ehrlich, V. W. Hughes, K. Kondo, M. S. Lubell, S. Miyashita, R. H. Miller, D. A. Palmer, W. Raith, N. Sasao, K. P. Schüler and P. A. Souder, Phys. Rev. Lett. **45**, 2000-2003 (1980).

188. *First Observation of the Ground-State Hyperfine-Structure Resonance of the Muonic Helium Atom.* H. Orth, K. -P. Arnold, P. O. Egan, M. Gladish, W. Jacobs, J. Vetter, W. Wahl, M. Wigand, V. W. Hughes and G. zu Putlitz, Phys. Rev. Lett. **45**, 1483-1486 (1980).

189. *Measurements of Parity Violation in the Scattering of Polarized Electrons from Protons.* P. A. Souder, V. W. Hughes, M. S. Lubell and S. Kowalski, **Future Directions in Electromagnetic Nuclear Physics**, 385-395 (1980).

190. *Limits on Neutrino Oscillations from Muon-Decay Neutrinos.* P. Némethy, S. E. Willis, V. W. Hughes, R. L. Burman, D. R. F. Cochran, J. S. Frank, R. P. Redwine, J. Duclos, H. Kaspar, C. K. Hargrove and U. Moser, Phys. Rev. **D23**, 262-264 (1981).

191. *Search for Long-Lived 2S Muonic Hydrogen in H_2 Gas.* P. O. Egan, S. Dhawan, V. W. Hughes, D. C. Lu, F. G. Mariam, P. A. Souder, J. Vetter, G. zu Putlitz, P. A. Thompson and A. B. Denison, Phys. Rev. **A23**, 1152-1163 (1981).

192. *Internal Spin Structure of the Proton from High Energy Polarized e-p Scattering.* V. W. Hughes, G. Baum, M. R. Bergström, P. R. Bolton, J. E. Clendenin, N. R. DeBotton, S. K. Dhawan, R. A. Fong-Tom, Y. -N. Guo, V. -R. Harsh, K. Kondo, M. S. Lubell, C. -L. Mao, R. H. Miller, S. Miyashita, K. Morimoto, U. F. Moser, I. Nakano, R. F. Oppenheim, D. A. Palmer, L. Panda, W. Raith, N. Sasao, K. P. Schüler, M. L. Seely, J. Sodja, P. A. Souder, S. J. St. Lorant, K. Takikawa and W. Werlen,

High-Energy Physics with Polarized Beams and Polarized Targets, ed. by C. Joseph and J. Soffer (Birkhäuser, Verlag, 1981) pp. 331-343.

193. *Measurements of the Polarization Parameter in K^+p Elastic Scattering at Low Energies.* B. R. Lovett, V. W. Hughes, M. Mishina, M. Zeller, D. M. Lazarus and I. Nakano, Phys. Rev. **D23**, 1924-1932 (1981).

194. *Experiments on the 2^3P State of Helium. III. Measurement of the $2^3P_0 - 2^3P_1$ Fine Structure Interval.* A. Kponou, V. W. Hughes, C. E. Johnson, S. A. Lewis and F. M. J. Pichanick, Phys. Rev. **A24**, 264-278 (1981).

195. *Experiments on the 2^3P State of Helium. IV. Measurement of the $2^3P_0 - 2^3P_2$ Fine Structure Interval.* W. Frieze, E. A. Hinds, V. W. Hughes and F. M. J. Pichanick, Phys. Rev. **A24**, 279-287 (1981).

196. *A Planned Experiment on Parity Violation for Atomic Hydrogen.* V. W. Hughes, **Weak Interactions as Probes of Unification**, ed. by G. B. Collins, L. N. Chang and J. R. Ficenec, (AIP, 1981) pp. 78-83.

197. *New Results on Polarized Electron-Proton Scattering at SLAC.* G. Baum, M. R. Bergström, P. R. Bolton, J. E. Clendenin, N. R. DeBotton, S. Dhawan, R. Fong-Tom, Y. Guo, V. Harsh, V. W. Hughes, K. Kondo, M. S. Lubell, R. Miller, S. Miyashita, K. Morimoto, U. Moser, I. Nakano, R. Oppenheim, D. Palmer, L. Panda, W. Raith, N. Sasao, K. P. Schüler, M. Seely, J. Sodja, P. A. Souder, S. St. Lorant, K. Takikawa and M. Werlen, **High Energy Physics - 1980**, ed. by L. Durand and L. G. Pondrom, (AIP, XX International Conference, Madison, Wisconsin, 1981) pp. 781-783.

198. *Dynamic Nuclear Polarization of Irradiated Targets.* M. L. Seely, M. R. Bergström, S. K. Dhawan, R. A. Fong-Tom, V. W. Hughes, R. K. Oppenheim, K. P. Schüler, P. A. Souder, K. Kondo, S. Miyashita, I. Nakano, S. J. St. Lorant, Y. -N. Guo and A. Winnacker, **Polarized Phenomena in Nuclear Physics - 1980**, ed. by G. G. Ohlsen, R. E. Brown, N. Jarmie, W. W. McNaughton and G. M. Hale, (AIP, 1981) pp. 933-935.

199. *Dynamic Nuclear Polarization of Irradiated Targets.* M. L. Seely, M. R. Bergström, S. K. Dhawan, V. W. Hughes, R. F. Oppenheim, K. P. Schüler, P. A. Souder, K. Kondo, S. Miyashita, S. J. St. Lorant and Y. -N. Guo, **High Energy Physics with Polarized Beams and Polarized Targets**, ed. by C. Joseph and J. Soffer (Birkhäuser, Verlag, 1981) pp. 453.

200. *Additive Versus Multiplicative Muon Conservation.* P. Némethy and V. W. Hughes, Comments on Nuclear and Particle Physics **10**, 147-153 (1981).

201. *Observation of Muonium in Vacuum.* P. R. Bolton, A. Badertscher, P. O. Egan, C. J. Gardner, M. Gladisch, V. W. Hughes, D. C. Lu, M. Ritter, P. A. Souder, J. Vetter, G. zu Putlitz, M. Eckhause and J. Kane, Phys. Rev. Lett. **47**, 1441-1444 (1981).

202. *Polarization Effects.* V. W. Hughes et al., **ISABELLE, Proceedings of the 1981 Summer Workshop**, (Brookhaven National Laboratory, 1981) pp. 601-617.

203. *Plans for Measurement of Parity Nonconservation in Elastic Scattering of Polarized Electrons by Nuclei at the Bates Linear Accelerator Center.* P. A. Souder, G. Cates, T. J. Gay, V. W. Hughes, D. C. Lu, C. W. Tu, S. Kowalski, W. Bertozzi, C. P. Sargent, W. Turchinetz, M. S. Lubell and R. Wilson, **High Energy Physics with Polarized Beams and Polarized Targets**, ed. by C. Joseph and J. Soffer, (Birkhäuser, Verlag, 1981) pp. 454-457.

204. *New Experimental Limit on the Muon-Neutrino Lifetime.* J. S. Frank, R. L. Burman, D. R. F. Cochran, P. Némethy, S. E. Willis, V. W. Hughes, R. P. Redwine, J. Duclos, H. Kaspar, C. K. Hargrove and U. Moser, Phys. Rev. **D24**, 2001-2003 (1981).

205. *Precise Measurement of the Hyperfine-Structure Interval and Zeeman Effect in the Muonic Helium Atom.* C. J. Gardner, A. Badertscher, W. Beer, P. R. Bolton, P. O. Egan, M. Gladisch, M. Greene, V. W. Hughes, D. C. Lu, F. G. Mariam, P. A. Souder, H. Orth, J. Vetter and G. zu Putlitz, Phys. Rev. Lett. **48**, 1168-1171 (1982).

206. *Measurement of Spin-Exchange Effects in Electron-Hydrogen Collisions: 90° Elastic Scattering from 4 to 30 eV.* G. D. Fletcher, M. J. Alguard, T. J. Gay, V. W. Hughes, C. W. Tu, P. F. Wainwright, M. S. Lubell, W. Raith and F. C. Tang, Phys. Rev. Lett. **48**, 1671-1674 (1982).

207. *Dynamic Nuclear Polarization of Irradiated Targets.* M. L. Seely, A. Amittay, M. R. Bergström, S. K. Dhawan, V. W. Hughes, R. F. Oppenheim, K. P. Schüler, P. A. Souder, K. Kondo, S. Miyashita, K. Morimoto, S. J. St. Lorant, Y. -N. Guo and A. Winnacker, Nucl. Inst. Meth. **201**, 303-308 (1982).

208. *Higher Precision Measurement of the hfs Interval of Muonium and of the Muon Magnetic Moment.* F. G. Mariam, W. Beer, P. R. Bolton, P. O. Egan, C. J. Gardner, V. W. Hughes, D. C. Lu, P. A. Souder, H. Orth, J. Vetter, U. Moser and G. zu Putlitz, Phys. Rev. Lett. **49**, 993-996 (1982).

209. *Theoretical Hyperfine Structure of the Muonic* ^3He *and* ^4He *Atoms.* K. N. Huang and V. W. Hughes, Phys. Rev. **A26**, 2330-2333 (1982).

210. *Summary Talk.* V. W. Hughes, **Polarized Proton Ion sources**, ed. by A. D. Krisch and A. T. M. Lin, (AIP, 1982) pp. 8-20.

211. *Measurement of Spin-Exchange Effects in Electron-Hydrogen Collisions: Further Studies of Impact Ionization.* T. J. Gay, G. D. Fletcher, M. J. Alguard, V. W. Hughes, P. F. Wainwright and M. S. Lubell, Phys. Rev. **A26**, 3664-3667 (1982).

212. *Workshop Report on Polarized Proton Ion Sources.* V. W. Hughes, **High Energy Spin Physics - 1982**, ed. by G. Bunce, (AIP, 1983) pp. 534-545.

213. *Polarized Electron Source for Parity Experiment at Bates.* P. Souder, A. Barber, W. Bertozzi, G. Cates, G. Dodson, T. J. Gay, M. Goodman, V. W. Hughes, S. Kowalski, M. S. Lubell, A. Magnon, C. P. Sargent, R. Schaefer, W. Turchinetz and R. Wilson, **High Energy Spin Physics - 1982**, ed. by G. Bunce, (AIP, 1983) pp. 574-579.

214. *Measurement of the Internal Spin Structure of the Proton.* R. Oppenheim, G. Baum, M. R. Bergström, P. R. Bolton, J. E. Clendenin, N. R. DeBotton, S. K. Dhawan, R. A. Fong-Tom, Y. -N. Guo, V. -R. Harsh, V. W. Hughes, K. Kondo, M. S. Lubell, C. -L. Mao, R. H. Miller, S. Miyashita, K. Morimoto, U. F. Moser, I. Nakano, D. A. Palmer, L. Panda, W. Raith, N. Sasao, K. P. Schüler, M. L. Seely, J. Sodja, P. A. Souder, S. J. St. Lorant, K. Takikawa and M. Werlen, **High Energy Spin Physics - 1982**, ed. by G. Bunce, (AIP, 1983) pp. 255-258.

215. *Status of the AGS Polarized* H^- *Source.* A. Kponou, K. P. Schüler and V. W. Hughes, **High Energy Spin Physics - 1982**, ed. by G. Bunce, (AIP, 1983) pp. 607-610.

216. *Dynamic Nuclear Polarization of Irradiated Targets.* M. L. Seely, A. Amittay, M. R. Bergström, S. K. Dhawan, V. W. Hughes, R. F. Oppenheim, K. P. Schüler, P. A. Souder, K. Kondo, S. Miyashita, K. Morimoto, S. J. St. Lorant, Y. -N. Guo and A. Winnacker, **High Energy Physics - 1982**, ed. by G. Bunce, (AIP, 1983) pp. 526-533.

217. *Internal Spin Structure of the Nucleon.* V. W. Hughes and J. Kuti, Ann. Rev. Nucl. Part. Sci. **33**, 611-644 (1983).

218. *New Measurement of Deep-Inelastic e-p Asymmetries.* G. Baum, M. R. Bergström, P. R. Bolton, J. E. Clendenin, N. R. DeBotton, S. K. Dhawan,

Y. -N. Guo, V. -R. Harsh, V. W. Hughes, K. Kondo, M. S. Lubell, Z. - L. Mao, R. H. Miller, S. Miyashita, K. Morimoto, U. F. Moser, I. Nakano, R. F. Oppenheim, D. A. Palmer, L. Panda, W. Raith, N. Sasao, K. P. Schüler, M. L. Seely, P. S. Souder, S. J. St. Lorant, K. Takikawa and M. Werlen, Phys. Rev. Lett. **51**, 1135-1138 (1983).

219. *Reply to "Direct Comparison Between the γ-Ray Fluxes from Proton Beam Dumps at LAMPF and SIN."* J. S. Frank, R. L. Burman, D. R. F. Cochran, P. Némethy, S. E. Willis, V. W. Hughes, R. P. Redwine, J. Duclos, H. Kaspar, C. K. Hargrove, U. Moser, Phys. Rev. **D28**, 1790-1792 (1983).

220. *Precision Exotic Atom Spectroscopy.* V. W. Hughes, **Precision Measurement and Fundamental Constants II**, ed. by B. N. Taylor and W. D. Phillips, (Natl. Bur. Stand, (U.S.), Spec. Publ. 617, 1984) pp. 237-248.

221. *Formation of Muonium in the 2S State and Observation of the Lamb Shift Transition.* A. Badertscher, S. Dhawan, P. O. Egan, V. W. Hughes, D. C. Lu, M. W. Ritter, K. A. Woodle, M. Gladisch, H. Orth, G. zu Putlitz, M. Eckhause, J. Kane, F. G. Mariam and J. Reidy, Phys. Rev. Lett. **52**, 914-917 (1984).

222. *Muonium Has Not Yet Decayed!* V. W. Hughes and G. zu Putlitz, Comments Nucl. Part. Phys. **12**, 259-272 (1984).

223. *Precision Determination of the Hyperfine-Structure Interval in the Ground State of Positronium. V.* M. W. Ritter, P. O. Egan, V. W. Hughes and K. A. Woodle, Phys. Rev. **30**, 1331-1338 (1984).

224. *A Possible Higher Precision Measurement of the Muon g-2 Value.* V. W. Hughes and G. T. Danby, **Intersections Between Particle and Nuclear Physics**, ed. by R. E. Mischke, (AIP 123, 1984) pp. 534-537.

225. *The Lamb Shift in Muonium.* A. Badertscher, V. W. Hughes, D. C. Lu, M. W. Ritter, K. A. Woodle, M. Gladisch, H. Orth, G. zu Putlitz, M. Eckhause, J. Kane and F. G. Mariam, **Atomic Physics 9**, ed. by R. S. van Dyck, Jr. and E. N. Fortson, (World Scientific, 1985) pp. 83-98.

226. *Development of "Subsurface" Positive Muon Beam at LAMPF.* A. Badertscher, P. O. Egan, M. Gladisch, M. Greene, V. W. Hughes, F. G. Mariam, D. C. Lu, G. zu Putlitz, M. W. Ritter, G. Sandars, P. A. Souder and R. Werbeck, Nucl. Instr. and Meth. **A238**, 200-205 (1985).

227. *The Anomalous Magnetic Moment of the Muon.* V. W. Hughes, and T. Kinoshita, Comments on Nucl. Partical Physics **14**, 341-360 (1985).

228. *The Muon and the Electron.* V. W. Hughes, Ann. Phys. Fr. **10**, 955-983 (1985).

229. *Experimental Study of Spin-Exchange Effects in Elastic and Ionizing Collisions of Polarized Electrons with Polarized Hydrogen Atoms.* G. D. Fletcher, M. J. Alguard, T. J. Gay, V. W. Hughes, P. F. Wainwright, M. S. Lubell and W. Raith, Phys. Rev. **A31**, 2854-2884 (1985).

230. *High-Energy Polarized Electrons and Muons as a Probe for Studying the Quark Structure of Hadrons.* V. W. Hughes, Proc. Sixth Int. Symp. Polar. Phenom. in Nucl. Phys., Osaka, 1985, J. Phys. Soc. Jpn. **55**, 327-343 (1986).

231. *The Muon Anomalous g-Value.* V. W. Hughes, **Proc. of the Workshop on Fundamental Muon Physics: Atoms, Nuclei, and Particles**, LA-10714-C, Los Alamos National Laboratory, (Los Alamos, New Mexico, May, 1986) pp. 87-98.

232. *Muonium.* H. Orth and V. W. Hughes, **Proc. of the Workshop on Fundamental Muon Physics: Atoms, Nuclei, and Particles**, LA-10714-C, Los Alamos National Laboratory, (Los Alamos, New Mexico, May, 1986) pp. 62-74.

233. *Storage Ring Magnet for a Proposed New Precision Measurement of the Muon Anomalous Magnetic Moment.* V. W. Hughes, G. Danby, J. Jackson, E. Kelly, A. Prodell, R. Shutt, W. Stokes, S. K. Dhawan, A. Disco, F. J. M. Farley, Y. Kuang, H. Orth, G. Vogel, W. Williams, F. Krienen, M. Lubell, P. Marston and J. Tarrh, **Intersections Between Particle and Nuclear Physics**, ed. by D. F. Geesaman, (AIP, New York, 1986) pp. 382-390.

234. *Workshop Summary.* V. W. Hughes, **Proceedings of the Parity Violation Workshop**, CEBAF (December, 1986) pp. 317-330.

235. *Atomic Physics and Fundamental Principles.* V. W. Hughes, **Atomic Physics 10**, ed. by H. Narumi and I. Shimamura (Elsevier Science Publishers, 1987) pp. 1-34.

236. *Atomic Physics and Fundamental Principles.* V. W. Hughes, Nucl. Phys. **A463**, 3c-36c (1987).

237. *The Muon Anomalous g-Value.* V. W. Hughes, **Fundamental Symmetries**, ed. by P. Bloch, P. Pavlopoulos and R. Klapisch, (Plenum Publishing Corp., 1987) pp. 271-285.

238. *Muonium.* V. W. Hughes, **Fundamental Symmetries**, ed. by P. Bloch, P. Pavlopoulos and R. Klapisch, (Plenum Publishing Corp., 1987) pp. 287-300.

239. *First Observation of the Negative Muonium Ion Produced by Electron Capture in a Beam-Foil Experiment.* Y. Kuang, K. -P. Arnold, F. Chmely, M. Eckhause, V. W. Hughes, J. R. Kane, S. Kettell, D. -H. Kim, K. Kumar, D. C. Lu, B. Ni, B. Matthias, H. Orth, G. zu Putlitz, H. R. Schaefer, P. A. Souder and K. Woodle, Phys. Rev. **A35**, 3172-3175 (1987).

240. *Search for Spontaneous Conversion of Muonium to Antimuonium.* B. Ni, K. -P. Arnold, F. Chmely, V. W. Hughes, S. H. Kettell, Y. Kuang, J. Markey, B. E. Matthias, H. Orth, H. R. Schaefer, K. Woodle, M. D. Cooper, C. M. Hoffman, G. E. Hogan, R. E. Mischke, L. E. Piilonen, R. A. Williams, M. Eckhause, P. Guss, J. Kane, J. Reidy and G. zu Putlitz, Phys. Rev. Lett. **59**, 2716-2719 (1987).

241. *Exclusive ρ^0 and ϕ Production in Deep Inelastic Muon Scattering.* EM Collaboration, J. Ashman, et al., Z. Phys. C **39**, 169-175 (1988).

242. *Search for Spontaneous Conversion of Muonium to Antimuonium.* B. Ni, K. -P. Arnold, F. Chmely, V. W. Hughes, S. H. Kettell, Y. Kuang, J. Markey, B. E. Matthias, H. Orth, H. R. Schaefer, K. Woodle, M. D. Cooper, C. M. Hoffmann, G. E. Hogan, R. E. Mischke, L. E. Piilonen, R. A. Williams, M. Eckhause, P. Guss, J. Kane, J. Reidy, and G. zu Putlitz, Nucl. Phys. **A478**, 757c-767c (1988).

243. *Ultraprecise Superferric Storage Ring Magnet for the Muon G-2 Experiment.* D. Brown, T. deWinter, F. Krienen, D. Loomba, D. Stassinopoulos, G. Cottingham, J. Cullen, G. Danby, J. Jackson, E. Kelly, S. Kuznetsov, M. May, I. Polk, A. Prodell, R. Shutt, W. Stokes, K. Endo, H. Hirabayashi, S. Kurokawa, Y. Yamamoto, P. G. Marston, J. M. Tarrh, K. Nagamine, J. M. Bailey, A. Disco, S. Dhawan, F. J. M. Farley, V. W. Hughes, Y. Kuang, and G. Vogel, IEEE Trans. Magn. **24**, 1381-1383 (1988).

244. *The Muon Anomalous Magnetic Moment.* V. W. Hughes, Physica Scripta **T22**, 111-118 (1988).

245. *A Measurement of the Spin Asymmetry and Determination of the Structure Function g_1 in Deep Inelastic Muon-Proton Scattering.* EM Collaboration, J. Ashman, et al., Phys. Lett. **B206**, 364-370 (1988).

246. *Measurement of the Polarization of Thermal Muonium in Vacuum.* K. A. Woodle, K. -P. Arnold, M. Gladisch, J. Hofmann, M. Janousch, K. P. Jungmann, H. -J. Mundinger, G. zu Putlitz, J. Rosenkranz, W. Schäfer, G. Schiff, W. Schwarz, V. W. Hughes and S. H. Kettell, Z. Phys. **D9**, 59-64 (1988).

247. *The Integral of the Spin-Dependent Structure Function g_1^p and the Ellis-Jaffe Sum Rule.* V. W. Hughes, V. Papavassiliou, R. Piegaia, K. P. Schüler and G. Baum, Phys. Lett. **212B**, 511-514 (1988).

248. *Measurement of the Ratios of Deep Inelastic Muon-Nucleus Cross Sections on Various Nuclei Compared to Deuterium.* EM Collaboration, J. Ashman, et al., Phys. Lett. **B202**, 603-610 (1988).

249. *Progress Report on the Bates Parity Experiment.* P. A. Souder, D. - H. Kim, K. S. Kumar, M. E. Schulze, M. S. Lubell, J. S. Patch, R. Wilson, G. W. Dodson, K. A. Dow, M. Farkhondeh, J. Flanz, K. Isakovich, S. Kowalski, C. P. Sargent, W. Turchinetz, G. D. Cates, V. W. Hughes, R. Michaels and H. R. Schaefer, **Intersections Between Particle and Nuclear Physics**, ed. by G. Bunce (AIP Conf. Proc. 176, 1988) pp. 543-548.

250. *The Electrical Neutrality of Atoms.* V. W. Hughes, L. J. Fraser and E. R. Carlson, Z. Phys. **D10**, 145-151 (1988).

251. *Measurements of Nucleon Spin-Dependent Structure Functions – Past and Future.* V. W. Hughes, **Proceedings of the Symposium on Future Polarization Physics at Fermilab**, (1988) pp. 19-36.

252. *Atoms, Molecules and I. I. Rabi (Post-World War II Period).* V. W. Hughes, **Atomic Physics 11**, ed. by S. Haroche, J. C. Gay and G. Grynberg, (World Scientific, 1989) pp. 15-35.

253. *Some Recent Advances in Muonium.* V. W. Hughes, **The Hydrogen Atom**, ed. by G. F. Bassani, M. Inguscio and T. W. Hänsch, (Springer-Verlag, Berlin, Heidelberg, 1989) pp. 171-181.

254. *Muon Anomalous Magnetic Moment.* V. W. Hughes, **Particles and Fields Series 37**, ed. by K. J. Heller (AIP Conf. Proc. 187, 1989) pp. 326-347.

255. *First Observation of the Free Pionium Atom in Vacuum.* H. -J. Mundinger, K. -P. Arnold, M. Gladisch, J. Hofmann, W. Jacobs, H. Orth, G. zu Putlitz, J. Rosenkranz, W. Schäfer, K. A. Woodle and V. W. Hughes, Europhys. Lett. **8**, 339-344 (1989).

256. *The Bates Polarized Electron Source.* G. D. Cates, V. W. Hughes, R. Michaels, H. R. Schaefer T. J. Gay, M. S. Lubell, R. Wilson, G. W. Dodson, K. A. Dow, S. B. Kowalski, K. Isakovich, K. S. Kumar, M. E. Schulze, P. A. Souder and D. H. Kim, Nucl. Inst. and Meth. **A278**, 293-317 (1989).

257. *Formation of the Negative Muonium Ion and Charge-Exchange Processes for Positive Muons Passing through Thin Metal Foils.* Y. Kuang, K. - P. Arnold, F. Chmely, M. Eckhause, V. W. Hughes, J. R. Kane, S. Kettell, D. -H. Kim, K. Kumar, D. C. Lu, B. Matthias, B. Ni, H. Orth, G. zu Putlitz, H. R. Schaefer, P. A. Souder and K. Woodle, Phys. Rev. **A39**, 6109-6123 (1989).

258. *Parity Violation in Electron Scattering from Carbon A Progress Report.* G. W. Dodson, K. A. Dow, M. Farkhondeh, J. Flanz, K. Isakovich, S. Kowalski, C. P. Sargent, W. Turchinetz, D. -H. Kim, K. S. Kumar, P. A. Souder, M. S. Lubell, J. S. Patch, R. Wilson, G. D. Cates, V. W. Hughes, R. Michaels and H. R. Schaefer, **Particles and Fields Series 37**, ed. by K. J. Heller, (AIP Conf. Proc. 187, 1989) pp. 486-492.

259. *Search for Spontaneous Conversion of Muonium to Antimuonium.* V. W. Hughes, B. E. Matthias, H. Ahn, A. Badertscher, F. Chmely, M. Eckhause, K. P. Jungmann, J. R. Kane, S. H. Kettell, Y. Kuang, H. J. Mundinger, B. Ni, H. Orth, G. zu Putlitz, H. R. Schaefer, M. T. Witkowski and K. A. Woodle, **Nuclear Weak Process and Nuclear Structure**, ed. by M. Morita, H. Ejiri, H. Ohtsubo and T. Sato, (World Scientific Pub. Co., 1989) pp. 157-163.

260. *Acceleration of Polarized Protons to 22 GeV/c and the Measurement of Spin-Spin Effects in* $p \uparrow + p \uparrow \rightarrow p + p$. F. Z. Khiari, P. R. Cameron, G. R. Court, D. G. Crabb, M. Fujisaki, I. Gialas, P. H. Hansen, M. E. Hejazifar, A. D. Krisch, A. M. T. Lin, S. L. Linn, D. C. Peaslee, R. S. Raymond, R. R. Raylman, T. Roser, T. Shima, K. M. Terwilliger, L. A. Ahrens, J. G. Alessi, H. N. Brown, K. A. Brown, E. D. Courant, G. T. Danby, S. Giordano, H. J. Halama, A. Kponou, R. Lambiase, S. Y. Lee, Y. Y. Lee, R. E. Lockey, Y. I. Makdisi, P. A. Montemurro, R. J. Nawrocky, L. G. Ratner, J. F. Skelly, T. J. Sluyters, A. Soukas, S. Tepikian, R. L. Witkover, J. B. Roberts, G. C. Phillips, V. W. Hughes, P. Schüler, J. A. Bywater, R. L. Martin, J. R. O'Fallon, T. S. Bhatia, L. C. Northcliffe and M. Simonius, Phys. Rev. **D39**, 45-85 (1989).

261. *An Investigation of the Spin Structure of the Proton in Deep Inelastic Scattering of Polarized Muons on Polarized Protons.* EM Collaboration, J. Ashman, et al., Nucl. Phys. **B328**, 1-35 (1989).

262. *Measurement of the Lamb Shift in the n=2 State of Muonium.* K. A. Woodle, A. Badertscher, V. W. Hughes, D. C. Lu, M. Ritter, M. Gladisch, H. Orth, G. zu Putlitz, M. Eckhause, J. Kane and F. G. Mariam, Phys. Rev. **41**, 94-105 (1990).

263. *High Energy Physics with Polarized Electrons and Muons.* V. W. Hughes, Nucl. Phys. **A518**, 371-388 (1990).

264. *Muonium.* V. W. Hughes and G. zu Putlitz, **Quantum Electrodynamics**, ed. by T. Kinoshita (World Scientific, 1990) pp. 822-904.

265. *Fine Structure in the 2^3P State of Helium.* F. M. J. Pichanick and V. W. Hughes, **Quantum Electrodynamics**, ed. by T. Kinoshita (World Scientific, 1990) pp. 905-936.

266. *Measurement of Parity Violation in the Elastic Scattering of Polarized Electrons from ^{12}C.* P. A. Souder, R. Holmes, D. -H. Kim, K. S. Kumar, M. E. Schulze, K. Isakovich, G. W. Dodson, K. A. Dow, M. Farkhondeh, S. Kowalski, M. S. Lubell, J. Bellanca, M. Goodman, S. Patch, R. Wilson, G. D. Cates, S. Dhawan, T. J. Gay, V. W. Hughes, A. Magnon, R. Michaels and R. Schaefer, Phys. Rev. Lett. **65**, 694-697 (1990).

267. *A Spectrometer for Muon Scattering at the Tevatron.* E665 Collaboration, M. R. Adams, et al., Nucl. Inst. Meth. **A291**, 533-551 (1990).

268. *Results from the Bates ^{12}C Parity Experiment.* P. A. Souder, J. Bellanca, G. D. Cates, G. W. Dodson, K. A. Dow, M. Farkhondeh, R. Holmes, V. W. Hughes, T. J. Gay, K. Isakovich, D. -H. Kim, S. Kowalski, K. S. Kumar, M. S. Lubell, R. Michaels, J. S. Patch, H. R. Schaefer, M. E. Schulze and R. Wilson, to be published in PANIC XII Conference (1990).

269. *The Anomalous Magnetic Moment of the Muon.* AGS 821 Collaboration, V. W. Hughes, et al., to be published in the Proceedings of the 9th International Symposium on High Energy Spin Physics, Bonn, 1990.

270. *The Anomalous Magnetic Moment of the Muon.* V. W. Hughes, to be published in the 19 INS Symposium, Japan 1990, (World Sci. Pub.).

271. *The Spin Dependent Structure Functions of the Nucleon.* V. W. Hughes, **Polarized Collider Workshop**, ed. by J. Collins, S. F. Heppelman, R. W. Robinett, (AIP Conf. Proc. 223, NY, 1991) pp. 51-64.

272. *New Search for the Spontaneous Conversion of Muonium to Antimuonium.* B. E. Matthias, H. E. Ahn, A. Badertscher, F. Chmely, M. Eckhause, V. W. Hughes, K. P. Jungmann, J. R. Kane, S. H. Kettell, Y. Kuang, H. -J. Mundinger, B. Ni, H. Orth, G. zu Putlitz, H. R. Schaefer, M. T. Witkowski and K. A. Woodle, Phys. Rev. Lett. **66**, 2716-2719 (1991).